Advances in
MICROBIAL ECOLOGY

Volume 8

ADVANCES IN MICROBIAL ECOLOGY

A Continuation Order Plan is available for this series. A continuation order will bring delivery of each new volume immediately upon publication. Volumes are billed only upon actual shipment. For further information please contact the publisher.

Advances in
MICROBIAL ECOLOGY

Volume 8

Edited by

K. C. Marshall

University of New South Wales
Kensington, New South Wales, Australia

PLENUM PRESS · NEW YORK AND LONDON

The Library of Congress cataloged the first volume of this title as follows:

Advances in microbial ecology. v. 1–
  New York, Plenum Press c1977–
  v. ill. 24 cm.
  Key title: Advances in microbial ecology, ISSN 0147-4863
  1. Microbial ecology — Collected works.
QR100.A36   576′.15   77-649698

Library of Congress Catalog Card Number 77-649698
ISBN 0-306-41877-0

© 1985 Plenum Press, New York
A Division of Plenum Publishing Corporation
233 Spring Street, New York, N.Y. 10013

Printed in the United States of America

Contributors

Martin Alexander, Laboratory of Soil Microbiology, Department of Agronomy, Cornell University, Ithaca, New York 14853

Trevor Duxbury, Department of Microbiology, University of Sydney, Sydney, New South Wales 2006, Australia

Karl-Erik Eriksson, Swedish Forest Products Research Laboratory, S-11486, Stockholm, Sweden

Hans van Gemerden, Laboratorium voor Microbiologie, University of Groningen, Haren, The Netherlands

J. Gijs Kuenen, Laboratorium voor Microbiologie, Delft University of Technology, Delft, The Netherlands

Adrian Lee, School of Microbiology, University of New South Wales, Kensington, New South Wales 2033, Australia

Lars G. Ljungdahl, Center for Biological Resource Recovery, Department of Biochemistry, University of Georgia, Athens, Georgia 30602

Lesley A. Robertson, Laboratorium voor Microbiologie, Delft University of Technology, Delft, The Netherlands

Joseph A. Robinson, The UpJohn Company, Kalamazoo, Michigan 49001

Preface

Advances in Microbial Ecology was established by the International Committee on Microbial Ecology (ICOME) as a vehicle for the publication of critical reviews selected to reflect current trends in the ever-expanding field of microbial ecology. Most of the chapters found in *Advances in Microbial Ecology* have been solicited by the Editorial Board. Individuals are encouraged, however, to submit outlines of unsolicited contributions to any member of the Editorial Board for consideration for inclusion in a subsequent volume of *Advances*. Contributions are expected to be in-depth, even provocative, reviews of topical interest relating to the ecology of microorganisms.

With the publication of Volume 8 of *Advances* we welcome to the panel of contributors Martin Alexander, the founding editor of this series, who discusses the range of natural constraints on nitrogen fixation in agricultural ecosystems. Ecological aspects of cellulose degradation are discussed by L. G. Ljungdahl and K.-E. Eriksson, and of heavy metal responses in microorganisms by T. Duxbury. In his chapter, A. Lee considers the gastrointestinal tract as an ecological system, and comments on the possibility of manipulating this system. The complex interactions among aerobic and anaerobic sulfur-oxidizing bacteria are discussed in terms of natural habitats and chemostat culture by J. G. Kuenen, L. Robertson, and H. van Gemerden. Finally, J. A. Robinson presents the advantages and limitations in the use of nonlinear regression analysis in determining microbial kinetic parameters in ecological situations.

K. C. Marshall, Editor
R. M. Atlas
B. B. Jørgensen
J. H. Slater

Contents

Chapter 2

Determining Microbial Kinetic Parameters Using Nonlinear Regression Analysis: Advantages and Limitations in Microbial Ecology

Joseph A. Robinson

Chapter 3

Neglected Niches: The Microbial Ecology of the Gastrointestinal Tract
Adrian Lee

Chapter 4

Ecological Constraints on Nitrogen Fixation in Agricultural Ecosystems

Martin Alexander

Chapter 5

Ecological Aspects of Heavy Metal Responses in Microorganisms

Trevor Duxbury

Chapter 6

Ecology of Microbial Cellulose Degradation

Lars G. Ljungdahl and Karl-Erik Eriksson

Microbial Interactions among Aerobic and Anaerobic Sulfur-Oxidizing Bacteria

J. GIJS KUENEN, LESLEY A. ROBERTSON, and
HANS VAN GEMERDEN

1. Introduction

Life on the earth is dependent on the balanced recycling of various elements. Well-known examples are the carbon, nitrogen, and sulfur cycles. Although these cycles are often discussed separately, they are, in fact, closely linked. This is due not only to the fact that organic matter contains these elements, but also to the role that nitrate or sulfate can play in replacing oxygen as an electron acceptor for the mineralization of organic compounds. Thus, during the breakdown of organic matter, nitrate is reduced to ammonia or nitrogen, and sulfate to sulfide. Both the ammonia and the sulfide can be reoxidized. Since this chapter is mainly concerned with the ecology of bacteria involved in the sulfur cycle, a brief discussion of this cycle is appropriate (Pfennig and Widdel, 1982; Kuenen, 1975; Trudinger, 1982). Sulfate serves as the sulfur source for the biosynthesis of organic sulfur compounds by plants and microorganisms using the process known as assimilatory sulfate reduction (Fig. 1). In biological materials, sulfur is usually present in its most reduced form (e.g., as sulfide in amino acids such as cysteine). During the decomposition of this material under aerobic conditions, the organic sulfide is initially oxidized and subsequently released as sulfate. Under anaerobic

J. GIJS KUENEN and LESLEY A. ROBERTSON • Laboratorium voor Microbiologie, Delft University of Technology, Delft, The Netherlands.   HANS VAN GEMERDEN • Laboratorium voor Microbiologie, University of Groningen, Haren, The Netherlands.

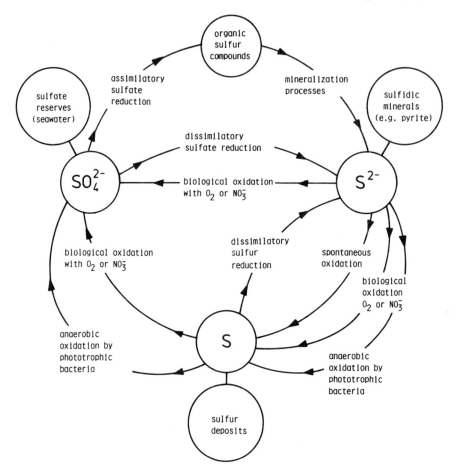

Figure 1. The sulfur cycle. [Adapted from Bos and Kuenen (1983).]

conditions, the sulfide is liberated as hydrogen sulfide. Another important source of hydrogen sulfide is from dissimilatory sulfate reduction, whereby sulfate-reducing bacteria can use sulfate as their electron acceptor for the oxidation of organic compounds or molecular hydrogen. Some sulfide precipitates as ferrous sulfide or forms pyrite, but much of it is reoxidized via elemental sulfur to sulfate (Bos and Kuenen, 1983). This can be done under anaerobic conditions in the light by phototrophic sulfur-oxidizing bacteria, which use the sulfide as an electron donor in the generation of reducing power for the reduction and assimilation of carbon dioxide (Trüper and Fischer, 1982). By this process, the electrons produced in mineralization are rechanneled into organic compounds.

Elemental sulfur, which may be formed as an intermediate, can also be reduced to sulfide by sulfur-reducing heterotrophs. Under aerobic or denitrifying conditions hydrogen sulfide can also be spontaneously oxidized by oxygen or nitrate if these substances interact at high enough concentrations. At lower concentrations of oxygen (or nitrate), sulfide is usually oxidized by the colorless sulfur bacteria to give sulfate (Kelly, 1982). During this process, the majority of the electrons from the sulfide are used to reduce oxygen or nitrate to give water or molecular nitrogen, respectively. Some electrons can be used for CO_2 reduction and be recycled into organic compounds. Thus, it is the complementary action of the two types of oxidative bacteria with the sulfur and sulfate-reducing bacteria that maintains most of the global sulfur cycle. However, it should be remembered that industrial pollution and geothermal processes also contribute substantially.

Both the phototrophic and colorless sulfur bacteria comprise large, heterogeneous groups of organisms. Whereas the difference between the two is obviously based on the possession or lack of photosynthetic pigments, the groups of different species can be further subdivided by their degree of physiological specialization or versatility, type of photosynthetic pigmentation, and other characteristics, for example, the ability to denitrify. These subdivisions are dealt with in Section 2, and are summarized in Tables I and II.

The sulfide-oxidizing bacteria are dependent on reduced sulfur compounds for growth, and therefore are found in environments where sulfate reduction occurs or where a geological source of sulfur compounds is available. Since most of the colorless sulfur bacteria are dependent on oxygen, they often live at the interface between aerobic and anaerobic zones where low concentrations of oxygen and sulfide can coexist. Examples of such environments are the aerobic surfaces of otherwise anaerobic freshwater marine sediments and the interface between the aerobic and anaerobic zones of stratified bodies of water. An example of such an interface is found in Solar Lake (Sinai), where sulfide and oxygen coexist over a depth of a few centimeters (Fig. 2). In sediments, this layer can be as narrow as 0.1 mm or less (Jørgensen, 1982). Under special circumstances, blooms of colorless sulfur bacteria (e.g., *Beggiatoa* mats) can be found on the surface of anaerobic sediments. When light can reach these interfaces, the phototrophic sulfur bacteria may thrive, and in many stratified lakes annual blooms of a variety of these phototrophs occur. These blooms can produce intensely colored green, brown, or red layers sometimes as thick as 1 m. Blooms of these organisms can also be found in the top layers of light-exposed sediments, where they can, for example, form a discrete layer in the complex microbial communities of algal mats.

In our laboratories we have been investigating the occurrence and

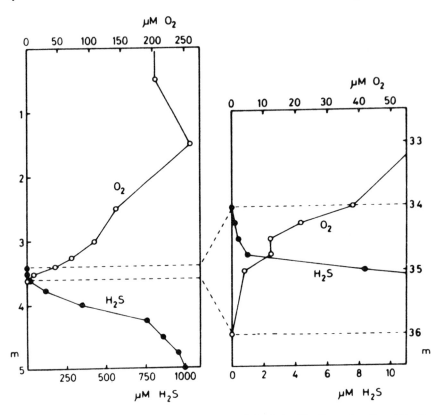

Figure 2. Concentrations of sulfide and oxygen in Solar Lake (Sinai) as a function of depth. Oxygen and sulfide can be seen to coexist at the interface. [Jørgensen *et al.* (1979).]

ecological niches of the seemingly endless variety of species among the colorless and phototrophic bacteria. In addressing these questions, we have tried to consider the abiotic environmental parameters together with some of the biotic variables, especially the interactions that occur among the phototrophic and colorless sulfur bacteria as well as between the two groups. Our approach has been through an ecophysiological study of pure cultures of these organisms under conditions that may be relevant to their existence in nature. As a second step, we have studied mixtures of the pure cultures of these organisms, and in a few cases we have also done some field studies to verify predictions made in the laboratory.

One of the most important environmental pressures imposed on microorganisms is that of nutrient or light limitation. An understanding of bacterial response to these limitations and their survival under such conditions is crucial to a better knowledge of microbial ecology. For the study of the ecophysiology and interactions of microorganisms under nutrient limitation in the laboratory, continuous cultivation in a chemostat has been an indispensable technique. In our investigation of the interactions between the different types of sulfur bacteria, this technique has allowed us to study the physiology of test organisms under a variety of nutrient limitations, and also to use the chemostat as a device by which the selective pressures exerted on species competing for limiting nutrients or light can be simulated and amplified.

In Section 4 we describe a selected number of examples of microbial competition involving the sulfur-oxidizing bacteria. It will become clear that many of the phototrophs and the colorless sulfur bacteria are very well suited for use as model organisms for the exploration of the basic principles that determine the survival value of different metabolic strategies in the struggle for existence.

2. The Types of Bacteria

2.1. The Colorless Sulfur Bacteria

The group of organisms known as the colorless sulfur bacteria make up a heterogeneous collection of Gram-negative bacteria, which includes intensively studied species, such as some members of the genus *Thiobacillus,* and others that have not been obtained in pure culture and have only been studied superficially, such as the genus *Thiobacterium* (Vishniac, 1974; la Rivière, 1974). The various physiological types represented within the group are shown in Table I. Colorless sulfur bacteria are found at a wide range of temperatures, pH values, and degrees of aerobiosis or anaerobiosis. Some are obligate chemolithoautotrophs (e.g., *T. neapolitanus*), some can only oxidize sulfur compounds if they are supplied with an organic carbon source (e.g., *T. perometabolis*), while others are capable of autotrophic, heterotrophic, or mixotrophic growth (e.g., *T. novellus*). The group is commonly subdivided on the basis of the degree of physiological specialization shown by the various species (Tables I and III).

2.1.1. The Obligate Chemolithoautotrophs

These highly specialized species can only grow autotrophically. They must use an inorganic source of energy and obtain cell carbon from carbon dioxide fixation via the Calvin cycle. Most, however, are able to utilize small amounts of exogenous organic carbon (Matin, 1978), which can serve as a source of carbon, but not energy. The citric acid cycle in these organisms is inoperative, its enzymes only serving for biosynthesis. With

Table I. Metabolic Types Found among Colorless Sulfur Bacteria

Metabolic definition	Genera	Representative species	Respiratory type	Special requirement
Obligate chemolithotroph	*Thiobacillus, Thiomicrospira*	*Thiobacillus neapolitanus*	O_2	—
		Thiobacillus ferrooxidans	O_2	Acidophilic
		Thiobacillus denitrificans	O_2/NO_3^-	—
		Thiomicrospira denitrificans	O_2/NO_3^-	Microaerophilic
Facultative chemolithotroph	*Thiobacillus, Sulfolobus,*	*Thiobacillus intermedius*	O_2	—
	Thermothrix, Paracoccus,	*Beggiatoa*	O_2	Microaerophilic
	Thiosphaera, Beggiatoa	*Thiobacillus acidophilus*	O_2	Acidophilic
		Sulfolobus acidocaldarius	O_2	Acidophilic, thermophilic
		Thiobacillus versutus	O_2/NO_3^-	Denitrifies heterotrophically
		Thiosphaera pantotropha	O_2/NO_3^-	—
Chemolithoheterotroph	*Thiobacillus, Pseudomonas*	*Thiobacillus perometabolis*	O_2	—
Heterotroph (oxidizes S^{2-}, but does not appear to gain energy)	*Beggiatoa, Pseudomonas*	*Beggiatoa*	O_2/S^0	—

Table II. Physiological Types Found among Phototrophic Bacteria

Pigmentation	Family	Photopigment	Metabolic definition	Electron donor
Green	Chlorobiaceae	Bacteriochlorophyll a and c, d, or e, chlorobactene	Obligate phototroph, facultative photoautotroph, external S^0 accumulates, obligate anaerobe	S^{2-}, $S_2O_3^{2-}$, S^0, H_2, organic acids
Green	Chloroflexaceae	Bacteriochlorophyll a and c, β- and γ-carotene	Facultative phototroph, facultative photoautotroph, external S^0 accumulates, thermophil	S^{2-}, organic acids
Purple	Chromatiaceae (formerly Thiorhodaceae)	Bacteriochlorophyll a or b	Obligate phototroph, facultative photoautotroph, internal S^0 accumulates, facultative aerobe	S^{2-}, $S_2O_3^{2-}$, S^0, H_2, organic acids
Purple	Rhodospirillaceae (formerly Athiorhodaceae)	Bacteriochlorophyll a or b	Facultative phototroph, facultative photoautotroph	S^{2-}, $S_2O_3^{2-}$, organic acids
Blue-green	Cyanobacteriaceae (kingdom)	Chlorophyll a, phycocyanin, and/or phycoerythrin, allophycocyanin, β-carotene	Phototroph, obligately photoautotroph	S^{2-}, H_2O

**Table III. Metabolic Definitions Used to Describe Colorless Sulfur
Bacteria**

	Energy source		Carbon source	
	Reduced sulfur compounds	Organic compounds	CO_2	Organic compounds
Obligate autotroph	+	−	+	−
Facultative autotroph	+	+	+	+
Chemolithoheterotroph	+	+	−	+

regard to their carbon metabolism, the obligate chemolithotrophs appear to be completely geared to autotrophic growth and, in this respect, they can be regarded as metabolically rigid. However, in other metabolic aspects they may exhibit a surprising degree of flexibility. Studies with *T. neapolitanus* have shown that its carbon and nitrogen metabolism can adapt in response to fluctuating environmental conditions. This includes the regulation of its Calvin cycle in response to the concentration of available carbon dioxide, the induction and repression of enzymes for nitrogen metabolism, and the formation of intracellular storage products, such as glycogen. This serves as an energy source, even under anaerobic conditions, by means of a heterolactic fermentation (Beudeker *et al.,* 1981). The phenomena have been extensively reviewed by Kuenen and Beudeker (1982). The representatives of the genus *Thiomicrospira* are all obligate chemolithotrophs, which are very similar to their *Thiobacillus* counterparts (Kuenen and Tuovinen, 1981).

2.1.2. The Facultative Chemolithotrophs

These physiologically versatile species are capable of autotrophic growth, but are also able to grow heterotrophically on various organic substrates. Many of them also grow mixotrophically (i.e., simultaneously using two different substrates, often by autotrophic and heterotrophic pathways of substrate utilization) under dual substrate limitation in the chemostat (Gottschal and Kuenen, 1980a; A. L. Smith *et al.,* 1980; Leefeldt and Matin, 1980). In batch culture, however, the enzymes of the Calvin cycle are usually repressed in the presence of organic substrates (Gottschal and Kuenen, 1981), and a form of diauxy may occur. One of the most studied organisms in this group is *Thiobacillus* A2, recently renamed *T. versutus* (Harrison, 1983).

This organism displays a remarkable versatility in its use of a variety of organic compounds as substrates. During continuous culture on limiting mixtures of organic compounds or on mixtures of organic compounds and reduced sulfur compounds, *T. versutus* adapts its metabolism to that required for the optimal use of the limiting substrates. Thus, during growth on either acetate or glucose with thiosulfate, its enzyme levels are adjusted to the required turnover rates. For example, when the supply of organic substrate is high relative to that of the reduced sulfur compounds, the enzymes of the Calvin cycle are reduced to the amounts required for supplementary carbon dioxide fixation or, if possible, completely repressed. The flexibility of this organism is such that not only can it grow mixotrophically, but it is also capable of adjusting its metabolism to suit strongly fluctuating nutrient supplies. With alternating 4-hr supplies of organic and reduced sulfur compounds in continuous flow experiments (Fig. 3), this organism is able to grow efficiently by switching its enzyme levels for the required rate of turnover (Gottschal *et al.*, 1981).

It is only in recent years, with the advent of the chemostat as a tool for ecological studies, that the importance of these versatile species has come to be appreciated.

A completely different group of colorless sulfur bacteria are the *Beggiatoa* species, some of which are able to grow autotrophically while using the Calvin cycle for carbon dioxide fixation. These *Beggiatoa* can also grow heterotrophically on acetate, and therefore should be considered facultative chemolithotrophs (Nelson and Jannasch, 1983). Very little is known of their autotrophic physiology, since they are extremely difficult to grow, and can only be cultivated autotrophically in sulfide/oxygen gradients. Other *Beggiatoa* species are chemolithoheterotrophs (see Section 2.1.3).

2.1.3. The Chemolithoheterotrophs

This is a little-studied group, of which *T. perometabolis* is the only recognized species in the genus *Thiobacillus* (Vishniac, 1974). These organisms can use inorganic energy sources such as reduced sulfur compounds, but cannot fix carbon dioxide and therefore require organic carbon. Their ecological niche probably lies in areas where a supply of reduced sulfur compounds is available to provide energy for growth, but where there is enough organic material present to make the operation of the Calvin cycle for carbon dioxide fixation energetically undesirable. Many of the *Beggiatoa* species that have been obtained in pure culture may also be chemolithoheterotrophs. Although their aerobic growth is often dependent on sulfide, it is not always clear whether sulfide serves as an energy source. Sulfide can also be used by these organisms to detoxify hydrogen peroxide formed under aerobic conditions. Güde *et al.* (1981) have obtained indications that sulfide may act as an ancillary

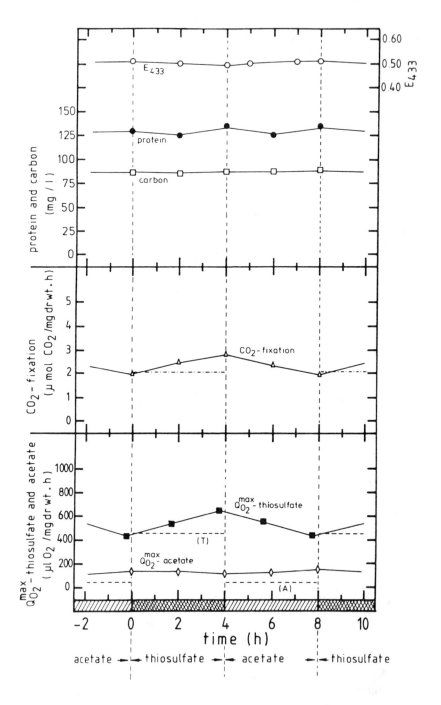

energy source during growth on acetate. For further details, the reader is referred to the review by Larkin and Strohl (1983).

2.1.4. The Denitrifying Sulfur Bacteria

All of the colorless sulfur bacteria described thus far are able to use oxygen as the terminal electron acceptor for respiration, although one species, *Thiomicrospira denitrificans,* can only do so under microaerophilic conditions. A few species are also capable of anaerobic growth if nitrate or another nitrogen oxide is available to serve as an electron acceptor. This is the process known as denitrification, the nitrate being reduced to nitrogen gas. The group includes several obligate autotrophs, such as *T. denitrificans* and *Thiomicrospira denitrificans* (Vishniac, 1974; Timmer-ten Hoor, 1975), versatile species such as *Thiosphaera pantotropha* (Robertson and Kuenen, 1983), and at least two species, *T. versutus* and *Paracoccus denitrificans* (Wood and Kelly, 1983; Friedrich and Mitringa, 1981), that are versatile aerobically but do not appear to be able to grow autotrophically on reduced sulfur compounds under anaerobic conditions.

2.2. The Phototrophic Sulfur Bacteria

The different types of phototrophic bacteria are listed in Table II. The Cyanobacteria (formerly known as the blue-green algae) differ from the other types of phototrophic bacteria in being able to use water as an electron donor for photosynthesis, with oxygen as the end product. The purple and green bacteria cannot carry out an oxygenic photosynthesis, and require anaerobic conditions for phototrophic growth. They are dependent on electron donors with a lower redox potential than water (e.g., reduced sulfur compounds, hydrogen, or small organic molecules). The green phototrophic bacteria fall into two families, the Chlorobiaceae and the Chloroflexaceae, both of which are considered sulfur bacteria. A recent report indicates that many Chloroflexaceae may be primarily heterotrophs (Castenholz, 1984). The purple bacteria also fall into two families, the Chromatiaceae and the Rhodospirillaceae, the latter previously being regarded as the purple nonsulfur bacteria. Some Cyanobacteria and

---

Figure 3. *Thiobacillus versutus* (formerly *Thiobacillus* A2) grown in a chemostat at a dilution rate of 0.05 hr^{-1} with alternating 4-hr periods of acetate (10 mM) and thiosulfate (40 mM). (O) Optical density at 433 nm; (●) protein as mg/liter; (□) organic carbon as mg/liter; (△) μmol carbon dioxide fixed per mg dry weight per hr; (■) $Q_{O_2}^{max}$ thiosulfate as μl oxygen per mg dry weight per hr; (♦) $Q_{acetate}^{max}$ as μl oxygen per mg dry weight per hr. [Gottschal *et al.* (1981).]

some of the Rhodospirillaceae use sulfide or thiosulfate as an electron donor (Cohen *et al.,* 1975; Hansen and van Gemerden, 1972), but they do not use sulfur and are numerically less important in this respect than the Chlorobiaceae and the Chromatiaceae. The Chloroflexaceae are a group of predominantly thermophilic, green, gliding bacteria about which comparatively little is known.

The Chlorobiaceae and the Chromatiaceae differ from each other in their ultrastructure, types of bacteriochlorophyll, and mode of carbon dioxide fixation, the Chlorobiaceae lacking a Calvin cycle. The general physiology of the phototrophic bacteria has been described in detail in several reviews (Pfennig, 1978; Trüper and Fischer, 1982). As with the colorless sulfur bacteria, the group can be subdivided by degree of physiological specialization with regard to their obligate or facultative requirement for light or for an inorganic electron donor.

2.2.1. The Obligate Phototrophs

These are phototrophic bacteria that grow either very slowly or not at all in the absence of sulfide. They have green or brown photosynthetic pigments and are classified as the family Chlorobiaceae. They require sulfide first as a source of sulfur, a process termed assimilatory sulfide utilization, and second as an electron donor in photosynthesis. These bacteria can be grown in sulfide/carbon dioxide media, but, unlike the Chromatiaceae, fix carbon dioxide by means of a reversed citric acid cycle rather than the Calvin cycle. The controversy over the presence or absence of citrate lyase, the key enzyme of the reversed citric acid cycle, was resolved when Ivanovsky *et al.* (1980) demonstrated that the enzyme is ATP dependent. ATP had not, as a rule, been included in the enzyme assay mixture. The Chlorobiaceae are able to assimilate a limited number of low-molecular-weight organic compounds, acetate being ecologically the most important. However, since their citric acid cycle is reversed, it cannot be used for the generation of reducing power from acetate to allow growth. The Chlorobiaceae therefore convert acetate to pyruvate via a reductive carboxylation mediated by reduced ferredoxin. Sulfide is required for the formation of the reduced ferredoxin, and therefore acetate cannot be assimilated in its absence (Sadler and Stanier, 1960).

These organisms are equipped with bacteriochlorophylls *c, d,* or *e* as light-harvesting pigments (Pfennig, 1978), but contain bacteriochlorophyll *a* in their reaction center (Olsen and Romano, 1962). The brown species, *Chlorobium phaeobacteroides* and *Chl. phaeovibroides,* have carotenoids as additional light-harvesting moieties. Like all other phototrophs, the Chlorobiaceae are able to adapt to low light intensities by increasing their pigment content (Broch-Due *et al.,* 1978).

The Chlorobiaceae are obligate phototrophs. They appear to be able to gain some ATP from glycogen degradation in the dark (Sirevåg and Ormerod, 1977), but this is merely a survival mechanism sufficient for dark periods of limited length rather than a means of growth. They are obligate anaerobes and cannot grow under even microaerophilic conditions (Kämpf and Pfennig, 1980). All of the Chlorobiaceae so far studied are specialists with a high reactivity, since their affinity for sulfide is very high, their maintenance requirements are low, and they grow well at reduced light intensities.

As already stated, the Cyanobacteria differ from all of the other phototrophic bacteria in being able to use water as an electron donor for photosynthesis, producing oxygen. In this, the Cyanobacteria resemble the higher plants. Most of the Cyanobacteria are metabolically very rigid and unable to develop on organic media. There is a strong analogy between their carbon and nitrogen metabolism and that of the obligate chemolithoautotrophs. The Cyanobacteria are mentioned in this chapter because some of them are able to use sulfide as an electron donor for photosynthesis, with sulfur as the end product, and in this they resemble some of the other phototrophic bacteria. Because of their facility for oxygenic photosynthesis, the Cyanobacteria exert great influence over the position of the sulfide/oxygen interface in stratified bodies of water and sediments.

The Cyanobacteria frequently form blooms (Walsby, 1978; van Liere, 1979). This appears to be due to their ability to grow at much lower light intensities than the green algae. Some species are capable of both oxygenic and anoxygenic photosynthesis. Field data support this (Jørgensen et al., 1979).

2.2.2. The Facultative Photolithotrophic Bacteria

With the exception of the Chlorobiaceae and most of the Cyanobacteria, the phototrophic bacteria have a certain degree of flexibility in their use of inorganic or organic electron donors, and for this reason may be termed facultative photolithotrophs. As with the thiobacilli, many of them are able to grow mixotrophically. Some only do this under growth-limiting conditions, but others exhibit mixotrophy even in batch culture. For example, the Chromatiaceae are classically grown on sulfide and carbon dioxide, but can also grow on mixtures of sulfide and acetate or another low-molecular-weight organic carbon source. Most species can grow on acetate alone, but a few require sulfide (e.g., *Chromatium okenii, Chr. weissei*), since they do not have adenosine phosphosulfate (APS) reductase and therefore cannot use the pathway of assimilatory sulfate reduction. Thus, the sulfide requirement of the latter group is different from that of the Chlorobiaceae, which need sulfide for the assimilation of any organic compound.

The Rhodospirillaceae are best characterized as photoorganotrophs, but now that some of them have been shown to use sulfide as an electron donor (Hansen and van Gemerden, 1972), they could also be regarded as facultative photolithotrophs. However, the importance of the Rhodospirillaceae, or purple nonsulfur bacteria, in the global sulfur cycle is not clear. Comparatively low concentrations of sulfide inhibit most of these species.

In comparison with the substrate range of the Rhodospirillaceae, the variety of organic molecules used by the Chromatiaceae is fairly restricted (e.g., they cannot grow on amino acids). Another difference between the two is that *Chr. vinosum* oxidizes sulfide equally rapidly whether grown on sulfide or acetate (van Gemerden and Beeftink, 1981). Since in the phototrophic sulfur bacteria the oxidation of sulfide in the absence of organic carbon is stoichiometrically linked to the fixation of carbon dioxide, this finding demonstrates the operation of the Calvin cycle under conditions where it is not required. This stoichiometric linkage does not, of course, occur in the colorless sulfur bacteria, since they transfer the majority of their electrons to oxygen or one of the nitrogen oxides rather than to carbon dioxide. The Rhodospirillaceae appear to adapt their enzyme levels more to the immediate need than do the Chromatiaceae (Hurlbert and Lascelles, 1963; Slater and Morris, 1973). It would thus appear that the Rhodospirillaceae are better adapted for life under a steady substrate supply, whereas the Chromatiaceae are more suited to fluctuating nutrient supplies. In nature diurnal fluctuations of light not only influence the photosynthetic processes directly, but also result in strong fluctuations of other nutrients, in particular those of sulfide and (excreted) organic compounds. As will be shown below, these fluctuations play a decisive role in the selection and competition of the phototrophs.

2.2.3. The Facultative Phototrophs

Some phototrophs are able to generate ATP by fermentation or respiration, as well as by light absorption. However, this is more of a survival mechanism than a provision for growth. The growth rates are very low and nothing is known about affinities under these conditions.

The Rhodospirillaceae are physiologically the most diverse of the phototrophs. They are able to use anoxygenic photosynthesis (some with sulfide as electron donor), and most species can grow under microaerophillic or fully aerobic conditions at the expense of organic substrates. This group is very well adapted to life under oxygen fluctuation.

3. Types of Microbial Interaction

As stated in the introduction, the purpose of this chapter is to discuss the interactions of the bacteria involved in the oxidative part of the sulfur cycle. At the same time, we will try to demonstrate that these studies have also been useful in elucidating some of the general principles governing microbial interactions. We begin with a short discussion of the definitions of the different types of microbial relationships.

It is often convenient to define these relationships by considering only the dominant and most apparent phenomenon (e.g., competition for a substrate), but it must be remembered that in a mixed population more than one type of interaction may be occurring. Additionally, many interactions require more than one definition. For example, methanogens compete with sulfate-reducing bacteria for acetate, but require sulfide produced by those same sulfate-reducers. The various interactions are summarized in Table IV, and are discussed briefly in the following sections. For more details, the reader should consult the excellent reviews by Frederickson (1977) and Bull and Slater (1982). An extensive and complex terminology for the description of microbial interactions has arisen over the years, and the terms used in this review will be those quoted by Bull and Slater (1982).

As stated in the introduction, the main emphasis of this review is on the competitive interactions among the phototrophic and colorless sulfur bacteria (Section 4). In addition, some of the few known examples of other interactions are discussed in Section 5.

Table IV. Definitions of the Terms Used to Describe the Various Microbial Interactions[a]

Term	Population A	Population B	Description
Neutralism	0	0	The coexistence of two populations that have no effect at all on each other
Mutualism	+	+	Mutual benefit of coexisting populations
Commensalism	0	+	Benefit of one population from another
	+	0	that itself is unaffected
Amensalism	0	−	The inhibition or restriction of one
	−	0	population by another, the second population remaining unaffected
Predation	+	−	The feeding of one population on
Parasitism	−	+	another; a predator is larger and a parasite smaller than the prey
Competition	−	−	Competition for a limiting nutrient or other factor

[a]+, Positive effect; −, negative effect; 0, no effect.

3.1. Neutralism

Neutralism is defined as a state in which two populations coexist without affecting each other in any way. Where actively growing cultures are concerned, some kind of physical separation is often implied, since in most cases any action on the part of one population must have some effect on a second population with which it is in contact (for example, reducing the pH or altering the redox level). In the artificial environment of a chemostat, it is possible to set up a system where all nutrients and other growth factors required are in excess except the energy source. A trivial example of this type can be created by growing two populations in a mixed chemostat culture while each is supplied with a substrate not used by the other (Kuenen and Gottschal, 1982). It has also been shown that if alternating supplies of reduced sulfur compounds and an organic substrate such as acetate are provided, it is possible for the obligate chemolithotroph *T. neapolitanus* and an obligate heterotroph known as spirillum G7 to coexist, each on an alternate substrate supply (Gottschal *et al.*, 1981). Other cases have been reported (Brock, 1966).

3.2. Mutualism

Mutualism is a condition in which both populations in a culture derive benefit from the presence of the other. This might take the form of cross-feeding of required nutrients such as vitamins or substrate precursors, or involve the removal of toxic excretion products. For a theoretical treatment of cross-feeding, the reader is referred to papers by de Freitas and Frederickson (1978) and Frederickson (1977). Specific aspects of mutualism have been allocated individual terms by various authors. For example, synergism, also known as protocooperation, has been described as a facultative and transient phenomenon and mutualism as an obligate interaction. Symbiosis is a mutualistic relationship involving specific partners and permitting them to achieve some development that would not be possible individually. In the interests of simplicity, the single term mutualism will be used here. Examples of this are dealt with in Section 5.

3.3. Commensalism

Commensalism is a relationship in which one population benefits from the presence of a second, while the second population remains unaffected by the first. It is likely that many systems presently considered to be commensal will, on closer study, prove to be mutualistic, since the "productive partner" in a commensal pair may actually benefit from the removal of its product by the second species. However, many instances of one species providing another with a vitamin while both are growing on separate substrates are known.

3.4. Amensalism

Amensalism, or antagonism, is said to occur when one population adversely affects the growth of others without being affected itself. Again, this definition is probably an oversimplification, since the first population may be said to benefit from the reduced competitive activity of the second.

3.5. Predation

Predation is the feeding of one species on another, to the detriment of the second species. Parasitism and predation are frequently separated, but the main difference is one of scale. In parasitism the benefiting species is normally smaller than the prey; in predation it is larger and the interaction is generally fatal to the prey.

3.6. Competition

Competition may be defined as the fight for survival between two species requiring a common nutrient, space, or light and is one of the strongest evolutionary pressures. Since the appearance of the chemostat, it has become one of the most easily tested interactions and has been studied with several sets of organisms under different conditions. This is dealt with extensively in Section 4.

4. Competition

The principles of competition for growth-limiting substrates have been reviewed in detail in a number of recent papers (Veldkamp and Jannasch, 1972; Harder *et al.,* 1977; Parkes, 1982; Kuenen and Robertson 1984a,b). The practical aspects of the design of continuous flow experiments for the study of competition for growth-limiting substrates have been discussed by Veldkamp and Kuenen (1973) and Kuenen *et al.* (1977). Before embarking on a detailed discussion of the chemostat studies that we have carried out to study competition among the sulfur bacteria, a few points must be made. If the substrate saturation curves (μ versus s) of any particular pair of organisms competing for the same substrate were known, at least in theory, the outcome of any competition could be predicted. However, the μ versus s curve may be dependent on the environmental conditions applied and only represent balanced growth. A practical aspect is that the concentration of growth-limiting substrates in the culture is frequently below the detection level for the substrate, and therefore cannot be measured. Therefore μ versus s curves cannot be made. In such cases, competition experiments will reveal the relative positions of the μ versus s curves of the organisms. If one organism outcompetes the other at a low dilution rate, but the other is more successful at higher dilution rates, their μ versus s curves must cross. The dominant organism at the lower dilution rate is said to have a higher "affinity" for the growth-limiting substrate. A rough approximation of the affinity can be obtained from the slope of the μ versus s curve at very low substrate concentrations (μ_{max}/K_s), and not by comparing K_s values (Healey, 1980).

There are many other reasons why competition experiments are necessary. Deviations from the theoretical predictions often occur, and these very deviations may deepen our understanding of the principles of competition, and possibly the influence of other interactions. As will become clear from the following paragraphs, the physiology of one organism can often be influenced by the presence of another to such an extent that predictions made from pure culture studies no longer apply. Finally, it must be stressed that the outcome of competition experiments under steady state conditions can often be predicted by means of mathematical modeling, which is relatively simple for this type of situation, but when fluctuations are introduced as an experimental variable, the complications are such that this becomes very difficult.

In the study of the ecology of the sulfur-oxidizing bacteria, it is necessary to have a clear picture of the important environmental stresses or limitations likely to occur in their habitats. A short outline of these habitats was given in Section 1. From this it is clear that both the phototrophic and the colorless sulfur bacteria are typical gradient organisms, which must compete for sulfide diffusing upward from the anaerobic zones where the sulfate bacteria are active, and also for oxygen or light, which penetrate from above. Bicarbonate can serve the autotrophs as a carbon source, but almost all of these bacteria can use some organic compounds that may also be produced within the habitat. These are often the products of fermentation, especially acetate, which is made by a variety of organisms, ranging from sugar-fermenting heterotrophs to the acetogenic bacteria, which reduce carbon dioxide to acetate. It will be shown in the following section that the presence of organic compounds can play a decisive role in the outcome of competition for sulfide. It will also become clear that the influence of these organic molecules, together with that of oxygen, light, and sulfide itself, plays an important role in maintaining the wide spectrum of metabolic types and the control of the interactions between the organisms involved in the oxidation of reduced sulfur compounds.

4.1. Competition Involving Chemolithotrophs

It is convenient to divide this section into two parts: (1) competition for growth-limiting substrates under a steady flow of substrates, and (2) the effect of substrate fluctuations on the outcome of competition.

4.1.1. Steady Nutrient Supply

Gottschal and co-workers (1979) studied competition in a chemostat among an obligate chemolithotroph, *T. neapolitanus,* a facultative chemolithotroph, *T. versutus,* and an obligate heterotroph, spirillum G7, with limiting single or double substrate supplies. Thiosulfate was used as a typical substrate for chemolithotrophic growth, and acetate as a representative organic substrate for heterotrophic growth. As expected, when either thiosulfate or acetate was supplied to the mixed continuous culture, the appropriate specialist outcompeted the versatile organism, not only at their maximum specific growth rates, but also at lower dilution rates. An example of the results obtained during competition between *T. neapolitanus* and *T. versutus* for thiosulfate is shown in Fig. 4. It should be noted that *T. versutus* is not completely washed out of the culture. This is due to the excretion by *T. neapolitanus* of glycollate, which can be used by *T. versutus.* This will be further discussed in Section 5. These experiments lead to the conclusion that the specific growth rates of the specialists lie above that of the versatile organism over a whole range of substrate concentrations provided that only one substrate is present. The specialists will therefore outcompete the nonspecialist under these conditions.

When the competition experiments were repeated using mixtures of acetate and thiosulfate, the picture changed. The addition of acetate or glycollate to the autotrophic cultures (Fig. 5) and thiosulfate to the heterotrophic cultures (not shown) resulted in an increase in the steady state number of *T. versutus,* which depended on the concentration of the additive. This eventually allowed the versatile species to outcompete either specialist.

In further experiments it was shown that the versatile species became dominant in cultures containing all three organisms when the concentrations of acetate and thiosulfate in the mixture were of the same order of magnitude, but that either of the specialists dominated when the concentration of one of the substrates greatly exceeded the other (Gottschal *et al.,* 1979).

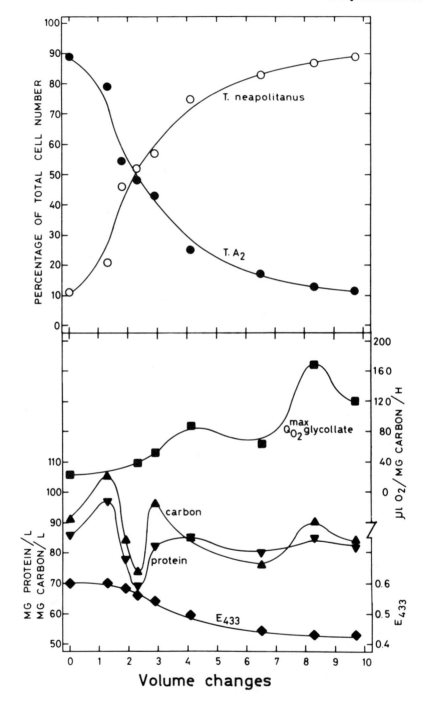

Mathematical modeling of mixed cultures of bacteria competing for a mixture of two or more substrates has been described (P. A. Taylor and Williams, 1974; Frederickson, 1977; Gottschal and Thingstad, 1982). All of these models predict that the maximum number of species that can coexist when competing for growth-limiting substrates can never be more than the number of substrates available. The results of experiments with two-species cultures can also be predicted from such models. A somewhat unexpected prediction derived from the model for two substrates is that, as long as two organisms are present in the culture, the specialist species will dictate the concentration of the common substrate. This implies that, in our case, the physiological state (i.e., enzyme levels, respiratory capacity, etc.) of *T. versutus* does not change with a varying acetate to thiosulfate ratio as long as the specialist is present. This constant physiological state is in contrast with the predictions made from pure culture studies of *T. versutus* (Gottschal and Kuenen, 1980a). An important general conclusion derived from this work is that the physiological state of a mixotroph can be influenced by the presence of other bacteria competing for the same substrate.

The mathematical modeling also predicted that in the acetate–thiosulfate medium, only two organisms could coexist. However, the coexistence of all three species was observed in experiments involving all three. This remains unexplained. It is possible that, rather than a steady state, a so-called pseudo steady state existed in which changes took so long that hundreds of volume changes would be required before a true steady state could be established.

For the survival of versatile organisms such as *T. versutus* in a chemostat, substrate limitation is essential, and since in nature nutrients are probably often limiting, this would indicate an explanation for the survival of versatile species in association with specialists in the wild. In the presence of excess substrate, such as is found in batch culture, the specialists have an advantage both because of their superior maximum specific growth rate and also because the versatile organism often cannot grow mixotrophically if the substrates are not limiting (the Calvin cycle enzymes are usually repressed in the presence of a nonlimiting organic source of carbon).

←_____

Figure 4. Competition in continuous culture at a dilution rate of 0.025 hr^{-1} between *T. versutus* (*T.* A2) and *T. neapolitanus* for growth-limiting thiosulfate (40 mM). The organisms were pregrown in separate chemostats ($D = 0.025$ hr^{-1}) and then mixed in a ratio of 9:1, respectively, at $t = 0$. (●) *T. versutus* and (○) *T. neapolitanus* percentage of total cell number; (■) Q_{O2}^{max} glycollate as μl oxygen per mg carbon; (▲) mg organic carbon per liter; (▼) mg protein per liter; (◆) optical density at 433 nm. [Gottschal *et al.* (1979).]

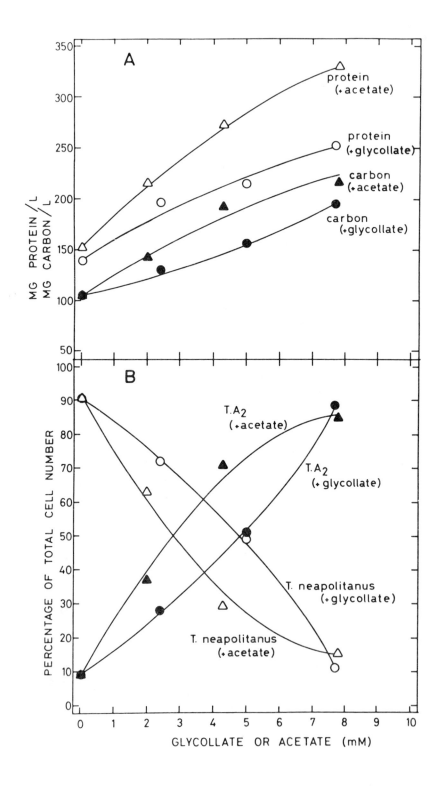

Another example of competition in a chemostat for a steady supply of a limiting nutrient was reported by Kuenen *et al.* (1977). Two species of obligate chemolithotroph, *T. thioparus* and *Thiomicrospira pelophila,* were isolated from the same intertidal mudflat. It was found that *Tms. pelophila* had a much higher tolerance for sulfide than *T. thioparus* and this led to speculation as to the niches occupied by the two species. The results of iron-limited experiments with the two organisms (Fig. 6) show that, at very low iron concentrations, *Tms. pelophila* was able to grow much faster than *T. thioparus,* but at higher iron concentrations the position was reversed. In the marine environment, at the interfaces where the sulfide concentrations are highest, iron may very well be limiting because of the low solubility of ferrous sulfide in water. An ability to grow rapidly at low iron concentrations and relatively high sulfide concentrations would thus give *Tms. pelophila* a selective advantage.

Additional competition experiments under thiosulfate limitation also showed that the pH optima of the two species were sufficiently different to play a possible role in the survival of the organism. Above pH 7.5, *Tms. pelophila* dominated, but below pH 6.5, *T. thioparus* was more successful. At intermediate pH values, variable results were obtained.

4.1.2. Fluctuating Substrate Supply

Although environments involving a steady supply of nutrients do occur in nature, fluctuations in the substrate supply or other conditions are more common. It was assumed that in an environment undergoing rapid or irregular fluctuations the ecological advantage would be possessed by a species that could adapt to the changing conditions and "scavenge" a discontinuous supply of very low concentrations of both inorganic and organic substrates rather than being dependent on one. In other words, the advantage should again be with the versatile species.

The results of competition in continuous culture between *T. versutus* and *T. neapolitanus* when thiosulfate and acetate were supplied in alternating 4-hr periods are illustrated in Fig. 7. In experiment A, the inoculum contained approximately equal numbers of each organism. Experi-

---

←——————————————————————————————————

Figure 5. The effect of different concentrations of organic substrate on the outcome of competition for thiosulfate (40 mM) between *T. versutus* (T. A2) and *T. neapolitanus* in a chemostat a dilution rate of 0.07 hr⁻¹. All measurements were taken after steady states had been established. (A) Protein concentration in the cultures limited by thiosulfate plus acetate (△) or glycollate (○). Organic carbon content of cultures limited by thiosulfate plus acetate (▲) or glycollate (●). (B) The percentage *T. versutus* cells with thiosulfate plus acetate (▲) or glycollate (●) in the feed, and the percentage of *T. neapolitanus* with thiosulfate plus acetate (△) or glycollate (○) in the feed. [Gottschal *et al.* (1979)].

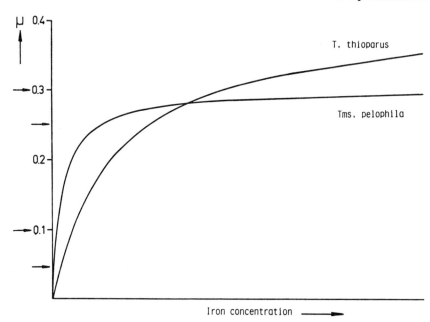

Figure 6. The specific growth rate μ of *Thiomicrospira pelophila* and *Thiobacillus thioparus* as a function of the iron concentration when growing in a chemostat at 25°C. The graph was constructed on the basis of competition experiments, actual iron concentrations not known. The arrows indicate the dilution rates at which competition experiments were carried out. [Kuenen *et al.* (1977).]

ment B was started with considerably more cells of *T. versutus* than of *T. neapolitanus*. In both cases, the end result was a community made up of roughly equal numbers of each. The most obvious explanation for this was that the specialist was able to grow autotrophically, claiming all of the substrate, during the thiosulfate period, and that the versatile species grew heterotrophically, obtaining its substrate during the acetate period. For example, *T. neapolitanus* maintained a very high thiosulfate-oxidizing capacity during the acetate supply period, which would amount to a starvation period for this organism. The *T. versutus*, while growing on acetate, repressed its thiosulfate-oxidizing potential. On the appearance of thiosulfate in the culture, *T. neapolitanus* was able to oxidize the thiosulfate at a much higher rate than the versatile species. In other words, *T. neapolitanus* was more "reactive" then *T. versutus* (Beudeker *et al.*, 1982). This interpretation was supported by the results of pure culture studies with *T. versutus* (Fig. 3), in which it was shown that *T. versutus* maintained just enough respiratory capacity for both acetate and thiosulfate to maintain growth at the required rate. The result of the higher reactivity of *T. neapolitanus* would be that at the onset of each thiosulfate

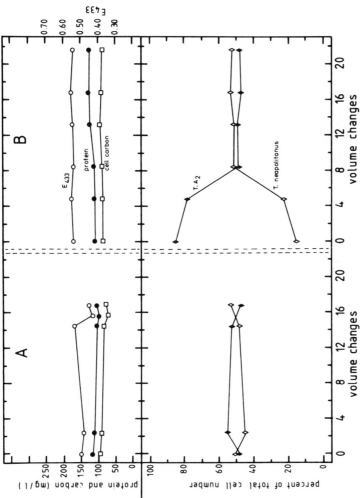

Figure 7. Competition in a chemostat at a dilution rate of 0.05 hr^{-1} between *T. versutus* (*T.* A2) and *T. neapolitanus* for growth-limiting thiosulfate (40 mM) or acetate (10 mM) supplied in alternating 4-hr periods. (A, B) Similar experiments started with different initial ratios of the two organisms. (○) Optical density at 433 nm; (●) mg protein per liter; (□) mg cell carbon per liter; (◆) percentage of *T. versutus*; (◇) percentage of *T. neapolitanus*. [Gottschal *et al.* (1981).]

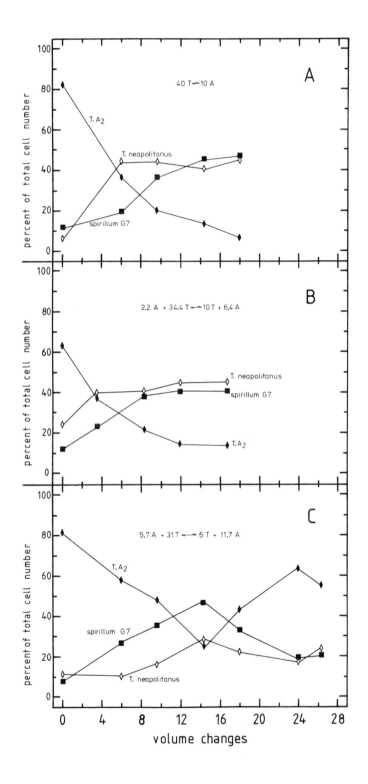

period, *T. versutus* would induce a lower level of thiosulfate-oxidizing capacity, eventually losing it completely. The specialist heterotroph (spirillum G7) responded to the alternating substrate supplies as the mirror image of *T. neapolitanus.*

The results obtained by repeating these experiments using all three organisms (*T. neapolitanus, T. versutus,* and spirillum G7) can be seen in Fig. 8A. When the feeds simply contained separate supplies of thiosulfate and acetate, both of the specialist species formed about half of the final culture and the versatile organism washed out. The explanation for this unexpected result lies in the adaptability of *T. versutus* and the rigidity of the two specialists. The specialist heterotroph (spirillum G7) could oxidize the acetate more rapidly than *T. versutus.* Since *T. versutus* tends to adapt gradually rather than to respond with full potential to a sudden change in the environment, it was outcompeted by both of the specialists. When small amounts of acetate were included in the thiosulfate supply, and thiosulfate in the acetate supply (Fig. 8B), *T. versutus* was able to maintain itself as a small proportion of the community. An increase in the proportions of these additives and a reduction of the main substrates (Fig. 8C) resulted in the versatile species regaining the advantage, since it was no longer obliged to switch between autotrophic and heterotrophic growth, but could grow mixotrophically throughout. A more general conclusion that may be derived from these experiments is that metabolic rigidity, as exemplified by *T. neapolitanus,* may have ecological advantages, since it provides the organism with the potential to respond to fluctuating nutrient supplies with full capacity.

Continuous enrichment cultures (see Section 6) made with alternating thiosulfate and acetate supplies resulted in the isolation of a facultative autotroph called *Thiobacillus* S (P. Spijkerman and J. G. Kuenen, unpublished results) rather than one or more specialists. *Thiobacillus* S differed from *T. versutus* since, although its capacity for thiosulfate utilization dropped during culture in the absence of thiosulfate, it was never completely lost. Rather than the rapid switches of metabolism shown by *T. versutus, Thiobacillus* S developed a different survival strategy. It stored significant quantities of poly-β-hydroxybutyrate during the periods

---

Figure 8. Competition in continuous culture among *T. versutus* (*T.* A2), *T. neapolitanus,* and spirillum G7 for thiosulfate and acetate as growth-limiting substrates. The dilution rate was 0.05 hr^{-1} with intermittent feeding of two media containing either thiosulfate or acetate, or different combinations of the two. These media were supplied alternately, to the culture in 4-hr periods. (♦) The *T. versutus,* (■) spirillum G7, and (◇) *T. neapolitanus* as the percentage of the total cell number present in the culture. (A) 10 mM acetate or 40 mM thiosulfate. (B) 2.2 mM acetate + 34.4 mM thiosulfate or 10 mM thiosulfate + 11.7 mM acetate. (C) 5.7 mM acetate + 31.0 mM thiosulfate or 6 mM thiosulfate. [Gottschal *et al.* (1981).]

of heterotrophic growth (Fig. 9). This storage product was then utilized as a supplementary carbon source during the initial stages of growth on thiosulfate, when its carbon dioxide-fixing capacity was low, allowing a gradual change of metabolic behavior. Thus, by showing a less dramatic response to changes in its environment than *T. versutus, Thiobacillus* S is able to retain enough "reactivity" to compete successfully with specialist organisms present in the enrichment inoculum.

4.2. Competition Involving Phototrophs

As previously stated, many phototrophic bacteria require sulfide, carbon dioxide, and light, and growth is often enhanced by organic molecules of low molecular weight. As with the colorless sulfur bacteria, the outcome of competition among the phototrophs is governed by the number and type of limiting substrates available, together with the occurrence and duration of environmental fluctuations such as those involving pH, redox potential, and, most important in this case, light.

4.2.1. Competition between Two Species of Chromatium

Many differences are apparent in the light absorption spectra of the different phototrophs (Fig. 10). This can often provide an explanation for

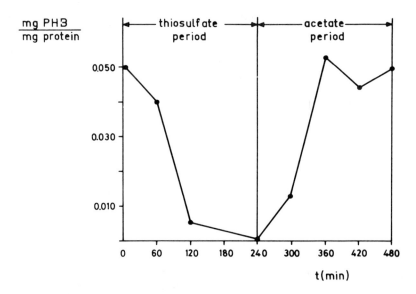

Figure 9. Fluctuation of the poly-β-hydroxybutyrate in *Thiobacillus* S during alternating 4-hr periods of thiosulfate and acetate in a chemostat ($D = 0.05$ hr^{-1}). [Kuenen and Robertson (1984a).]

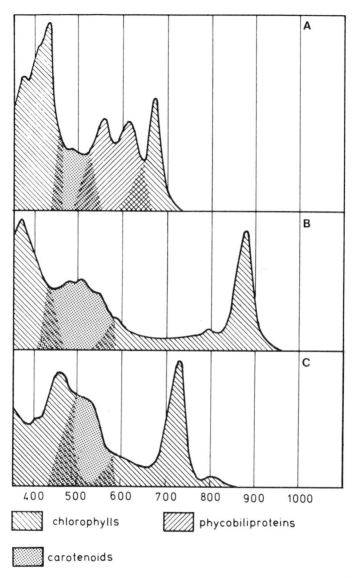

400   500   600   700   800   900   1000

chlorophylls   phycobiliproteins

carotenoids

Figure 10. The *in vivo* spectra of some types of phototrophic bacteria. (A) Cyanobacteria (chlorophyll *a*); (B) Chromatiaceae and Rhodospirillaceae (bacteriochlorophyll *a*); (C) brown Chlorobiaceae (bacteriochlorophylls *e* and *a*; green species lack the high carotenoid content). [Adapted from R. Stanier *et al.* (1976), p. 538, with permission.]

the vertical distribution of different types of phototrophs in natural environments. However, such spectral differences have a decisive influence under light-limiting conditions only, and are not necessarily essential when growth is limited by other environmental factors. An example of this is found with the relatively large *Chromatium weissei,* frequently the dominant population in blooms, and with *Chr. vinosum,* a smaller organism, which is often isolated from the same habitats. Laboratory studies in batch culture revealed that the maximum specific growth rate of *Chr. vinosum* was always higher than that of the larger organism. This observation posed the question of why *Chr. weissei* was able to produce blooms while competing with *Chr. vinosum.* Further studies with pure cultures of the organisms in a chemostat showed that at all, including inhibitory, concentrations, the specific growth rate of *Chr. weissei* was lower than that of *Chr. vinosum.* The explanation for the ecological success of *Chr. weissei* does not, therefore, lie in its higher specific growth rate under sulfide limitation.

In the natural habitat of phototrophic bacteria, the sulfide concentration has been reported to show daily fluctuations (Sorokin, 1970; van Gemerden, 1967). Basically this can be explained by the fact that during the night sulfate reduction continues and phototrophic sulfide consumption stops (Jørgensen *et al.,* 1979; Jørgensen, 1982). For this reason, the physiologies of the two species were studied under fluctuating sulfide supplies. When the organisms were supplied with 0.5 mM sulfide, both species oxidized it via intracellular sulfur to sulfate. However, at identical culture densities *Chr. vinosum* depleted the medium of sulfide at a much lower rate (Fig. 11). This was because *Chr. weissei* was able to rapidly store intracellular sulfur, whereas *Chr. vinosum* oxidized most of the sulfide directly to sulfate. The specific rate of glycogen synthesis in both species was very similar, showing that the rate of oxidation (number of electrons obtained per unit of time) of the sulfur compounds was the same in both organisms. Thus, the sulfur-hoarding capacity of *Chr. weissei* was considerably larger than that of *Chr. vinosum.* From these results, it was predicted that during a transient accumulation of sulfide, *Chr. weissei* would have an ecological advantage, since it would be able to claim more of any available sulfide than would *Chr. vinosum.* Once the sulfide had been depleted, *Chr. weissei* would then be able to continue growing on its intracellular sulfur for a longer period than the other organism. Clearly under a continuous sulfide supply, this advantage would disappear, since the sulfide storage capacity of *Chr. weissei* is finite.

Competition experiments between the two species confirmed these predictions. When supplied with growth-limiting sulfide in continuous light, *Chr. vinosum* was the more successful species, but when the culture was subjected to a light/dark regimen, sulfide accumulated during the

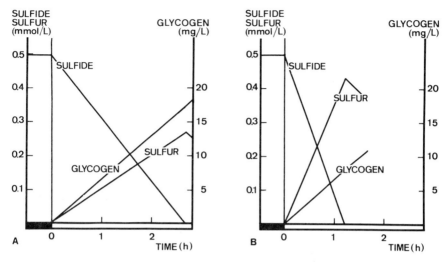

Figure 11. Substrate oxidation and product formation in pure cultures of *Chromatium vinosum* (A) and *Chromatium weissei* (B). Cultures were supplied with sulfide in the dark. During subsequent illumination, the time courses of sulfide, sulfur, glycogen, and protein were followed. No significant growth was observed during the observation time shown. For ease of comparison, the data have been standardized to identical population densities of 10 mg protein per liter. [Modified from van Gemerden (1974).]

dark phase and *Chr. weissei* was able to profit from its higher hoarding capacity. The *Chr. weissei* was able to coexist with and even dominate the smaller organism during light/dark regimes (Table V).

The general conclusion from these results is that the ability to rapidly store intracellular sulfur, an ability probably shared by all of the large

Table V. ***Chromatium vinosum* and *Chromatium weissei* in Balanced Coexistence in Sulfide-Limited Continuous Cultures under Various Light Regimens**[a]

	Relative abundance during steady state (%)	
Illumination regimen	*C. vinosum*	*C. weissei*
Continuous light	100	0
18 hr light, 6 hr dark	100	0
6 hr light, 6 hr dark	63	37
4 hr light, 6 hr dark	30	70

[a]After van Gemerden (1974). The dilution rate was 0.01 hr^{-1} throughout.

Chromatiaceae, may well explain the success of these species in nature. An interesting parallel has already been discussed in a preceding section, where we analyzed the success of a poly-β-hydroxybutyrate-accumulating *Thiobacillus* species grown under fluctuating supplies of acetate and thiosulfate. Clearly a capacity for hoarding is an important factor in microbial competition.

4.2.2. Competition between a Species of Chromatium and a Species of Rhodopseudomonas

The general differences between the Chromatiaceae and the Rhodospirillaceae were briefly discussed in Section 2. In spite of their high metabolic versatility and flexibility, blooms of the Rhodospirillaceae have never been found. It has recently been demonstrated (Hansen and van Gemerden, 1972; A. F. Dijkstra, D. J. Wijbenga, and H. van Gemerden, in preparation) that although these organisms are generally less tolerant of high sulfide concentrations, some are able to grow autotrophically in minerals-sulfide media at rates comparable with those of *Chromatium* species while exibiting a high affinity for the growth-limiting sulfide. One such organism is *Rhodopseudomonas capsulata.* However, it cannot oxidize sulfide further than elemental sulfur, which is deposited extracellularly (Hansen, 1974). Since *Chromatium* species can use this extracellular sulfur, this implies that in competition for sulfide, the *Chromatium* species would be able to claim at least 75% of the electrons available from the sulfide, even if the affinity of the *Rhodopseudomonas* for sulfide was higher than that of the *Chromatium* species. Coexistence of the two species might thus be possible. This is, indeed, the case for *R. capsulata* and *Chr. vinosum* competing for growth-limiting sulfide in the chemostat. The *Rhodopseudomonas* population was, however, only 4–5% of the total community, despite its higher affinity for sulfide. The explanation can be found by drawing a parallel with the competition for mixtures of thiosulfate and acetate as discussed above.

Where the purple sulfur bacteria are concerned, sulfur and sulfide function as mixed substrates. The *Chromatium* is able to consume all of the elemental sulfur and thus build up a high population. Once this population is established, it will also be able to consume a significant amount of the sulfide, regardless of its lower affinity. The main difference between this example and that discussed for acetate and thiosulfate supplying the *Thiobacillus* species, of course, is that the sulfur and sulfide concentrations in this experiment are linked, and do not vary independently. Thus, one would expect that the *Rhodopseudomonas* would never be eliminated from the culture unless an external source of elemental sulfur is available to increase the population density of the *Chromatium*.

Since both *Chromatium* and *Rhodopseudomonas* species can oxidize acetate, it was thought that this organic compound, which is often produced in parallel with sulfide by the sulfate-reducing bacteria, might affect the outcome of competition between these species. During acetate limitation in continuous culture, *R. capsulata* outcompeted *Chr. vinosum* by virtue of its higher affinity for acetate. If the culture was provided with an equimolar mixture of sulfide and acetate, the proportion of *R. capsulata* was about 20% of the total community, despite its higher affinity for both sulfide and acetate. As explained above, this is due to the inability of *R. capsulata* to oxidize elemental sulfur.

When acetate and sulfide were alternately supplied to the mixed culture, a situation reminiscent of the results obtained with competing versatile and specialist *Thiobacillus* species was created. In this case, *R. capsulata* may be regarded as the versatile organism able to adapt its acetate- and sulfide-oxidizing potentials to suit the available substrates. The *Chr. vinosum* can therefore be regarded as the specialist that maintained constitutive levels of the enzymes required for acetate and sulfide metabolism. It might thus be expected that *Chr. vinosum* would possess a much higher "reactivity" than *R. capsulata* and therefore should outcompete the latter when supplied with alternating feeds of acetate and sulfide. Indeed, D. J. Wijbenga and H. van Gemerden (unpublished observations) were able to show that *Chr. vinosum* did outcompete the *Rhodopseudomonas* almost completely under this substrate regime. Whether this finding can also be generally applied to those *Rhodopseudomonas* species that are able to oxidize sulfide to sulfate (e.g., *R. sulfidophila*) remains to be investigated.

4.2.3. Competition between the Green and Purple Sulfur Bacteria under Sulfide Limitation

The difference in niches between the Chlorobiaceae and the Chromatiaceae can in part be explained by their different absorption spectra, since these are partially complementary (Fig. 10). As pointed out in Section 4.2.1, this is particularly important under light-limiting conditions. Under saturating light conditions other differences come into play in the competition between these groups of bacteria.

Such an example can be found in the competition of *Chromatium* and *Chlorobium* species for sulfide.

Both *Chromatium* and *Chlorobium* oxidize sulfide to sulfate, with the temporary accumulation of elemental sulfur, the difference being that the latter deposits the sulfur outside the cells, whereas *Chromatium* stores the sulfur intracellularly. This might put the *Chlorobium* at a disadvantage, since the *Chromatium* can also grow at the expense of the oxidation of extracellular sulfur to sulfate. When these data are considered together with the abilities of both species to metabolize small organic molecules, it becomes clear that there should be many possibilities for the coexistence of the two types of phototrophs. With this in mind, van Gemerden and Beeftink (1981) studied competition between green and purple sulfur bacteria. It was found that the *Chlorobium* species *(Chl. limicola f. thiosulfatophilum)* had a three to four times higher affinity for sulfide than the *Chromatium* species *(Chr. vinosum)* studied. Competition experiments performed in continuous cultures illuminated with saturating light intensities revealed that at the lower dilution rates the *Chlorobium* dominated, whereas at higher dilution rates the *Chromatium* had the advantage (Table VI). It is important to stress that, under all conditions, stable coexistence of both participants was observed.

It appeared unlikely that the sulfur produced by the *Chlorobium* was freely available to the *Chromatium,* because the sulfur adheres to the *Chlorobium* cells. At the lower growth rates, the *Chlorobium* was able to claim 90% of the available reducing power, indicating that, whatever the affinities for sulfur involved, the *Chromatium* was unable to claim any significant part of the excreted sulfur. However, at a higher dilution rate more extracellular sulfur would be present and a certain amount of this elemental sulfur would dissolve. At first sight, therefore, the results of Table VI might be interpreted to imply that the *Chromatium* would have a higher affinity for soluble sulfur than the *Chlorobium,* since otherwise the *Chromatium* would have been completely outcompeted. Although this may be the case, another possibility might be that the *Chlorobium* was excreting organic molecules (Sirevåg and Ormerod, 1977), which were then assimilated by the *Chromatium.* Other complicating factors might be the excretion of thiosulfate by the *Chlorobium* (Schedel, 1978)

Table VI. *Chlorobium limicola f. thiosulfatophilum* and
Chromatium vinosum in Balanced Coexistence in
Sulfide-Limited Continuous Cultures at Various Dilution
Rates[a]

| Dilution rate (hr^{-1}) | Relative abundance during steady state (%) | | | |
| | Chlorobium | | Chromatium | |
	Biovolume	Protein	Biovolume	Protein
0.02	90	89	10	11
0.05	87	91	13	9
0.08	50	49	50	51
0.10	5	2	95	98

[a]After van Gemerden and Beeftink (1981). The pH was 6.8 throughout.

or the formation of polysulfides from sulfide and sulfur (van Gemerden, 1983).

Another important factor that would affect the outcome of competition between the green and purple sulfur bacteria is the effect of acetate available in the natural environment. Experiments using a brown *Chlorobium* and a species of *Thiocapsa* (a close relative of *Chromatium*) indicate that the *Thiocapsa* had a higher affinity for acetate, but a lower affinity for sulfide. If this observation can be extrapolated to situations involving other species of *Chlorobium,* it would imply that the outcome of competition at lower growth rates shown by the results in Table VI might be reversed by the presence of acetate or other assimilable organic substrates. It should be stressed, however, that this expectation is confined to saturating light intensities.

Finally, it is known that, in contrast to the Chlorobiaceae, the Chromatiaceae possess the pathway of assimilatory sulfate reduction. Furthermore, the green sulfur bacteria are rapidly killed by oxygen, whereas most of the purple bacteria are not. The temporary absence of sulfide and the appearance of oxygen in their habitat might therefore present a considerable disadvantage to the *Chlorobium* species (see Section 4.2.5).

4.2.4. Competition between Green and Purple Sulfur Bacteria under Light Limitation

As indicated in Section 4.2.3, the Chlorobiaceae and Chromatiaceae have photosynthetic pigments with complementary absorption spectra (see Fig. 10). This is particularly true of the green, but not the brown, *Chlorobium* species. Some brown-pigmented *Chlorobium* species (for example, *Chl. phaevibroides*) differ from the green in having a high carotenoid content, and thus seem better equipped for competition with the Chromatiaceae. In fact, blooms containing green *Chlorobium* species together with a brown *Chlorobium* in place of the expected *Chromatium* have been found. Frequently, Chlorobiaceae are found just below a bloom of purple sulfur bacteria (see Section 4.4.3).

An example in which light limitation played a crucial role in the competition between brown Chlorobiaceae and purple sulfur bacteria was studied in connection with their simultaneous presence near the oxic/anoxic interface of Lake Kinneret (Israel). It was shown that the blooming *Chl. phaeobacteroides* has an advantage over the accompanying *Thiocapsa roseopersicina* because of its ability to grow relatively rapidly under light limitation, as occurred in this lake (Fig. 12). Studies with pure and mixed cultures of the two organisms demonstrated that at very low light intensities the *Chlorobium* entirely outcompeted the *Thiocapsa*. At somewhat higher, but still limiting, light levels the two species were able to

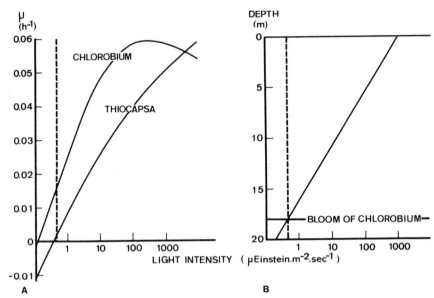

Figure 12. Growth responses of *Chlorobium phaeobacteroides* and *Thiocapsa roseopersicina* (both isolates from Lake Kinneret) as a function of (A) the intensity of light and (B) the relation between depth and light intensity in Lake Kinneret as related to the level of a bloom of the *Chlorobium* sp. The vertical dotted lines A and B indicate the light intensity at which the *Thiocapsa* sp. can no longer grow. Wind, blowing each day for some time, creates an internal seiche in the lake. The bloom thus moves up and down with the thermocline. For ease of comparison, the impact of the internal seiche on the light conditions in the lake is neglected. [Modified from van Gemerden (1983).]

coexist because of the availability of two substrates, sulfide and acetate. The former was used preferentially by the *Chlorobium* and the latter by the *Thiocapsa*. When light was supplied in excess, both organisms should have been able to grow mixotrophically on both substrates, but instead of coexisting, the *Thiocapsa* completely outcompeted the other species (van Gemerden, 1983).

The relatively high growth rate of the brown *Chlorobium* under light limitation is linked to its requirement for maintenance energy, which is tenfold lower than that of the *Thiocapsa* species (Fig. 12). At the light intensities available at Lake Kinneret, only minimal growth of the *Thiocapsa* was possible, whereas the *Chlorobium* was able to double its numbers once a day. The data seem to indicate that these differences in growth rate under light limitation are general features of the Chlorobiaceae and the Chromatiaceae, and this would be consistent with their natural distribution relative to light.

An important observation was recently made in Lake Cisó (Spain), where it was observed that competition between green and brown *Chlo-*

robium species seemed to be controlled by the presence or absence of an overlaying bloom of Chromatium species (Montesinos et al., 1982). It is known that at the depth in this lake where the phototrophs grow, only light wavelengths between 300 and 600 nm can penetrate. The maximum light intensity lies between 450 and 500 nm. Therefore, the accessory pigments rather than the absorption peaks of the bacteriochlorphyll in the far red will be important for light harvesting. Since it was known that the absorption spectrum of the brown Chlorobium overlapped more with that of Chromatium than did that of the green, it was postulated that the Chromatium acted as a biological light filter, thereby favoring growth of the green Chlorobium beneath it. Indeed, when fluorescent light lacking the red wavelengths was used, the two Chlorobium species grew at approximately the same rate. However, when a filter designed to mimic the absorption of light by 10 m of water plus the Chromatium bloom was used, the green Chlorobium became dominant (Montesinos et al., 1982). This provides support for the hypothesis formulated above.

4.2.5. Competition Involving Cyanobacteria

Thus far, we have only considered competition between the anaerobic, sulfide-utilizing phototrophs. As mentioned before, it was discovered some years ago that some of the Cyanobacteria are capable of an anaerobic, sulfide-dependent photosynthesis in addition to their normal oxygenic photosynthesis. The immediate cause of this discovery was a bloom of Cyanobacteria at a depth of 4 m in a small, hypersaline, stratified lake in Sinai called Solar Lake (Fig. 13). At this depth there was a concentration of almost 1 mM sulfide. The cyanobacterium involved turned out to be Oscillatoria limnetica, an organism capable of both oxygenic and anoxygenic photosynthesis (Cohen et al., 1975). This and related organisms were also found to bloom at the sulfide/oxygen interface in the lake, a position normally occupied by blooms of Chromatium species. Detailed studies of the diurnal variations of sulfide, oxygen, and oxygenic versus anoxygenic photosynthesis at the interface (Jørgensen et al., 1979) yielded a possible explanation. At the interface the cyanobacterial bloom carried out a sulfide-dependent photosynthesis during the early hours of the day for as long as sulfide was present. After depletion of the sulfide, the activity of the population shifted to oxygenic photosynthesis. This flexibility, allowing the switching from one form of photosynthesis to another, allowed the cyanobacteria to remain at the depth providing optimal light conditions. This may give them a competitive advantage over the purple sulfur bacteria, which must move up and down with the diurnal variation of the sulfide/oxygen interface (Jørgensen et al., 1979). It is obvious that growth of the obligate anaerobic green sulfur bacteria is strongly inhibited under these conditions.

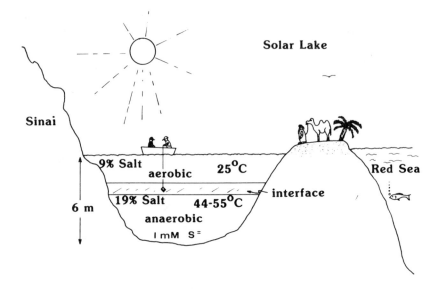

Figure 13. Section of Solar Lake (Sinai). The lake is stratified for 7–9 months of the year, but during the summer months complete mixing takes place. During the summer, the rate of evaporation exceeds the flow of seawater seeping in from the Red Sea, but in the winter this is reversed. The less concentrated seawater forms a layer above the more concentrated hypolimnion. The upper layer insulates the lower layer, thus creating a greenhouse effect.

4.3. Other Selective Pressures

Although competition for growth-limiting substrates and light is one of the strongest selective pressures, as already mentioned, other factors, such as pH, temperature, or oxygen availability, may influence the outcome of such competition. Some examples of this influence will be dealt with in this section.

4.3.1. pH

During the competition experiments with *T. thioparus* and *Tms. pelophila* described in Section 4.1.1, Kuenen *et al.* (1977) also looked at the effect of pH on the outcome of competition for growth-limiting thiosulfate. They found that above pH 7.5 *Tms. pelophila* outgrew *T. thioparus,* but that below 6.5 the positions were reversed. This observation gave some indication of the niches occupied by the two obligate chemolithotrophs in their natural environment, the mud of the Waddenzee. Under normal conditions, the pH in the mud can rise as high as 8.0, favoring the *Thiomicrospira*. As sulfuric acid is produced by the aerobic oxidation of sulfur and reduced sulfur compounds, however, the pH will fall and *T. thioparus* might gain an advantage. A. L. Smith and Kelly (1979) investigated the effect of pH on the competition for growth-limiting thiosulfate between *T. versutus* and *T. neapolitanus,* since it is known that the two organisms have different pH optima (7.5–8.0 and 6.5–7.0, respectively). It would be expected that the obligate chemolithotroph would outcompete *T. versutus* (see Section 4.1), but at pH values above the optimum of the specialist, the versatile species dominated. At pH values approaching its optimum, *T. neapolitanus* dominated. Again, the picture changed when an organic compound, in this case glucose, was included in the medium, giving *T. versutus* an advantage.

In the competition for sulfide and sulfur between *Chlorobium* and *Chromatium* (see Section 4.2.3), pH had a marked effect on the outcome of the experiments. At a dilution rate of 0.05 hr^{-1} and a pH of 6.8 (optimal for *Chlorobium*) the *Chlorobium* dominated by 91% (Table VI), whereas at a pH of 7.5 (optimal for *Chromatium*) the two organisms coexisted in equal numbers (van Gemerden and Beeftink, 1981).

In all of these examples, the optimum growth pH of the organisms was a complicating factor, and both cases serve to emphasize that bacterial relationships are a product of many factors, rather than just an obvious few.

4.3.2. Oxygen

Another selective factor can be oxygen tension. During competition studies with the denitrifying obligate chemolithotrophs *T. denitrificans* and *Tms. denitrificans,* Timmer-ten Hoor (1977) found that, although their maximum specific growth rates in pure culture were similar, in anaerobic competition experiments *Tms. denitrificans* dominated. If oxygen was not rigorously excluded from the chemostat, however, *T. denitrificans* gained the advantage. The *Tms. denitrificans* has constitutive denitrifying enzymes, whereas *T. denitrificans* will only induce production of such enzymes under anaerobiosis in the presence of nitrate. This probably explains the ability of *Tms. denitrificans* to establish itself in the anaerobic cultures. However, *Tms. denitrificans* is only capable of microaerophilic growth and it would appear that it cannot compete successfully with aerobic sulfide-oxidizing bacteria as long as a substantial turnover of oxygen is possible.

Again, this last example emphasizes the need for careful control of the experimental conditions during competition experiments. Any extra introduced variable can influence the outcome of the experiment and make interpretation of the results difficult.

4.3.3. Oxygen and the Vertical Distribution of Phototrophs

Very often in natural ecosystems the Chlorobiaceae are found below a layer of Chromatiaceae (Caldwell, 1977; Caldwell and Tiedje, 1975; Caumette, 1982; Kuznetsov and Gorlenko, 1973). The affinity of *Chl. limicola* for sulfide is higher than that of *Chr. vinosum* (van Gemerden and Beeftink, 1981; van Gemerden, 1974) and there are indications that the same is true for other representatives from the two genera. Since the concentration of sulfide declines as it approaches the surface of natural waters, the vertical distribution of the two genera might be expected to be the other way around. However, it has been reported that the purple sulfur bacteria are able to grow chemolithotrophically (Kämpf and Pfennig, 1980). The sulfide concentrations found in nature exibit diurnal fluctuations, and the lower sulfide concentrations are often found in contact with low oxygen levels (Jørgensen *et al.,* 1979). The Chlorobiaceae are obligate anaerobes (Kämpf and Pfennig, 1980), but the Chromatiaceae are not. It therefore seems likely that the positions of the species are governed by the oxygen profile rather than the sulfide profile. It is probable that the purple sulfur bacteria prevent the penetration of oxygen to the lower layers, enabling their potential competitors to develop beneath them. Conceivably, the Chlorobiaceae can grow at greater depth because of the lower light requirement generally observed in these organisms.

4.4. Competition between Colorless and Phototrophic Sulfur Bacteria

Habitats where the total turnover of inorganic and organic compounds is heavily dominated by the sulfur cycle are called sulfureta. They are often characterized by heavy blooms of phototrophic bacteria, and, particularly when the bloom occurs on the surface of a sediment, blooms of colorless sulfur bacteria, especially *Beggiatoa,* have been found. These blooms do not usually consist of single species, but rather of layers of species with different niches that together make up the sulfuretum. Such a community has recently been investigated by Jørgensen (1982), who used microelectrodes to measure the changes in the sulfide/dissolved oxygen interface throughout a diurnal cycle. These changes, together with their effect on the community of the sulfuretum, a mixture of dominating populations of *Beggiatoa, Chromatium,* and *Oscillatoria,* are shown in Fig. 14. It can be seen that the position of the *Chromatium* is governed by the sulfide concentration, and this in turn is affected by the oxygen produced by the cyanobacterium during oxygenic photosynthesis. The only time at which *Chromatium* was found at the surface of the sediment was during the night, when the oxygen in the sediment became depleted and the sulfide boundary moved upward. Just after sunrise, dissolved

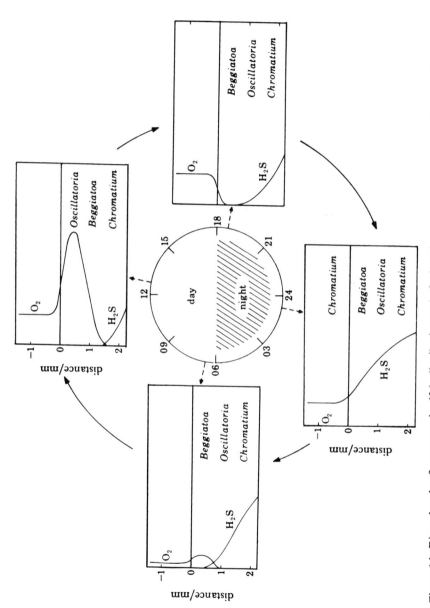

Figure 14. Diurnal cycle of oxygen and sulfide distribution and of microbial zonation in a marine sulfuretum. [Reprinted from Jørgensen (1982).]

oxygen began building up and the *Chromatium* followed the sulfide boundary down. Similarly, the position of the *Beggiatoa* within the layer was determined by the position of the sulfide/oxygen interface, although the fact that these organisms are only motile by a gliding action means that they are restricted to the solid phase and that during the night they were unable to maintain their customary position at the sulfide/oxygen boundary, since this then occurred above the surface of the sediment.

These observations, in which the movements of the different populations are so clearly defined as to change the surface color of the sediment, demonstrate how well the physiological reactions and chemotactic behavior of these bacteria are adjusted to the fluctuating nature of their habitat.

The movement of the *Chromatium* into the water above the sediment might, at first glance, seem unexpected. However it should be realized that although no measurable oxygen was present in the water, oxygen would rapidly diffuse in from the air and chemolithotrophic growth would be possible at the air/water interface. It is therefore interesting to remember that a number of the Chromatiaceae are able to grow, be it very slowly, chemolithoautotrophically on sulfide and oxygen (Kämpf and Pfennig, 1980). Clearly, the very dense population of *Chromatium* near the interface would present serious competition for smaller populations of colorless sulfur bacteria such as the thiobacilli. It may be assumed that in the combined presence of sulfide and oxygen the affinity of the thiobacilli for both substrates is better than that of the Chromatiaceae, but the population effect would annihilate this advantage entirely. However, it is clear that more experimental work must be carried out before further speculation is justified. As a first step, competition experiments between phototrophic and colorless sulfur bacteria should be carried out.

When generalizing about the possible competition of blooms of Chromatiaceae with smaller populations of colorless sulfur bacteria, it should be realized that, at any sulfide/oxygen interface to which light can penetrate, competition between the phototrophic and colorless sulfur bacteria must be expected.

Nevertheless, in most cases a distinct peak of dark CO_2 fixation, indicative of the activity of colorless sulfur bacteria, is observed at the interface just above the blooms of phototrophic bacteria. (Dark carbon dioxide fixation is measured by incubating samples from a given depth in dark bottles *in situ*.) In those cases where such a peak is not found, it might be speculated that this is due to the fact that during the night the phototrophs would rise and compete with the colorless sulfur bacteria, thus keeping the nonphototrophic population at a low level. Another possibility is that it is partially due to the inadequacy of the experimental methods, which may be illustrated by the examples below.

When investigating the oxygen/sulfide interface of the Solar Lake, Jørgensen *et al.* (1979) found that, in order to analyze the diurnal cycle of oxygen and sulfide concentrations, the resolving power of their sampling had to be reduced to 2½-cm intervals. This was accomplished by employing a specially designed pump. Using this method, it was possible to demonstrate a complex diurnal rhythm of dark carbon dioxide fixation, which, with more usual sampling methods, might have appeared only as an indeterminate band.

An even more pronounced case was observed during an investigation of a small stratified lake in Norway (Y. Børsheim, J. C. Gottschal, I. Dundas, and J. G. Kuenen, unpublished observations). A bloom of phototrophic bacteria was observed below the oxygen/sulfide interface, and, with the usual water samplers (which take a 40-cm column), no dark carbon dioxide fixation was found. However, when the more accurate sampling technique was used, a small peak extending over only 2½–5 cm was detected distinctly above the layer of phototrophic organisms (Fig. 15). Had this activity been present in the larger sample, it would have disappeared in the background. The presence of the discrete carbon dioxide fixation peak indicated that the dark carbon dioxide fixation was not due to dark activity of the phototrophs, but was indeed due to the presence of colorless sulfur bacteria. The finding of these discrete layers indicates

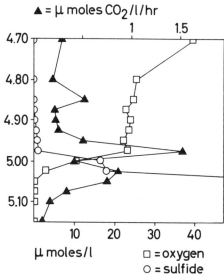

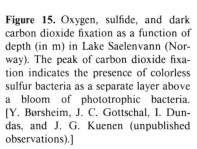

Figure 15. Oxygen, sulfide, and dark carbon dioxide fixation as a function of depth (in m) in Lake Saelenvann (Norway). The peak of carbon dioxide fixation indicates the presence of colorless sulfur bacteria as a separate layer above a bloom of phototrophic bacteria. [Y. Børsheim, J. C. Gottschal, I. Dundas, and J. G. Kuenen (unpublished observations).]

that, at least during the day, competition between the two types of sulfur bacteria did not occur, although the situation may have changed during the night.

A general conclusion from this work is that the outcome of competition between phototrophic and nonphototrophic bacteria for sulfide is very dependent on the population density of the phototroph and thus on the available light at the interface. As the available light decreases, the importance of the role of the phototrophs will also decrease. In environments where high sulfide and oxygen fluxes are available in the dark, dominant populations of the colorless sulfur bacteria will be found. For example, a bloom of *Beggiatoa* has been found on the surface of a sediment 15 m down in brackish water at Limfjorden in Denmark. In turbid water, 15 m is well below the level to which photosynthetically useful light can penetrate. A second example can be found in the ecology of the hydrothermal vents. At the enormous depths at which these vents occur (approximately 2500 m), light never penetrates and large numbers of free-living colorless sulfur bacteria that utilize the sulfides coming out of the vent have been found (Jannasch and Wirsen, 1981).

A final example of a situation where the colorless sulfur bacteria have a distinct, if unexpected, selective advantage is in the extremely hot springs and pools that are the terrestial equivalent of the marine vents. Brock (1978) has reported that different forms of life have distinct upper temperature limits. For example, fish are not found above 38°C, protozoa above 56°C, and photosyntheic bacteria above 70–73°C. However, some colorless sulfur bacteria have been found growing in water at 95°C. Indeed, it has been observed that at very high temperatures the niches that would normally be occupied by the phototrophs are occupied by at least one type of colorless sulfur bacterium, *Sulfolobus* (Brock, 1978).

5. Other Interactions Involving the Sulfide-Oxidizing Bacteria

5.1. Examples among the Chemolithotrophs

Relatively little is known of the interactions among the sulfur bacteria other than competition. The examples described in this section are merely case histories, which illustrate the nearly endless possible interactions occurring in nature. We discuss three cases for the chemolithotrophs, namely an interaction between an obligate chemolithotroph and a heterotroph that is mutually beneficial, the concerted action of two chemolithotrophs on one common substrate, and growth of a satellite population of one organism on excretion products of another, leading to decreased competitiveness of the latter.

Interactions of the symbiotic colorless sulfur bacteria that live in higher animals, for example, in the hydrothermal vents, will not be discussed and the reader is referred to the relevant literature (Jannasch and Wirsen, 1981; Felbeck *et al.*, 1981; Cavanaugh *et al.*, 1982; Cavanaugh, 1983).

We discuss two examples involving the phototrophic sulfur bacteria. The first illustrates the role of phototrophic bacteria as prey organisms in laminated (layered) ecosystems. The second example focuses on interactions between nonphototrophic, sulfide-producing, heterotrophic bacteria and sulfide-oxidizing phototrophs, and shows that a lower affinity for a substrate is not necessarily a disadvantage.

5.1.1. "Heterotrophic" T. ferrooxidans

There have been several reports that different *T. ferrooxidans* cultures, normally considered to be obligately chemolithotrophic, are able to metabolize glucose and various other organic substrates, frequently after an adaptation period (Shafia and Wilkinson, 1969; Lûndgren *et al.*, 1964). However, it has now been shown (Mackintosh, 1978; Harrison *et al.*, 1980) that many of the cultures of *T. ferrooxidans* available in culture collections are contaminated with acidophilic heterotrophs, which grow either on organic excretion products supplied by the autotroph or on traces of organic compounds in the media and possibly on the culture vessel. *Thiobacillus acidophilus,* a facultatively chemolithotrophic, acidophilic species able to use reduced sulfur compounds but not ferrous iron, was isolated from a supposedly pure culture of *T. ferrooxidans,* as was *Acidiphilium cryptum,* a heterotroph. The close associations between the different types of bacteria have been shown to be beneficial to both organisms. A recent experiment (A. Harrison Jr., personal communication) has shown that if a lawn of *T. ferrooxidans* is spread onto agar-containing ferrous sulfate and mineral salts, large, rust-colored colonies occur only where spot inoculations of various acidophilic heterotrophs (e.g., *Acidiphilium cryptum* and *T. acidophilus*) had been made. These colonies proved to be a mixture of *T. ferrooxidans* and the heterotroph. This may be a truly mutualistic relationship, since the *T. ferrooxidans* benefits by the removal of inhibitory organic compounds, and the heterotroph is probably supplied by *T. ferrooxidans* with a substrate, although the trace organic materials in the medium may be sufficient to support growth.

It has been reported that the natural bacterial communities active in industrial processes such as copper leaching are also mixtures of a dominant population of *T. ferrooxidans* with satellite populations of heterotrophs. These mixtures are more effective in the leaching process than pure cultures of *T. ferrooxidans.* Similarly, mixed cultures of *T. ferroox-*

idans and acidophilic hetrotrophs have been found to be more effective in the desulfurization (depyritization) of coal (Kelly, 1978). A variety of explanations have been put forward to explain these phenomena, such as the removal of toxic organic compounds or fixation of molecular nitrogen by the heterotrophs, but these remain to be substantiated.

5.1.2. *Leptospirillum ferrooxidans and Acidophilic Thiobacilli*

It has been found that although neither organism can individually degrade pyrite or chalcopyrite, a mixed culture of *Leptospirillum ferrooxidans* and *T. organoparus* can grow well on either mineral. *Leptospirillum ferrooxidans* was originally isolated from copper deposits in Armenia (Balashova *et al.*, 1974). It is able to use ferrous iron as an energy source, but cannot use sulfur or reduced sulfur compounds. *Thiobacillus organoparus* is a facultatively chemolithotrophic bacterium that can utilize reduced sulfur compounds, but not ferrous iron (Markosyan, 1973). It has since been shown that if *Leptospirillum ferrooxidans* is mixed with other sulfide-oxidizing bacteria, such as *T. thiooxidans* or *T. acidophilus,* the rapid degradation of pyrite occurs (Kelly, 1978; Norris and Kelly, 1978). Since *T. thiooxidans* is itself an obligate chemolithotroph, the stimulation is unlikely, in this case, to be due to the removal by the secondary population of organic excretory products. It has been suggested that the role of *T. thiooxidans* might be the removal of a coating of elemental sulfur from the pyrite crystals.

5.1.3. *Thiobacillus versutus/Thiobacillus neapolitanus*

An apparently commensal relationship was reported by Gottschal *et al.* (1979) and was briefly mentioned in Section 4.1. It was found that when *T. neapolitanus,* an obligate chemolithotroph, and *T. versutus,* a facultative chemolithotroph, were grown together in a chemostat with thiosulfate as the growth-limiting substrate, *T. neapolitanus* became the dominant species and *T. versutus* did not, as expected, wash out (Fig. 4). It has been found that pure cultures of *T. neapolitanus* excreted glycollate at relatively high concentrations of oxygen. At low concentrations (10–20% of air saturation), little glycollate was excreted. Glycollate cannot be metabolized by *T. neapolitanus* and it was assumed that in mixed culture the *T. versutus* was growing on the glycollate. In concord with this hypothesis, a sample of the mixed culture of *T. neapolitanus* and *T. versutus* could respire glycollate at a rate much higher than that of *T. versutus* when grown alone on thiosulfate. Furthermore, the number of *T. versutus* cells in the mixed culture dropped from 10% to 2% when the dissolved oxygen concentration was reduced from 80% to 10% of air saturation (J. C. Gottschal and J. G. Kuenen, unpublished results). This apparently commensal relationship changed, however, when the dilution rate was severely reduced (i.e., to about 1% of the maximum specific growth rate of *T. neapolitanus*). The viability of the obligate chemolithotroph fell, and *T. versutus* became the dominant organism. It might be speculated that *T. neapolitanus* was still, as at higher dilution rates, producing glycollate. If this was the case, the ability of *T. versutus* to grow mixotrophically on the glycollate and thiosulfate might be regarded as amensalism. This relationship is a clear example of the difficulty of defining any interbacterial relationship by a single term.

5.2. Examples among the Phototrophs

5.2.1. The Trophic Role of Phototrophic Bacteria in Laminated, or Stratified, Ecosystems

As discussed in the preceding sections, phototrophic bacteria, especially the purple and green types, often form blooms. The organisms in these blooms grow at the expense of nutrients in the anaerobic zone, which of course are inaccessible to aerobic species. In this zone the productivity of the phototrophic bacteria is, on average, 100–200 mg carbon per m^3 per day. This accounts for about one-third of the total primary production for the whole water column [see Biebl and Pfennig (1979) for a compilation]. Just above such blooms, at depths where oxygen is still present, high populations of predators, including protozoa and rotifers, are frequently observed. That they are feeding on the phototrophic bacteria can be demonstrated by the presence of photopigments in the digestive systems of these planktonic predators. Although these predators might be expected to feed preferentially on the phototrophic microorganisms readily available in their own aerobic layer, this may not always be the case. Laboratory experiments reported by Gophen *et al.* (1974) showed that a common predator in Lake Kinneret preferred the brown *Chl. phaeobacteroides* (grown on [^{3}H]-acetate) to algae (grown on [^{14}C]-carbon dioxide). This would imply that the products of anaerobic photosynthesis are mineralized under aerobic conditions. These data illustrate the importance of phototrophic bacteria as a link between the nutrient supplies in the anaerobic areas of aqueous ecosystems and production in the aerobic part.

5.2.2. Interactions between Sulfide-Producing Heterotrophic Bacteria and Sulfide-Oxidizing Phototrophs

In a mixed culture involving a green sulfur bacterium *(Chl. limicola)* and *D. desulfuricans,* competition for extracellular sulfur produced by *Chl. limicola* can arise. The *Chl. limicola* is able to oxidize sulfur to sulfate, and *D. desulfuricans* in turn reduces the sulfate to sulfide (Biebl and Pfennig, 1978). However, if *Chl. limicola* can only obtain carbon dioxide, the outcome of competition for sulfur is irrelevant, since both organisms use the product made by the other (i.e., sulfide or sulfate). The *Chl. limicola* gains eight electrons from the oxidation of one molecule of sulfide to sulfate, but only two if the oxidation proceeds only to sulfur, with *D. desulfuricans* then reducing all of the sulfur produced. However, this would mean that *D. desulfuricans* produces four times as much sulfide. The factor governing the final growth yield of the mixture is the initial concentration of the organic electron donor of *D. desulfuricans*. If *Chl. limicola* is provided with a carbon source other than carbon dioxide—for example, acetate, which cannot be used by *D. desulfuricans*—the position changes and it becomes advantageous for the phototroph to lose the competition for sulfur because it requires the presence of sulfide in order to assimilate acetate (Sadler and Stanier, 1960). Increased sulfide production by *D. desulfuricans* would thus allow *Chl. limicola* to increase its acetate uptake. This relationship may be regarded as commensal, since *D. desulfuricans* remains unaffected by the presence of *Chl. limicola*. If the acetate is replaced by ethanol, which cannot be assimilated by *Chl. limicola,* the phototroph becomes doubly dependent on *D. desulfuricans,* requiring sulfide and also benefitting from the acetate produced by the sulfate-reducer from the ethanol (Fig. 16A).

When *Desulfuromonas acetooxidans* is the sulfide-producing partner instead of *Desulfovibrio,* it becomes vitally important that the phototroph loses the competition for extracellular sulfur, because *Ds. acetooxidans* cannot reduce sulfate, but requires sulfur for sulfide production (Biebl and Pfennig, 1978). Any sulfide oxidized to sulfate cannot be reduced again, and this ultimately results in the end of sulfide production by *Ds. acetooxidans* and acetate uptake by *Chl. limicola* (Fig. 16B). Since the two species grow well together on very small amounts of sulfur, it is obvious that this does not occur and extensive recycling of the sulfur must take place. Apparently the *Desulfuromonas* is able to outcompete *Chl. limicola* for any extracellular sulfur produced by the phototroph, which is to the benefit of both organisms.

Another example of syntrophic growth is that of *"Chloropseudomonas ethylica."* This organism, thought to be a motile, green bacterium, was investigated by Gray *et al.* (1973) and found to be a mixture of two bacteria. One proved to be the sulfide-oxidizing phototroph *Chlorobium limicola,* and the other was a sulfate-reducing bacterium that they did not fully characterize, but was probably a *Desulfovibrio* sp. The mutualistic relationship existed in a cycling of sulfide/sulfate to provide the *Chlorobium* with an electron donor and the *Desulfovibrio* with an electron acceptor.

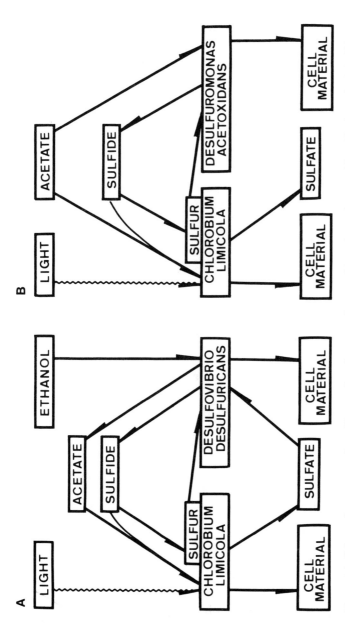

Figure 16. Schematic representation of interactions between phototrophic and nonphototrophic sulfur bacteria. (A) *Chlorobium limicola* and *Desulfovibrio desulfuricans* in media supplemented with ethanol. (B) *Chlorobium limicola* and *Desulfuromonas acetooxidans* in media supplemented with acetate. [Based on data from Biebl and Pfennig (1978), and modified from van Gemerden (1983).]

6. Enrichment Studies

Conditions in batch culture where the substrate is present in excess for the greatest part of the growth phase favor the isolation of specialist species able to grow rapidly on the substrate supplied (see Section 4). Versatile species that have a lower substrate affinity and a lower specific growth rate will be outcompeted. However, when more than one substrate is supplied at growth-limiting concentrations, as in a chemostat, the increased affinities of versatile species for these multiple substrates allow them to outcompete specialists dependent on only one of the substrates. The principles of competition as demonstrated by these examples can also be applied to continuous flow enrichments for organisms with mixotrophic abilities, the techniques being the same as used previously by Veldkamp and others (see, for example, Veldkamp and Jannasch, 1972) for the chemostat enrichment of organisms competing for a single growth-limiting substrate.

6.1. Aerobic Enrichments of Colorless Sulfur Bacteria

Table VII shows the dominant populations obtained by continuous enrichment under different conditions. In each case the outcome could have been predicted from the principles established in the competition studies described above. Versatile species were obtained when two substrates were in equivalent proportions, and a chemolithoheterotroph when only a relatively small amount of thiosulfate was used. Although it seems that these enrichments can be successfully and reproducibly applied to freshwater samples, enrichments for mixotrophs from marine environments using thiosulfate–acetate media were unsuccessful in our hands. In all cases, mixtures of obligate chemolithotrophs and heterotrophs were obtained. As expected, when using thiosulfate as the only growth-limiting substrate, specialists were enriched for under both aerobic and anaerobic conditions (Kuenen, 1972; Timmer-ten Hoor, 1977), but the unpredictability of continuous enrichments inoculated from marine sources is further demonstrated by the enrichment of a facultatively chemolithotrophic marine strain of *T. intermedius* in thiosulfate-limited medium (D. W. Smith and Finazzo, 1981).

6.2. Anaerobic Enrichments of Sulfide-Oxidizing Bacteria

Various types of colorless sulfur bacteria can be enriched for under aerobic conditions. The successful enrichment of a denitrifying obligate chemolithotroph, *Thiomicrospira denitrificans,* in a nitrate-limited, thiosulfate-fed chemostat was reported by Timmer-ten Hoor (1977).

Table VII. The Results of Aerobic and Anaerobic Continuous Flow Enrichment Cultures with Different Substrate Supplies

Inoculum source	Culture conditions	Substrate (mM)		Dominant population	Predominant secondary population	Reference
		S^{2-} or $S_2O_3^{2-}$	Acetate			
Freshwater ditch	Aerobic, constant feed, mixed substrate	20	5	Mixotroph	Mixotroph	Gottschal and Kuenen (1980b)
		20	10	Mixotroph	Heterotroph	Gottschal and Kuenen (1980b)
		10	15	Chemolithoheterotroph	Heterotroph	Gottschal and Kuenen (1980b)
	Aerobic, alternate feeds (4 hr each), single substrates	40	10	Mixotroph	Chemolithoheterotroph	Gottschal et al. (1981)
Effluent treatment system	Anaerobic, denitrifying, constant feed, single substrate	10	10	Mixotroph	Heterotroph	Robertson and Kuenen (1983)

   Batch enrichments for thiosulfate-utilizing denitrifiers almost invariably result in the enrichment of the obligately chemolithotrophic *Thiobacillus denitrificans* (Taylor *et al.*, 1971). If the principles of Gottschal *et al.* (see Section 4) are also applied to anaerobic systems, however, the isolation of facultative organisms should be possible by using mixtures of thiosulfate and organic substrates. Indeed, by using anaerobic chemostat enrichments fed with a mixture of nitrate and growth-limiting concentrations of thiosulfate and acetate, and using an inoculum from a sulfide-oxidizing, denitrifying waste water treatment system, we were able to isolate such a species (Table VII). This has now been described (Robertson and Kuenen, 1983). The new isolate, *Thiosphaera pantotropha,* has since been reisolated from samples directly streaked from the column. It has proved unusually versatile and can denitrify or grow aerobically on a wide range of substrates, including sulfide, thiosulfate, hydrogen, and many organic compounds. Preliminary results indicate that other isolates from the same source are equally versatile.

   A system more complex than a chemostat running on synthetic medium is currently under study at the microbiology laboratory in Delft. It consists of an anaerobic fluidized bed column fed from a methanogenic reactor and a nitrifying column (patent number EP 005188A1). It receives roughly equivalent concentrations of small organic molecules (e.g., acetate, butyrate) and reduced inorganic sulfur compounds (mostly sulfide). For this reason it might be regarded as a system similar to the chemostats designed for the selective culture of versatile species. However, the similarity is not complete, since the bulk of the biomass in the column is attached to sand particles. The problem is further complicated by the fact that the column receives from the preceding methane reactor a "protein soup" formed from lysing methanogens. However, no specialist autotrophs have been isolated from the system, and the versatile sulfide-oxidizers make up a significant proportion of the total community (Table VIII). This system was the source of the inoculum for the enrichment experiments.

   These results indicate that the concept of the ecological niche for versatile sulfide-oxidizing bacteria as developed from simple laboratory studies can successfully be applied to more complex ecosystems.

7. Conclusion

   It can be seen from the examples discussed in this chapter that many of the sulfur-oxidizing bacteria are ideal for the study of microbial interactions, offering a spectrum of physiological types for testing the role and importance of a variety of selective pressures, such as light quantity and

Table VIII. The Proportion of Each Metabolic Type Found among the Isolates from the Desulfurizing Denitrification Column

Metabolic type	Percentage of the total community
Obligate autotroph	0
Facultative autotroph	29
Chemolithoheterotroph	1
Heterotroph	65
Miscellaneous isolates (each constituting less than 1% of the population)	5

quality, pH, single or mixed substrate limitation, and the influence of other organisms. These studies have yielded information both about the physiology of individual species and the way in which the differences between these species serve to fit them for the different niches available in the community involved in the maintenance of the sulfur cycle.

For the occurrence and distribution of the many different physiological types among the colorless sulfur bacteria, it has been postulated that the relative turnover rates of organic compounds determine the success of specialized versus versatile species in nature. Strong fluctuations in the supply of inorganic sulfur compounds and organic compounds favor more specialized species, since they are more reactive than versatile organisms. Measurements in the field to check this hypothesis are lacking, but observations made on the complex community of a waste water purification system strongly support it.

In competition among the phototrophic bacteria, analogous principles of selection can be observed. Again, reactivity, that is, the ability to respond instantaneously to sudden increases in the nutrient supply, is of importance. Thus, the Rhodospirillaceae, being metabolically versatile, are less reactive than the Chromatiaceae and Chlorobiaceae, which contain constitutive levels of sulfide-oxidizing enzymes. The niche of the Rhodospirillaceae, therefore, may lie in environments where trace amounts of sulfide are continuously available in addition to organic compounds. On the other hand, the Chromatiaceae often form blooms, whereas the Rhodospirillaceae are ubiquitous but never occur at high densities in nature.

Although few other systems have been studied in such detail, it would appear from a survey of the literature that the conclusions derived from the work described here can be more generally applied [for a review see Kuenen and Gottschal (1982)]. The techniques of selective enrichment on mixed substrates have already been successfully applied in the study of clostridia isolated from waste water (Laanbroek et al., 1979) and the methylotrophs (W. Harder, unpublished results, quoted in Kuenen and Gottschal, 1982). Enrichments in continuous flow systems for phototrophic bacteria in the presence of mixed substrates have not yet been attempted, but should prove rewarding.

There is a striking resemblance between the sulfur and nitrogen cycles, and also between the different types of organisms involved in the cycles. This is especially true for the chemolithotrophs, and it is interesting to speculate whether the principles controlling sulfide oxidation might also apply to the oxidation of ammonia. Although heterotrophic nitrification does not appear to supply the cells with energy, as does heterotrophic sulfide oxidation, it is a little-understood phenomenon (Witzel and Overbeek, 1979). By use of the techniques and approach described in this chapter, it might be possible to determine the reasons for heterotrophic nitrification and the impact of heterotrophic nitrifiers on the ecology of the chemolithotrophic ammonia-oxidizing bacteria.

As well as being of academic interest, the studies discussed here would be of some importance in fields such as waste water purification. In this area, it is the very competition of microbes that produces the characteristics of the treatment plant. As modern treatment plants become more sophisticated and complex, a clearer picture of the effect that variations in the mixture of sulfides, ammonia, organic compounds, and other materials have on the composition of the microbial community will be necessary in order to understand the performance, (over)capacity, and resilience of the plant with respect to sulfide removal or nitrification. Hopefully, a better knowledge of the interactions among the different metabolic types will make their behavior more predictable than at present.

References

Balashova, V. V., Vedenina, I. Ya., Markosyan, G. E., and Zavarzin, G. A., 1974, The auxotrophic growth of *Leptospirillum ferrooxidans, Microbiology* **43**:491–494.

Beudeker, R. F., de Boer, W., and Kuenen, J. G., 1981, Heterolactic fermentation of intracellular polyglucose by the obligate chemolithotroph *Thiobacillus neapolitanus* under anaerobic conditions, *FEMS Microbiol. Lett.* **12**:337–342.

Beudeker, R. F., Gottschal, J. C., and Kuenen, J. G., 1982, Reactivity versus flexibility in *Thiobacilli, Antonie Leeuwenhoek J. Microbiol. Serol.* **48**:39–51.

Biebl, H., and Pfennig, N. P., 1978, Growth yields of green sulfur bacteria in mixed culture with sulfur and sulfate reducing bacteria, *Arch. Microbiol.* **117**:9–16.

Biebl, H., and Pfennig, N. P., 1979, Anaerobic CO_2 uptake by phototrophic bacteria. A review, *Arch. Hydrobiol. Beitr.* **12**:48–58.

Bos, P., and Kuenen, J. G., 1983, Microbiology of sulphur oxidizing bacteria, in: *Microbial Corrosion*, pp. 18–27, The Metals Society, London.

Broch-Due, M., Ormerod, J. G., and Fjerdingen, B. S., 1978, Effect of light intensity on the vesicle formation in *Chlorobium, Arch. Microbiol.* **116:**269–274.

Brock, T. D., 1966, *Principles of Microbial Ecology,* Prentice-Hall, Englewood Cliffs, New Jersey.

Brock, T. D., 1978, *Thermophilic Microorganisms and Life at High Temperatures,* Springer-Verlag, Berlin.

Bull, A. T., and Slater, J. H., 1982, Microbial interactions and community structure, in: *Microbial Interactions and Communities,* (A. T. Bull and J. H. Slater, eds.), pp. 13–44, Academic Press, London.

Caldwell, D. E., 1977, The planktonic microflora of lakes, *CRC Crit. Rev. Microbiol.* **5:**305–370.

Caldwell, D. E., and Tiedje, J. M., 1975, The structure of anaerobic bacterial communities in the hypolimnia of several Michigan lakes, *Can. J. Microbiol.* **21:**377–385.

Castenholz, R. W., 1984, Habitats of *Chloroflexus* and related organisms, in: *Current Perspectives in Microbial Ecology* (M. J. Klug and C. A. Reddy, eds.), pp. 196–200, American Society for Microbiology, Washington, D. C.

Caumette, P., 1982, Contribution of phototrophic bacteria to the food chain in a stratified tropical lagoon, in: *4th International Symposium on Photosynthetic Prokaryotes* (R. Y. Stanier, ed.), Abstract A9, Institut Pasteur, Lyon.

Cavanaugh, C. M., 1983, Symbiotic chemoautotrophic bacteria in marine invertebrates from sulphide rich habitats, *Nature* **302:**58–61.

Cavanaugh, C. M., Gardiner, S. L., Jones, L. M., Jannasch, H. W., and Waterbury, J. B., 1982, Prokaryotic cells in the hydrothermal vent tube worm *Riftia pachyptila* (Jones): Possible chemoautotrophic symbionts, *Science* **213:**340–342.

Cohen, Y., Padan, E., and Shilo, M., 1975, Facultative bacteria-like photosynthesis in the blue-green alga *Oscillatoria limnetica, J. Bacteriol.* **123:**855–861.

De Freitas, M. J., and Frederickson, A. G., 1978, Inhibition as a factor in the maintenance of diversity of microbial ecosystems, *J. Gen. Microbiol.* **106:**307–320.

Felbeck, H., Childress, J. J., and Somero, G. N., 1981, Calvin–Benson cycle and sulphide oxidation enzymes in animals from sulphide rich habitats, *Nature* **293:**291–293.

Frederickson, A. G., 1977, Behavior of mixed cultures of microorganisms, *Annu. Rev. Microbiol.* **31:**63–87.

Friedrich, C. G., and Mitringa, G., 1981, Oxidation of thiosulphate by *Paracoccus denitrificans* and other hydrogen bacteria, *FEMS Microbiol. Lett.* **10:**209–212.

Gophen, M., Cavari, B. Z., and Bernan, T., 1974, Zooplankton feeding on differentially labelled algae and bacteria, *Nature* **247:**393–394.

Gottschal, J. C., and Kuenen, J. G., 1980a, Mixotrophic growth of *Thiobacillus* A2 on acetate and thiosulphate as growth limiting substrates in the chemostat, *Arch. Microbiol.* **126:**33–42.

Gottschal, J. C., and Kuenen, J. G., 1980b, Selective enrichment of facultatively chemolithotrophic thiobacilli and related organisms in the chemostat, *FEMS Microbiol. Lett.* **7:**241–247.

Gottschal, J. C., and Kuenen, J. G., 1981, Physiological and ecological significance of facultative chemolithotrophy and mixotrophy in chemolithotrophic bacteria, in: *Microbial Growth on Cl-Compounds* (H. Dalton, ed.), pp. 92–104, Heyden, London.

Gottschal, J. C., and Thingstad, T. F., 1982, Mathematical description of competition between two and three bacterial species under dual substrate limitation in the chemostat: A comparison with experimental data, *Biotechnol. Bioeng.* **24:**1403–1418.

Gottschal, J. C., de Vries, S., and Kuenen, J. G., 1979, Competition between the facultatively chemolithotrophic *Thiobacillus* A2, an obligately chemolithotrophic *Thiobacil-*

lus and a heterotrophic spirillum for inorganic and organic substrates, *Arch. Microbiol.* **121**:241–249.

Gottschal, J. C., Nanninga, H., and Kuenen, J. G., 1981, Growth of *Thibacillus* A2 under alternating growth conditions in the chemostat, *J. Gen. Microbiol.* **126**:23–28.

Gray, B. H., Fowler, C. F., Nugent, N. A., Rigopoulos, N., and Fuller, R. C., 1973, Reevaluation of *Chloropseudomonas ethylica* strain 2K, *Int. J. Syst. Bacteriol.* **23**:256–264.

Güde, H., Strohl, R., and Larkin, J. M., 1981, Mixotrophic and heterotrophic growth of *Beggiatoa alba* in continuous culture, *Arch. Microbiol.* **129**:357–361.

Hansen, T. A., 1974, Sulfide als electrondonor voor *Rhodospirillaceae*, Ph. D. thesis, University of Groningen.

Hansen, T. A., and van Gemerden, H., 1972, Sulfide utilization by purple nonsulfur bacteria, *Arch. Microbiol.* **86**:49–56.

Harder, W., Kuenen, J. G., and Matin, A., 1977, Microbial selection in continuous culture, *J. Appl. Bacteriol.* **43**:1–24.

Harrison, A. P., Jr., 1983, Genomic and physiological comparisons between heterotrophic *Thiobacilli* and *Acidiphilium cryptum*, *Thiobacillus versutus* sp. nov. and *Thiobacillus acidophilus* nom. rev., *Int. J. Syst. Bacteriol.* **33**:211–217.

Harrison, A. P. Jr., Jarvis, B. W., and Johnson, J. L., 1980, Heterotrophic bacteria from continuous cultures of autotrophic *Thiobacillus ferrooxidans:* Relationships as studied by means of deoxyribonuceic acid homology, *J. Bacteriol.* **143**:448–454.

Healey, F. P., 1980, Slopes of the Monod equations as an indicator of advantage in nutrient competition, *Microb. Ecol.* **5**:281–286.

Hurlbert, R. E., and Lascelles, J., 1963, Ribulose diphosphate carboxylase in *Thiorhodaceae*, *J. Gen. Microbiol.* **33**:445–458.

Ivanovsky, R. N., Sintsov, N. V., and Kondratieva, E. N., 1980, ATP linked citrate lyase activity in the green sulfur bacterium *Chlorobium limicola* forma *thiosulfatophilum*, *Arch. Microbiol.* **128**:239–241.

Jannasch, H. W., and Wirsen, C. O., 1981, Morphological survey of microbial mats near deep sea thermal vents, *Appl. Environ. Microbiol.* **41**:528–538.

Jørgensen, B. B., 1982, Ecology of the bacteria of the sulphur cycle with special reference to anoxic–oxic interface environments, *Philos. Trans. R. Soc. Lond. B* **298**:543–561.

Jørgensen, B. B., Kuenen, J. G., and Cohen, Y., 1979, Microbial transformations of sulfur compounds in a stratified lake (Solar Lake, Sinai), *Limnol. Oceanogr.* **24**:799–822.

Kämpf, C., and Pfennig, N., 1980, Capacity of *Chromatiaceae* for chemotrophic growth. Specific respiration rates of *Thiocystis violaceae* and *Chromatium vinosum*, *Arch. Microbiol.* **127**:125–135.

Kelly, D. P., 1978, Microbial ecology, in: *The Oil Industry and Microbial Ecosystems* (K. W. A. Chater, H. J. Somerville, and H. J. Heyden, eds.), pp. 12–27, Institute of Petroleum, London.

Kelly, D. P., 1982, Biochemistry of the chemolithotrophic oxidation of inorganic sulphur, *Philos. Trans. R. Soc. Lond. B* **298**:499–528.

Kuenen, J. G., 1972, Een studie van kleurloze zwavelbacteriën uit het Groninger wad, Ph. D. thesis, University of Groningen.

Kuenen, J. G., 1975, Colourless sulfur bacteria and their role in the sulphur cycle, *Plant Soil* **43**:49–76.

Kuenen, J. G., and Beudeker, R. F., 1982, Microbiology of *Thiobacilli* and other sulphur-oxidizing autotrophs, mixotrophs and heterotrophs, *Philos. Trans. R. Soc. Lond. B* **298**:473–497.

Kuenen, J. G., and Gottschal, J. C., 1982, Competition among chemolithotrophs and methylotrophs and their interactions with heterotrophic bacteria, in: *Microbial Interactions and Communities* (A. T. Bull and J. H. Slater, eds.), pp. 153–188, Academic Press, London.

Kuenen, J. G., and Robertson, L. A., 1984a, Interactions between obligately and facultatively chemolithotrophic sulphur bacteria, in: *Continuous Culture Vol. 8: Biotechnology, Medicine, and the Environment* (A. C. R. Dean, D. C. Ellwood, and C. G. T. Evans, eds.), pp. 139–158, Ellis Horwood, Chichester.

Kuenen, J. G., and Robertson, L. A., 1984b, Competition among chemolithotrophic bacteria under aerobic and anaerobic conditions, in: *Current Perspectives in Microbial Ecology* (M. J. Klug and C. A. Reddy, eds.), pp. 306–313, American Society for Microbiology, Washington, D.C.

Kuenen, J. G., and Tuovinen, O. H., 1981, The genera *Thiobacillus* and *Thiomicrospira,* in: *The Prokaryotes* (M. P. Starr, H. Stolp, H. G. Truper, A. Balows, and H. G. Schlegel, eds.), pp. 1023–1036, Springer-Verlag, Berlin.

Kuenen, J. G., Boonstra, J., Schröder, H. G. H., and Veldkamp, H., 1977, Competition for inorganic substrates among chemoorganotrophic and chemolithotrophic bacteria, *Microb. Ecol.* **3:**119–130.

Kutznetsov, S. I., and Gorlenko, V. M., 1973, Limnologischen and mikrobiologischen eigenschaften von karstseen der A. S. R. Mari, *Arch. Hydrobiol.* **71:**475–486.

Laanbroek, H. J., Smit, A. J., Klein-Nulend, G., and Veldkamp, H., 1979, Competition for L-glutamate between specialised and versatile *Clostridium* species, *Arch. Microbiol.* **120:**61–67.

Larkin, J. M., and Ströhl, W. R., 1983, *Beggiatoa, Thiothrix,* and *Thioploca, Annu. Rev. Microbiol.* **37:**341–367.

Leefeldt, R. H., and Matin, A., 1980, Growth and physiology of *Thiobacillus novellus* under nutrient limited mixotrophic conditions, *J. Bacteriol.* **142:**645–650.

La Rivière, J. W. M., 1974, The genus *Thiobacterium,* in: *Bergey's Manual of Determinative Bacteriology,* 8th ed. (R. E. Buchanen and N. E. Gibbons, eds.), p. 462, Williams and Williams, Baltimore.

Lündgren, D. G., Andersen, K. J., Penson, C. C., and Mahony, R. P., 1964, Culture structure and physiology of the chemoautotroph *Ferrobacillus ferrooxidans, J. Gen. Microbiol.* **105:**215–218.

Mackintosh, M. E., 1978, Nitrogen fixation by *Thiobacillus ferrooxidans, Dev. Ind. Microbiol.* **6:**250–259.

Markosyan, G. E., 1973, A new mixotrophic sulphur bacterium developing in acid media. *Thiobacillus organoparus* sp. n., *Dok. Akad. Nauk SSSR* **211:**1205–1208 [*Sov. Phys. Dok.* **211:**318–320 (1973)].

Matin, A., 1978, Organic nutrition of chemolithotrophic bacteria, *Annu. Rev. Microbiol.* **32:**433–469.

Montesinos, E., Estev, I., Abella, C., and Guerrero, R., 1982, Ecology and physiology of the competition for light between *Chlorobium limicola* and *Chlorobium phaeobacteroides* in natural habitats, in *4th International Symposium on Photosynthetic Prokaryotes* (R. Y. Stanier, ed.), Abstract A19, Institut Pasteur, Lyon.

Nelson, D. C., and Jannasch, H. W., 1983, Chemoautotrophic growth of a marine *Beggiatoa* in sulfide-gradient cultures, *Arch. Microbiol.* **136:**262–269.

Norris, P. R., and Kelly, D. P., 1978, Dissolution of pyrite (FeS$_2$) by pure and mixed cultures of some acidophilic bacteria, *FEMS Microbiol. Lett.* **4:**143–146.

Olsen, J. M., and Romano, C. A., 1962, A new chlorophyll from green bacteria, *Biochim. Biophys. Acta* **59:**726–728.

Parkes, R. J., 1982, Methods for enriching, isolating and analysing microbial communities in laboratory systems, in: *Microbial Interactions and Communities* (A. T. Bull and J. H. Slater, eds.), pp. 45–102, Academic Press, London.

Pfennig, N., 1978, General physiology and ecology of photosynthetic bacteria, in: *The Photosynthetic Bacteria* (R. I. C. Clayton and V. R. Sistrom, eds.), pp. 3–18, Plenum Press, New York.

Pfennig, N., and Widdel, F., 1982, The bacteria of the sulphur cycle, *Philos. Trans. R. Soc. Lond. B* **298**:433–441.

Robertson, L. A., and Kuenen, J. G., 1983, *Thiosphaera pantotropha* gen. nov. sp. nov., a facultatively anaerobic, facultatively autotrophic sulphur bacterium, *J. Gen. Microbiol.* **129**:2847–2855.

Sadler, W. R., and Stanier, R. Y., 1960, The function of acetate in photosynthesis by green bacteria, *Proc. Natl. Acad. Sci. USA* **46**:1328–1334.

Schedel, M., 1978, Untersuchungen zur anaeroben oxidation reduzierter Schwefelverbindungen durch *Thiobacillus denitrificans, Chromatium vinosum* und *Chlorobium limicola,* Ph. D. Thesis, University of Bonn.

Shafia, F., and Wilkinson, R. F., 1969, Growth of *Ferrobacillus ferrooxidans* on organic matter, *J. Bacteriol.* **97**:251–260.

Sirevåg, R., and Ormerod, J. G., 1977, Synthesis, storage and degradation of polyglucose in *Chlorobium thiosulfatophilum, Arch. Microbiol.* **111**:239–244.

Slater, J. H., and Morris, I., 1973, The pathway of carbon dioxide assimilation in *Rhodospirillum rubrum* grown in turbidostat continuous flow culture, *Arch. Microbiol.* **92**:235–244.

Smith, A. L., and Kelly, D. P., 1979, Competition in the chemostat between an obligately and a facultatively chemolithotrophic *Thiobacillus, J. Gen. Microbiol.* **115**:377–384.

Smith, A. L., Kelly, D. P., and Wood, A. P., 1980, Metabolism of *Thiobacillus* A2 grown under autotrophic, mixotrophic and heterotrophic conditions in chemostat cultures, *J. Gen. Microbiol.* **121**:127–138.

Smith, D. W., and Finazzo, S. F., 1981, Salinity requirements of a marine *Thiobacillus intermedius, Arch. Microbiol.* **129**:199–203.

Sorokin, Y., 1970, Interrelations between the sulphur and carbon turnover in leromictic (sic) lakes, *Arch. Hydrobiol.* **66**:391–446.

Stanier, R. Y., Adelberg, E. E., and Ingraham, J. L., 1976, *The Microbial World,* pp. 527–563, Prentice Hall, Englewood Cliffs, New Jersey.

Taylor, B. F., Hoare, D. S., and Hoare, S. L., 1971, *Thiobacillus denitrificans* as an obligate chemolithotroph. I: Isolation and growth studies, *Arch. Mikrobiol.* **88**:285–298.

Taylor, P. A., and Williams, P. J. LeB., 1974, Theoretical studies on the coexistence of competing species under continuous flow conditions, *Can. J. Microbiol.* **21**:90–98.

Timmer-ten Hoor, A., 1975, A new type of thiosulphate oxidizing nitrate-reducing microorganism: *Thiomicrospira denitrificans* sp.nov., *Neth. J. Sea Res.* **9**:344–350.

Timmer-ten Hoor, A., 1977, Denitrificerende kleurloze zwavelbacteriën, Ph. D. thesis, University of Groningen.

Trudinger, P. A., 1982, Geological significance of sulphur oxidoreduction by bacteria, *Philos. Trans. R. Soc. London. B* **298**:563–581.

Trüper, H. G., and Fischer, U., 1982, Anaerobic oxidation of sulphur compounds as electron donors for bacterial photosynthesis, *Philos. Trans. R. Soc. Lond. B* **298**:529–542.

Van Gemerden, H., 1967, On the bacterial sulfur cycle of inland waters, Ph. D. thesis, University of Leiden.

Van Gemerden, H., 1974, Coexistence of organisms competing for the same substrate: An example among the purple sulfur bacteria, *Microb. Ecol.* **1**:104–119.

Van Gemerden, H., 1980, Survival of *Chromatium vinosum* at low light intensities, *Arch. Microbiol.* **125**:115–121.

Van Gemerden, H., 1983, Physiological ecology of purple and green bacteria, *Ann. Microbiol. (Inst. Pasteur)* **134B**:73–92.

Van Gemerden, H., and Beeftink, H. H., 1981, Coexistence of *Chlorobium* and *Chromatium* in a sulfide-limited chemostat, *Arch. Microbiol.* **129:**32–34.

Van Liere, E., 1979, On *Oscillatoria agardhii* Gomont, experimental ecology and physiology of a nuisance bloom-forming cyanobacterium, Ph.D. thesis, University of Amsterdam.

Veldkamp, H., and Jannasch, H. W., 1972, Mixed culture studies with the chemostat, *J. Appl. Chem. Biotechnol.* **22:**105–123.

Veldkamp, H., and Kuenen, J. G., 1973, The chemostat as a model system for ecological studies, in: *Modern Methods in the Study of Microbial Ecology* (T. Rosswall, ed.), Bulletins from the Ecological Research Committee (Stockholm), Vol. 17, pp. 347–355.

Vishniac, W. V., 1974, Organisms metabolizing sulphur and sulphur compounds. The genus *Thiobacillus,* in: *Bergey's Manual of Determinative Bacteriology,* 8th ed. (R. E. Buchanen and N. E. Gibbons, eds.), pp. 456–461, Williams and Williams, Baltimore.

Walsby, A. E., 1978, The gas vesicles of aquatic prokaryotes, in: *Relations between Structure and Function in the Prokaryotic Cell,* 28th Symposium of the Society for General Microbiology (R. Y. Stanier, H. V. Rogers, and J. B. Ward, eds.), pp. 327–358, Cambridge University Press.

Witzel, K. P., and Overbeek, J. G., 1979, Heterotrophic nitrification by *Arthrobacter* sp. (strain 9006) as influenced by different cultural conditions, growth state and acetate metabolism, *Arch. Microbiol.* **122:**137–143.

Wood, A. P., and Kelly, D. P., 1983, Autotrophic, mixotrophic and heterotrophic growth with denitrification by *Thiobacillus* A2 under anaerobic conditions, *FEMS Microbiol. Lett.* **16:**363–370.

2

Determining Microbial Kinetic Parameters Using Nonlinear Regression Analysis
Advantages and Limitations in Microbial Ecology

JOSEPH A. ROBINSON

1. Introduction

Microbial ecologists, and biologists in general, have come to appreciate the power of the quantitative approach in their research. It is no longer enough to describe the organisms that occupy a given habitat; the rates at which they carry out metabolic functions of ecological importance must be estimated. Only when quantitative information of metabolic activities is coupled with knowledge of organismal types can our understanding of the concerted actions of the members of a community be considered complete.

The quantitative approach in microbial ecology involves the estimation of constants or parameters in equations chosen to represent the process under study, such as substrate depletion, growth, or surface colonization. In many important practical situations, the functions that best represent biological behavior are nonlinear with respect to their parameters. Nonlinear parameter estimation (NPE), also referred to as nonlinear regression analysis or nonlinear optimization, deals in part with the estimation of parameters of nonlinear equations.

NPE methods provide tools for solving "inverse problems," i.e., esti-

JOSEPH A. ROBINSON ● The UpJohn Company, Kalamazoo, Michigan 49001.

mating the parameters of nonlinear models chosen to represent given processes. But NPE methods also allow an investigator to (1) determine *a priori* whether or not it is possible to obtain unique estimates of parameters for a given nonlinear model, (2) determine which of two (or more) competing models best describes a particular data set, and (3) design optimal experiments for the estimation of parameters. This last subject is particularly useful to investigators who must optimally allocate their experimental resources.

The impetus for this review stems from the need by microbial ecologists of a survey of NPE that is intermediate between publications that merely acknowledge its existence (Zar, 1974; Cooper and Weekes, 1983) and more advanced monographs written by statisticians and engineers (Bard, 1974; Beck and Arnold, 1977; Jennrich and Ralston, 1979; Huber, 1981). For the novice, techniques described within the advanced literature may be difficult to decipher, let alone implement. This review bridges the gap between some of the advanced literature on NPE and the correct use of a few basic NPE methods. Much of the cited work comes from the statistical and enzymological literature, since NPE is not generally used to estimate microbial kinetic parameters.

It might seem odd that a review of this subject is necessary, since main-frame computers typically have numerical software packages that include subroutines for fitting data to nonlinear models, e.g., the P3R subroutine of the BMDP package (Dixon *et al.,* 1981). However, other than knowing how to input the data, enter the chosen nonlinear model, and interpret the output, it is unnecessary for an experimenter to understand how his or her data are used by the computer to arrive at estimates of the model parameters. The appropriate use of NPE requires some familiarity with the mathematical elements and limitations of this statistical tool. Considering the limitations of NPE methods once data collection is over ultimately can lead to a significant waste of experimental resources.

NPE does not require the computational speed of a main-frame system for many models of microbiological interest, e.g., the Michaelis–Menten, Wright–Hobbie, Monod, and simple colonization models. I have implemented most of the techniques described in this review using microcomputers. Many models of microbiological interest, including the ones mentioned above, have 2–4 parameters, and for these the computational speed of a large computer is not required.

2. Fundamental Definitions

To begin, a distinction has to be made between variables and parameters. Further, the difference between models that are linear in their parameters versus those that are nonlinear with respect to their parameters must be understood.

2.1. Parameters versus Variables

In the following model Y and X are dependent and independent variables, respectively:

$$Y = AX/(B + X) \qquad (1)$$

The quantities A and B in Eq. (1) are the parameters to be estimated, given a Y versus X data set. The parameters A and B are considered to be constant for the given data set, but of course could themselves depend on other variables, such as temperature and pH.

2.2. Linear versus Nonlinear Models

Deciding whether or not a model is nonlinear in its parameters is easy if its sensitivity equations are examined. A sensitivity equation mathematically describes how sensitive a model is, in terms of changes in the dependent variable [e.g., Y in Eq. (1)], to changes in the parameters of the model [A or B for Eq. (1)]. A sensitivity equation is defined as the first derivative of the dependent variable with respect to a parameter of the chosen nonlinear model. There are two parameters in Eq. (1); hence this model has two sensitivity equations,

$$dY/dA = X/(B + X) \qquad (2)$$

and

$$dY/dB = -AX/(B + X)^2 \qquad (3)$$

In the derivation of Eq. (2), X and B are considered to be constants, whereas A and X are held constant for the derivation of dY/dB.

A model that is nonlinear in its parameters is defined as one whose sensitivity equations depend on one or more of the model parameters (Beck and Arnold, 1977; Draper and Smith, 1981). For Eq. (1), the sensitivity equations are functions of one (dY/dA) or both (dY/dB) of the parameters. Equation (1), therefore, is a model that is nonlinear in its parameters. If only one of the sensitivity equations of a given model depends on a single parameter, then this is sufficient for the model to be nonlinear with respect to its parameters.

In contrast, a model that is linear in its parameters is one whose sensitivity equations are all independent of the parameters. Such a model is $Y = BX + A$, for which dY/dA and dY/dB equal one and X, respectively.

To further complicate matters, models may be linear or nonlinear with respect to their parameters and with respect to the independent variable. As pointed out above, $Y = BX + A$ is linear in A and B. This equation is also linear with respect to the independent variable, X. In contrast, a polynomial equation like $Y = CX^2 + BX + A$ is nonlinear in X but linear in its parameters, since dY/dC, dY/dB, and dY/dA do not depend on A, B, or C. There are also examples of models that are nonlinear in their parameters but linear with respect to the independent variable. To avoid confusion from this point on, the terms "linear" and "nonlinear" will be reserved for how a model behaves with respect to its parameters and not the independent variable.

3. Use of Linearized Forms of Nonlinear Models

3.1. Rationale

A common practice is to transform a nonlinear model (e.g., the differential form of the Michaelis–Menten equation) into a linearized form (e.g., the Lineweaver–Burk expression) and to fit transformed data to the linearized form. The rationale for doing this is that estimates of the parameters may be obtained using linear least squares methods (actually, estimates of the parameters are calculated from new parameters appearing in the linearized version of the original nonlinear model). This technique is useful because of its simplicity, but it has several statistical faults.

Ease of analysis favors fitting data to linearized forms of nonlinear models. Linear least squares methods do not require provisional estimates of the parameters or the aid of a computer. In contrast, fitting data to a model using NPE methods requires initial estimates of the parameters and the computational speed of a computer. The latter is necessary because NPE is an iterative or recursive technique in which initial parameter estimates are sequentially improved until the "best" estimates (i.e., those that minimize differences between the observed and predicted values of the dependent variable) are calculated. This involves many arithmetic operations that would be prohibitive in practice without the aid of at least a microcomputer.

3.2. Limitations

3.2.1. Error Transformation

Transforming data and fitting them to a linearized version of a nonlinear model has several drawbacks. When data are transformed, the measurement errors are also transformed. Since *a priori* we do not know the magnitudes or types of errors associated with the data, there is no way of knowing how the errors are transformed. Indeed, different linearized forms of the same nonlinear model typically yield dissimilar estimates of the same parameters because each linearization transforms the measurement errors differently (Dowd and Riggs, 1965; Robinson and Characklis, 1984). Enzymologists and microbial kineticists have repeatedly pointed out the unreliability of certain linearized forms of Michaelis–Menten expressions for the estimation of V_{max} and K_m (Garfinkel *et al.*, 1977; Cornish-Bowden, 1979; Robinson and Tiedje, 1982; Robinson and Characklis, 1984).

3.2.2. Violation of Fundamental Assumptions

3.2.2a. Normally Distributed Errors in the Dependent Variable. In using standard (unweighted) least squares methods it is assumed that errors in the dependent variable are normally distributed, have a constant variance equal to one, and a mean of zero (Beck and Arnold, 1977; Draper and Smith, 1981). However, these methods retain much of their desirable properties when this assumption is violated (Beck and Arnold, 1977). The assumption of normally distributed measurement errors may be violated when data are transformed and fitted to a linearized form of a nonlinear model. For example, the measurement errors in the dependent variable of the Lineweaver–Burk equation ($1/v$) cannot be normally distributed, assuming those in v (the untransformed dependent variable) follow a normal distribution (Garfinkel *et al.*, 1977). Analysis of data containing errors that are not normally distributed produces parameter estimates that on average do not equal the population means of the parameters (Beck and Arnold, 1977). The degree of this bias depends on the type of linearization used and is dissimilar for different linearizations of the same nonlinear model.

3.2.2b. Lack of Error in the Independent Variable. Least squares analysis also assumes that the independent variable is free of error (Bard, 1974; Beck and Arnold, 1977; Draper and Smith, 1981), and fitting data to certain linearized forms of nonlinear models violates this condition. As pointed out by Beck and Arnold (1977), calculating the best parameter estimates for a linear model where the independent variable is not free from error is actually a nonlinear least squares problem. In other words, errors in the independent variable make the correct estimation of parameters in even a linear model a difficult proposition. In practice, the above assumption is sufficiently met if errors in the independent variable are at least tenfold less than errors in the dependent variable. Linearized forms of nonlinear equations in which the dependent variable appears on both sides of the equation are forms where the independent variable is not free from error. Examples of these forms include the Eadie–Hofstee linearization of the Michaelis–Menten equation, where v is regressed against v/S, and all of the linearized versions (Cornish-Bowden, 1979) of the integrated Michaelis–Menten equation.

3.2.3. Assessing Precision of the Estimated Parameters

It is difficult to assess the precision with which nonlinear parameters are determined when the parameters are actually estimated using a linearized form of the model. Equation (1) can be linearized by inverting both sides to obtain

$$1/Y = (B/A)/X + (1/A) \qquad (4)$$

[Actually, Eqs. (1) and (4) are analogous to the differential forms of the Michaelis–Menten and Lineweaver–Burk expressions, respectively.] If we consider the new parameters B/A and $1/A$, then Eq. (4) is a linear model. However, the actual parameters of interest are A and B, not B/A and $1/A$. Hence, although B and A may be estimated using Eq. (4), they are derived quantities and assessing their precision (i.e., calculating their standard errors) is difficult. On the other hand, various NPE methods allow for direct estimation of the standard errors of the parameters.

3.2.4. Number of Data Points Required for Determining Model Correctness

Linearized forms of nonlinear equations require that more data points be obtained for parameter estimation than if NPE techniques are used (Garfinkel *et al.,* 1977). Eight to ten data points are usually sufficient for parameter estimation when data are directly fitted to a nonlinear model (generally true for two and three-parameter models) (Cornish-Bowden, 1979). However, if a linearized form of the nonlinear model is used, then more data points are required and the spacing of the data points becomes important (Garfinkel *et al.,* 1977). NPE is more efficient in that fewer data points suffice to obtain parameter estimates and the spacing of the data points is less critical. For estimation of parameters via NPE, increasing the number of data points typically does not alter the parameter estimates themselves, although the precision with which they are determined may be increased (Garfinkel *et al.,* 1977). This contrasts with fitting transformed data to linearized models, where changing the number of data points can significantly influence estimates of the parameters (Garfinkel *et al.,* 1977).

3.2.5. Nonexistence of Linearized Forms of Nonlinear Models

There are many models of biological interest that cannot be algebraically converted into linearized forms. The following equation describes first-order appearance of a product:

$$P = S_0(1 - e^{-kt}) \tag{5}$$

where P is the product concentration, S_0 is the initial substrate concentration, t is the time, and k is a first-order rate coefficient. A reasonable approach to linearizing Eq. (5) would be to take the natural logarithm of both sides, but if this is done, then S_0 cancels out on the right-hand side. Thus, this linearization may not be used to estimate S_0 and k, assuming that S_0 is an unknown quantity of interest. In contrast, NPE methods allow simultaneous estimation of S_0 and k from P versus t data.

3.2.6. Identifiability

Linearized forms of nonlinear models also cannot be used to evaluate the uniqueness of the parameter estimates. Indeed, different linearized versions of the same nonlinear model yield different parameter estimates since they dissimilarly transform the measurement errors. The nonlinear form of the model must be used to determine whether or not unique parameter estimates can be calculated from data (Beck and Arnold, 1977).

Unique estimates of the parameters of a nonlinear model cannot be obtained (i.e., the model is not identifiable) if the sensitivity equations are multiples of one another (Beck and Arnold, 1977). For example, the parameters V_{max} and K_m in the differential form of the Michaelis–Menten model [Eq. (1) where Y is the initial velocity v, X is the substrate concentration, $A = V_{max}$, and $B = K_m$] cannot be uniquely identified if the dependence of the initial velocity on the substrate concentration is examined only in the first-order region. This is because dS/dV_{max} [Eq. (2)] and dS/dK_m [Eq. (3)] are multiples of one another when v is first order with respect to S (Fig. 1). Yet a linearized form of this model [e.g., Eq. (4)] could still be used to calculate values for these parameters, although the sensitivity equations show that the V_{max} and K_m estimates so obtained would not be unique. Robinson and Tiedje (1982) discuss this limitation of the Lineweaver–Burk linearization for the estimation of V_{max} and K_m for gaseous substrates. On the other hand, V_{max} and K_m can be uniquely estimated if the dependence of v on S is measured in the first-order through the zero-order regions (Fig. 2).

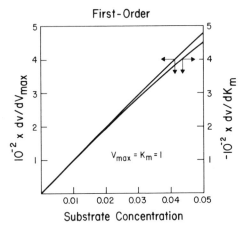

Figure 1. The sensitivity equations for V_{max} and K_m, derived from the differential form of the Michaelis–Menten equation [Eq. (1)], for the first-order region. Note that the two curves are multiples of one another (i.e., they are proportional).

The easiest way to check the identifiability of a model is to plot its sensitivity equations. This can and should be done before the investigator generates data to be fitted to the chosen nonlinear expression. Visual inspection of the curves described by the sensitivity equations can usually be relied on for assessing the identifiability of a nonlinear model.

For many models of interest to the microbial ecologist, the sensitivity equations are nearly proportional, resulting in parameter estimates that are highly correlated. This condition is undesirable, since it implies that several combinations of parameters may describe the same data set (Draper and Smith, 1981). This problem can be mitigated for some nonlinear models by measuring the dependent variable at values of the inde-

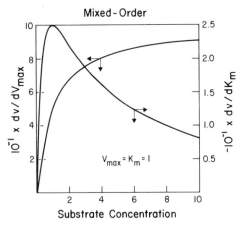

Figure 2. The sensitivity equations for V_{max} and K_m, derived from the differential form of the Michaelis–Menten equation [Eq. (1)], for the first-, mixed-, and zero-order regions. In this case the curves are not multiples of one another.

pendent variable where the sensitivity equations are maximal (Box and Lucas, 1959; Bard, 1974; Beck and Arnold, 1977). However, for some models [e.g., the integrated Monod equation (Holmberg, 1982; Robinson and Tiedje, 1983)] even optimal experimental designs will not significantly decrease the correlation among the parameter estimates.

3.2.7. *Appropriate Application of Linearized Forms of Nonlinear Models*

Linearized forms of nonlinear models do have their uses, both in parameter estimation and as diagnostic aids. NPE requires initial estimates (or guesses) of the parameters, which are updated until the best estimates are obtained. For some nonlinear models, the initial values must be close to the best parameter estimates and linearized forms can be used to satisfy this requirement. In addition, enzymologists use linearized forms of Michaelis–Menten models for mechanistic investigations of the effects of inhibitors on enzymatic activity. Determining the influence of competitive, uncompetitive, and noncompetitive inhibitors on the slope and intercept of Lineweaver–Burk plots is a legitimate application of a linearized model (Roberts, 1977). Such an application provides visual insight into how variables other than the independent variable affect the parameters. For this reason alone, linearized forms of nonlinear models should be retained.

4. Parameter Estimation Strategies

4.1. The Function to Be Minimized

Choice of the function to be minimized (i.e., the objective function) is one of the first problems to be solved in fitting data to a particular model (Bard, 1974; Beck and Arnold, 1977), whether it be linear or nonlinear. The objective function of choice is usually the sum of the squared deviations, given by

$$\text{RSS} = \text{SUM}(Y_{\text{obs}} - Y_{\text{pred}})^2 \tag{6}$$

where RSS is the sum of squares of the deviations, Y_{obs} represents the observed values of the dependent variable, and Y_{pred} represents the predicted Y values. The Y_{pred} values are calculated by solving the chosen model for the parameter estimates and the values of the independent variable at which Y_{obs} values were obtained. In words, RSS represents all of the information contained in the Y_{obs} values not explained by fitting the data to the chosen model.

Although Eq. (6) is typically the objective function chosen for minimization, other possibilities exist (Bard, 1974; Beck and Arnold, 1977; Arthenari and Dodge, 1981). For example, instead of Eq. (6) we could minimize

$$RSAD = SUM\ ABS(Y_{obs} - Y_{pred}) \tag{7}$$

where RSAD is the residual sum of the absolute differences, and ABS indicates the absolute value of the difference between Y_{obs} and Y_{pred}. As microbial ecologists, our choice of Eq. (6) (and its variants) over Eq. (7) stems from the overwhelming use of the least squares function for parameter estimation. Thus, our choice of Eq. (6) is somewhat arbitrary, although this objective function has mathematical properties that make it a good choice (Bard, 1974).

4.2. Least Squares Estimation

4.2.1. Linear Models

For linear models, explicit functions for the parameters that guarantee a minimum RSS can be readily derived. These equations can be found in most undergraduate textbooks on statistics (e.g., Zar, 1974; Meyer, 1975; Seber, 1977; Draper and Smith, 1981).

4.2.2. Nonlinear Models

In contrast to linear models, explicit functions giving the best parameter estimates do not exist for nonlinear models (Draper and Smith, 1981). Values for the parameters that minimize RSS must be recursively obtained. In words, an initial set of parameter estimates is determined either by using a linearized form of the chosen nonlinear model or by guesswork. An initial RSS value is calculated and then a new set of parameter values is calculated in some fashion. The new RSS is compared with the RSS for the initial parameter estimates, and if the former is less than the RSS for the initial estimates, then the second set of parameter estimates replaces the first. This process continues until the RSS reaches a minimum, at which point the best parameter values [i.e., those that minimize Eq. (6)] have been located.

There is no universal method for determining the path to be taken from the initial parameter estimates to the values that minimize Eq. (6). As I indicate below, many techniques for solving nonlinear least squares problems exist; they range from methods in which parameter improvement is largely guesswork to gradient methods where mathematical relations dictate the path to be taken for the improvement of the provisional

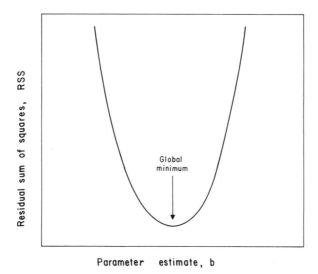

Figure 3. The residual sum of squares (RSS) curve for a hypothetical linear model with a single parameter. Note there is one local minimum, which is also the global minimum.

parameter estimates. Further, some NPE methods work better than others for determining parameters of nonlinear models currently used by microbial ecologists.

Finding best parameter estimates for some nonlinear models is no trivial matter, since markedly dissimilar estimates of the parameters may minimize Eq. (6). For linear models, the curve described by Eq. (6) is parabolic and only one set of parameter estimates corresponds to a minimal value of RSS (the "global" minimum). For the model $Y = BX$, there exists only one value of B at which $d\text{RSS}/dB = 0$ (Fig. 3). For nonlinear models there may be several points along the RSS curve (or RSS surface, if the model has two parameters) where the slope equals zero (Fig. 4). For models of microbiological interest, there may be one set of parameter estimates that is meaningful, corresponding to the global minimum, and other parameter values that yield "local" RSS minima. The latter represent cases where the parameter (for a one-parameter model) might be negative or have an unrealistically high value.

4.3. Methods of Minimizing Equation (6) for Nonlinear Models

4.3.1. Trial and Error Techniques

One of the simplest methods of minimizing RSS is by trial and error, although it is inefficient. A trial and error method begins with an initial

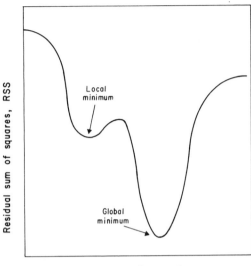

Parameter estimate, b

Figure 4. The residual sum of squares (RSS) curve for a hypothetical one-parameter nonlinear model. For this model, there are two local minima, one of which equals the global minimum.

set of parameter estimates, for which the RSS is calculated. Another set of parameter values is chosen, more or less arbitrarily, and the next RSS calculated. If the second RSS is less than the initial RSS, then the second set of parameter estimates replaces the initial parameter values for the next round. This process continues until the RSS changes by less than some value, the "convergence criterion" (e.g., 1×10^{-8}). Trial and error methods are inefficient, since there are no rules for predicting how the parameters of a nonlinear model should be changed to reduce the RSS from its initial or current value.

Trial and error methods have limited practical value for models with three or more parameters. For such equations, all parameters except one are typically fixed and this one is varied in either a systematic or random way until a relative RSS minimum is located. Then this parameter is fixed and another one varied, and so on. Although the entire process may be done interactively with a computer, this approach can require many changes to the parameter values before the RSS minimum is located. For this reason, trial and error techniques should be avoided (Beck and Arnold, 1977).

Trial and error methods have been used with some success despite their limitations. Koeppe and Hamann (1980) described a BASIC program (R'EVOL), written for a microcomputer, that fits data to nonlinear models using what they termed "the strategy of evolution." For a given model, their program randomly picks the parameter to be varied and also randomly chooses a new value for this parameter. The new value for the parameter is constrained to vary between high and low values, specified by the user before program execution. If the new parameter value reduces the previous RSS, then the initial value of the parameter is replaced with the new estimate and the process continued. Using R'EVOL, Koeppe and Hamann (1980) estimated parameters of several models that describe the fate of pharmaceuticals in mammalian tissues.

There is an additional limitation of trial and error search methods besides their computational inefficiency. These techniques do not require *a priori* knowledge of how sensitive the chosen model is to changes in its parameters. Thus, an examination of the sensitivity equations is not required for the application of trial and error methods, although these equations might predict that, for the chosen model, unique parameter estimates cannot be determined. In the R'EVOL program described by Koeppe and Hamann (1980) an attempt was made to assess the uniqueness of the parameters they obtained using the "radex error." This quantity, proposed by Goldberg (1968), is defined as the absolute amount by which a parameter can be changed without increasing the RSS by 100% of its minimum value. This is a useful concept, but the authors could have assessed the identifiability of the models they were interested in prior to data collection and analysis.

Unlike a random trial and error search, an exhaustive search entails systematically changing parameters in a model one at a time and calculating the RSS. Like a random search, an exhaustive search does not efficiently locate the best parameter estimates. But an exhaustive search can be instructive, since the shape of the RSS function, as dependent upon the parameter being varied, emerges from the analysis (Figs. 3 and 4). Further, this type of trial and error method gives a feeling for the difficulty of minimizing Eq. (6) for a nonlinear model.

4.3.2. The Gaussian Technique

4.3.2a. Mathematical Formulation. In contrast to trial and error searches, most NPE methods specify the direction (i.e., increasing or decreasing) and the magnitude of changes to be made to the parameter estimates during the recursive process. NPE methods seek an optimal path (i.e., the path with the fewest steps or iterations) to the parameter values that minimize the objective function (e.g., the least squares equation). There are several specific methods by which initial parameter estimates may be updated, of which the Gaussian method is probably the simplest. Further, the Gaussian method is useful for a wide range of NPE problems (Beck and Arnold, 1977; Draper and Smith, 1981). Many of the other techniques, such as the Levenberg–Marquardt method, are modified versions of the Gaussian method (Beck and Arnold, 1977).

The mathematical elements of the Gaussian method are derived through the application of a Taylor series expansion (Burden *et al.,* 1978). This expansion essentially linearizes the nonlinear RSS function in the neighborhood of the best parameter estimates. This is a reasonable practice for many nonlinear models, although for some that are strongly nonlinear, modifications must be made to ensure convergence.

If we consider a one-parameter nonlinear model, a Taylor series expansion about the best-parameter estimate B given an initial value b is described by

$$\text{RSS}(B) = \text{RSS}(b) + (B - b)\, d\text{RSS}(b)/db \qquad (8)$$

where RSS(B) and RSS(b) define the RSS curve as a function of B and b, respectively. The term $d\text{RSS}(b)/db$ is the first derivative of the RSS function with respect to the provisional estimate of B, namely b. Note that this derivative describes in mathematical terms the sensitivity of the RSS function to changes in b. Actually, Eq. (8) represents a truncated Taylor series, since only the first (linear) term is retained. For models with more than one parameter, Eq. (8) is amended to include (1) the derivatives for the other parameters and (2) the differences between the provisional estimates of these additional parameters and the values that define an RSS minimum.

Equation (8) can be rewritten to give

$$Y_{\text{obs}} - Y_{\text{pred}} = C\, dY_{\text{obs}}/db \qquad (9)$$

where now dY_{obs}/db equals the sensitivity equation for the model parameter B, evaluated for b, and C is a correction term that equals the difference between B and b. For models with more than one parameter, the right-hand side of Eq. (9) must be modified to include the sensitivity equations and correction terms for each parameter in the chosen nonlinear model. Equation (9) is the fundamental expression needed for implementation of the Gaussian method.

4.3.2b. An Example. To illustrate the application of the Gaussian method, I will estimate the parameters in the surface colonization equation proposed by Caldwell *et al.* (1981) for the data set and initial estimates of u and A given in Table I. This nonlinear model has the form

$$N = (A/u)e^{ut} - (A/u) \qquad (10)$$

where N is the number of cells as a function of time t, u is the growth rate, and A is the colonization rate. Note that Eq. (10) is another example

Table I. Estimation of u (growth rate) and A (colonization rate) of Equation (10) via the Gaussian Method [Equation (15)][a]

t	N_{obs}	dN/du	dN/dA	Residuals $N_{obs} - N_{pred}$
1	1.51	0.950287	1.1973	−0.286004
2	3.7	4.84744	2.8964	−0.644654
4	12	32.1071	8.7291	−1.093714
6	28.6	122.238	20.4748	−2.112157
8	65	374.698	44.1276	−1.191343
10	141	1025.98	91.7584	3.362349

Initial estimates: $u = 0.35$ and $A = 1.5$
Initial RSS: 18.8795

$$SM = \begin{bmatrix} 10872.0891 & 113475.383 \\ 113475.383 & 1209037.86 \end{bmatrix} \quad RV = \begin{bmatrix} 200.950024 \\ 2706.62009 \end{bmatrix}$$

$$CM = \begin{bmatrix} 4.509571 \times 10^{-3} & -4.232500 \times 10^{-4} \\ -4.232500 \times 10^{-4} & 4.0551632 \times 10^{-5} \end{bmatrix} \quad CTV = \begin{bmatrix} -0.2393786 \\ 0.0247058 \end{bmatrix}$$

First update for u: $0.35 + 0.025 = 0.37$
First update for A: $1.5 - 0.24 = 1.26$
New RSS: 4.00

[a]The above error-containing N versus t data were generated using a Monte Carlo simulation (Harbaugh and Bonham-Carter, 1970) for u and A values of 0.38 and 1.24, respectively.

of a nonlinear model that cannot be linearized by taking the natural logarithm of both sides of the equation; if this is done, then A/u cancels out.

To fit data to Eq. (10) using the Gaussian method, the sensitivity equations for the parameters are needed; these are

$$dN/du = (A/u)te^{ut} + (A/u^2)(1 - e^{ut}) \qquad (11)$$

and

$$dN/dA = (e^{ut} - 1)/u \qquad (12)$$

The problem of finding the best estimates of A and u using the Gaussian method is simplified if matrix notation is used. Three matrices are required, the first of which is the sensitivity matrix (SM). The SM is a square matrix where the number of rows and columns equals the number of parameters of interest. It contains the uncorrected sums of squares and

uncorrected sums of cross-products of the sensitivity equations. The SM
for Eq. (10) is

$$\text{SM} = \begin{bmatrix} \text{SM}_{11} & \text{SM}_{12} \\ \text{SM}_{21} & \text{SM}_{22} \end{bmatrix} \tag{13}$$

where $\text{SM}_{11} = \text{SUM}[(dN/du) \times (dN/du)]$, $\text{SM}_{22} = \text{SUM}[(dN/dA) \times (dN/dA)]$, $\text{SM}_{12} = \text{SUM}[(dN/du) \times (dN/dA)]$, and $\text{SM}_{21} = \text{SM}_{12}$. The first and
second subscripts of the elements in the SM give the respective row and
column numbers of these elements. Note that the sum of the uncorrected
cross-products is the same whether evaluated as $(dN/du)(dN/dA)$ or $(dN/dA)(dN/du)$, making the SM a symmetric matrix (Thomas, 1972; Burden
et al., 1978). The elements of the SM for Eq. (10) for the initial estimates
of u and A appear in Table I.

The other two matrices needed are actually column vectors, i.e.,
matrices with only one column (Thomas, 1972). The first of these vectors
contains the uncorrected sums of cross-products of the residuals and sensitivity equations, and is termed the residuals vector (RV). For equation
(10), the RV is

$$\text{RV} = \begin{bmatrix} \text{RV}_{11} \\ \text{RV}_{21} \end{bmatrix} \tag{14}$$

where $\text{RV}_{11} = \text{SUM}[(N_{\text{obs}} - N_{\text{pred}}) \times dN/du]$ and $\text{RV}_{21} = \text{SUM}[(N_{\text{obs}} - N_{\text{pred}}) \times dN/dA]$. In Eq. (14), N_{obs} and N_{pred} equal the observed (Y_{obs}) and
predicted (Y_{pred}) values of N, respectively. The latter values of the depen-
dent variable are calculated using Eq. (10), given the initial u and A estimates (Table I).

The last matrix (second column vector) needed is the correction
terms vector (CTV). The CTV has two elements: the corrections to be
made to the initial estimates of u and A. The CTV is found by solving
the following equation:

$$\text{CTV} = (\text{SM})^{-1} \times \text{RV} \tag{15}$$

As Eq. (15) indicates, the CTV is found by inverting the sensitivity
matrix (SM) and then multiplying the elements of this new symmetric
matrix, termed the covariance matrix (CM), by the elements of the residuals vector (RV). The inverse of the SM is

$$\text{CM} = \begin{bmatrix} \text{CM}_{11} & \text{CM}_{12} \\ \text{CM}_{21} & \text{CM}_{22} \end{bmatrix} \tag{16}$$

where $CM_{11} = SM_{22}/DET$, $CM_{22} = SM_{11}/DET$, $CM_{12} = -SM_{12}/DET$, and $CM_{21} = CM_{12}$. Here DET is the determinant of the above 2×2 SM, and it equals $SM_{11} \times SM_{22} - SM_{12}^2$.

The CTV according to Eq. (15) is then

$$CTV = \begin{bmatrix} RV_{12} \times CM_{11} + RV_{21} \times CM_{12} \\ RV_{12} \times CM_{12} + RV_{21} \times CM_{22} \end{bmatrix} \tag{17}$$

Once the CTV is calculated, the first element of this vector is added to u and the second element of the CTV is added to A, completing the first iteration (Table I). The new estimates of u and A are then used to calculate new elements for the SM and RV, and once more the CTV is found according to Eq. (15). The above steps are the mathematical elements of the Gaussian recursion process, whereby provisional estimates of parameters of nonlinear models are updated until the RSS is minimized. The analyst must choose a convergence criterion (e.g., when the correction terms are less than 1×10^{-4}); otherwise the updating process would cycle indefinitely.

The steps for updating u and A of Eq. (10) via the Gaussian technique are applicable to any two-parameter nonlinear model. The calculations are more cumbersome for models with three or more parameters, but they are analogous to those done for the colonization equation of Caldwell *et al.* (1981). For nonlinear models having more than two parameters, the inverse of the SM can be approximated using a numerical technique, such as Crout reduction (Burden *et al.*, 1978). The steps for multiplying matrices and column vectors having more than two rows can be found in most undergraduate math books (e.g., Thomas, 1972).

Implementation of the Gaussian method largely requires a knowledge of algebra. Only a few elementary rules of differentiation are needed to derive the sensitivity equations, and actually these equations can be approximated using algebraic techniques (Bard, 1974; Beck and Arnold, 1977). Hence the primary restraint to the application of this NPE method by microbial ecologists should not be a lack of the requisite mathematical knowledge; most microbiologists are exposed to the algebraic techniques needed for NPE at the undergraduate level.

The Gaussian method (or any NPE method) can be used to estimate the parameters of linear models. For a linear model, the elements of the SM are independent of the parameters. As a result, the Gaussian method converges after a single iteration regardless of the initial parameter values. This contrasts with the updating process for nonlinear models, which may require five or more iterations before the best parameter estimates are calculated. For linear models, the use of explicit functions that give the best parameter estimates (Zar, 1974; Draper and Smith, 1981) is much more efficient than estimating these parameters via NPE techniques.

4.3.2c. Modifications to Ensure Convergence. Modifications to the Gaussian method are necessary for models that are "ill-conditioned" (Bard, 1974; Beck and Arnold, 1977; Jennrich and Ralston, 1979). An ill-conditioned model is one whose sensitivity equations are nearly proportional, and for these models the unmodified Gaussian method can fail to locate the global RSS minimum, even when the initial parameter estimates are close to the best values. For ill-conditioned models, the minimum RSS given by the best parameter estimates is not well defined, and modifications are necessary to prevent the correction terms from diverging to meaningless values. Most of these modifications attempt to produce parameter searches in which the RSS is reduced with each iteration. Bard (1974) argues that an efficient NPE method must ensure that the RSS is lower after each iteration.

The simplest improvement to the Gaussian method is the "halving-doubling" technique (Beck and Arnold, 1977; Jennrich and Ralston, 1979), in which the correction terms are either halved or doubled, depending on how they affect the current RSS value. If a correction term produces a new RSS value greater than the previous RSS value, then the correction term is halved. On the other hand, if a correction term reduces the RSS, then it can be doubled and subsequently used to estimate another RSS. In this way, the RSS is forced to decrease with each iteration. This process is generally referred to as a "line search" and more sophisticated variations exist [see references in Jennrich and Ralston (1979)]. Line search methods are particularly useful for models that exhibit a high degree of nonlinearity, since for these models the RSS function may be poorly approximated by the truncated Taylor series in the neighborhood of the best parameter estimates (Bard, 1974; Beck and Arnold, 1977; Jennrich and Ralston, 1979).

One of the more commonly used variations of the Gaussian method is the Levenberg–Marquardt modification (Beck and Arnold, 1977). This technique has certain fundamental differences with the Gaussian method, but it can be treated as a variation of the Gaussian approximation (Beck and Arnold, 1977). The Levenberg–Marquardt method alters both the step size and direction taken by the Gaussian technique, attempting to ensure that the RSS is sequentially reduced. It works well when the sensitivity equations are nearly proportional (Beck and Arnold, 1977), such as can occur when the initial parameter estimates are poor. The specific equations for implementation of this NPE method can be found in Bard (1974), Beck and Arnold (1977), and Jennrich and Ralston (1979).

Another modification of the basic Gaussian method is the Box–

Kanemasu method and its variants (Box and Kanemasu, 1972; Beck and Arnold, 1977). The goal of the Box–Kanemasu modification is the same as those of other methods, namely, to ensure that changes taken in the parameter search do not widely oscillate and diverge away from the parameter value(s) defining the minimum RSS. The Box–Kanemasu modification approximates the RSS function at each iteration by a second-degree polynomial. The coefficients of this polynomial are determined and then used to alter the step size calculated via the Gaussian method. Thus, in contrast to the Levenberg–Marquardt method, the Box–Kanemasu modification only changes the step size of the Gaussian method. In their description of the Box–Kanemasu technique, Beck and Arnold (1977) recommend that a check be included in this method to ensure that the RSS continually decreases during the recursive process.

None of the above parameter estimation methods yield optimal results for all nonlinear models. The unmodified Gaussian, halving–doubling, Levenberg–Marquardt, and Box–Kanemasu methods each have strengths and weaknesses; which one is best partly depends on the nonlinear model of interest. In a survey of various techniques, Bard (1970) concluded that the Levenberg–Marquardt method performed best, whereas Box and Kanemasu (1972) found that the Levenberg–Marquardt method was not superior to the Box–Kanemasu modification for the nonlinear models they investigated. Davies and Whitting (1972) found in their comparison of several NPE methods that the unmodified Gaussian method performed quite well and was inferior to the Levenberg–Marquardt method for the models they used only when the initial parameter estimates were poor. Finally, Beck and Arnold (1977) point out that the unmodified Gaussian method is very competitive with its many variants, failing only when the sensitivity equations are approximately proportional.

Since no NPE algorithm is universally applicable, new algorithms (e.g., De Villiers and Glasser, 1981) continue to be developed and evaluated by theoreticians. For this reason, reports of parameters determined using an NPE technique should include mention of the NPE method employed.

4.3.2d. Approximation of Standard Errors of Parameters Using the CM. Point estimates of microbial parameters are never satisfactory alone. A microbial parameter (or any other, for that matter) can never be known with certainty. Hence, any report of a parameter should include an estimate of the precision with which the parameter was determined. The most commonly used measure of the precision of a parameter is the standard error (SE). The SE of a parameter can be approximated from replicate experiments, but it can also be estimated from a single Y versus

X data set. Regardless of how it is obtained, the SE can be used to construct an interval estimate (e.g., 95% confidence interval), which is more meaningful than a mere point estimate of a parameter.

Approximate SEs of parameters in nonlinear models can be calculated using elements from the main diagonal of the CM and the residual mean square (RMS). The RMS is estimated using the following equation:

$$RMS = SUM(Y_{obs} - Y_{pred})^2/(n - p) \qquad (18)$$

where n and p equal the number of data points and parameters, respectively. The Y_{pred} values are calculated using the best estimates of the parameters determined via the Gaussian method or one of its variants. The RMS and the element of the covariance matrix along the main diagonal that corresponds to the parameter for which the SE is to be estimated are used in the following expression:

$$SE_i = SQRT(RMS \times C_{ii}) \qquad (19)$$

where SE_i is the standard error of the ith parameter and C_{ii} is the ith element of the CM along the main diagonal. SQRT denotes the square root of the product of RMS and c_{ii}.

The appropriate use of Eq. (19) requires that (1) the measurement errors be normally distributed with a mean of zero and standard deviation of one and (2) the measurement errors be statistically independent quantities. The greater the degree to which these assumptions are violated, the greater the incorrectness of the SEs calculated via Eq. (19).

The SEs calculated using Eq. (19) are statistically optimistic. This results because a nonlinear model is treated as a linear model in the neighborhood of the best parameter estimates in the updating process. The degree of optimism, and hence the extent to which the SEs are underestimated, depends on how closely the linear Taylor series expansion approximates the RSS function near the RSS minimum. The SEs of parameters of a highly nonlinear model estimated using Eq. (19) will be more unrealistic than SEs calculated for a model that is nonlinear to a lesser degree. A more reliable method for calculating the SEs of nonlinear parameters is given in Section 4.7.

4.3.2e. Approximate Parameter Correlation Matrix. The parameter correlation matrix (PCM) can be used to assess the uniqueness of parameters estimated for a particular nonlinear model (Beck and Arnold, 1977). The PCM is a symmetric square matrix with dimensions set by the number of parameters of the nonlinear model [e.g., the PCM for Eq. (10) has two rows and two columns.] The elements of the PCM range from -1 to $+1$. The elements along the main diagonal all equal one, since any parameter correlates perfectly with itself. Elements off the main diagonal give the correlations for all pairwise combinations of the parameters. In this context, a high correlation (either positive or negative) is undesirable, since it indicates that the parameter estimates are not statistically independent.

For Eq. (10) (and for other two-parameter nonlinear models), there is only one element of the PCM of interest: the one that gives the correlation between the first and second parameters. For the model of Caldwell *et al.* (1981), the relevant element of the PCM is

$$CORR(u, A) = CM_{12}/SQRT(CM_{11} \times CM_{22}) \qquad (20)$$

where $CORR(u, A)$ denotes the correlation between the first (u) and second (A) parameters.

If all elements of the PCM, excluding the main diagonal, have absolute values greater than 0.9, then the parameter estimates are highly correlated and tend to be inaccurate (Beck and Arnold, 1977). This can be true even when the SEs of the parameters are small relative to the parameter estimates themselves, suggesting that the level of precision is high. For well-conditioned nonlinear models, and given an optimal experimental design, the parameter correlation coefficients are all less than 0.9 in absolute magnitude. Thus, the PCM is useful, perhaps necessary, in assessing the statistical significance of parameters estimated via NPE.

Parameters of nonlinear models may be more or less correlated, depending on (1) the values of the independent variable chosen at which to measure the dependent variable and (2) the nature of the nonlinear model itself. If the experimental design is such that the sensitivity equations are nearly proportional (e.g., in the first-order region for models incorporating saturation kinetic terms), then the parameter estimates will always be unreliable. For some nonlinear models (e.g., the integrated Monod equation), even an optimal experimental design cannot eliminate the high degree of correlation exhibited by estimates of the parameters. Hence, the choice of a particular nonlinear model should be based not only on how well experimental data fit the model, but also on the statistical reliability of the parameter estimates, as indicated by the PCM.

4.4. Weighted Least Squares (WLS) Analysis

4.4.1. When is WLS Necessary?

Parameter estimation using Eq. (6), as illustrated for a nonlinear model in Section 4.3.2b, may be termed ordinary least squares (OLS) analysis (Beck and Arnold, 1977). An assumption of OLS is that the variance of the dependent variable is constant over the range in which it is measured (Bard, 1974; Beck and Arnold, 1977; Draper and Smith, 1981). OLS is appropriate if nothing is known about the measurement errors, but if the experimenter knows that the measurement errors are not constant, then OLS is not an optimal NPE strategy (Beck and Arnold, 1977). OLS is also not appropriate when the dependent and independent variables differ substantially in magnitude (Bard, 1974). Regressing a dependent variable against an independent variable where the former is several orders of magnitude smaller than the latter makes little sense, because the residuals of the independent variable will dominate those of the dependent variable. In cases like the above, WLS should be used to minimize the RSS function.

In order to use WLS, weights must be assigned to the Y_{obs} values indicating the "degree of trust" to be placed in these values. Each observation should be weighted by an amount inversely proportional to its variance. If estimates of the variance are not available, then some weighting function is chosen to calculate the weights. For example, calculating weights as $1/Y_{obs}$ is appropriate when the variance of Y_{obs} is proportional to the magnitude of Y_{obs}.

WLS entails minimizing the following variation of Eq. (6):

$$\text{WRSS} = \text{SUM } w_i(Y_{obs} - Y_{pred})^2 \qquad (21)$$

where WRSS is the weighted residual sum of squares and w_i is the weight assigned to a given Y_{obs} value. As was true for OLS, minimization of Eq. (21) for a nonlinear model must be done using the Gaussian method or some other recursive procedure.

For WLS analysis using the Gaussian method, changes must be made to the elements of the SM [Eq. (13)] and RV [Eq. (14)]. For the surface colonization model [Eq. (10)], the new elements of the SM are

$$\text{SM}_{11} = \text{SUM } w_i(dN/du)(dN/du)$$
$$\text{SM}_{22} = \text{SUM } w_i(dN/dA)(dN/dA)$$
$$\text{SM}_{12} = \text{SUM } w_i(dN/du)(dN/dA)$$

and $\text{SM}_{21} = \text{SM}_{12}$. The new elements of the RV are

$$\text{RV}_{11} = \text{SUM } w_i(N_{obs} - N_{pred})(dN/du)$$
$$\text{RV}_{21} = \text{SUM } w_i(N_{obs} - N_{pred})(dN/dA)$$

For WLS NPE, the SEs of the parameters, as well as the elements of the approximate PCM, may be estimated using the weighted RMS and the appropriate elements of the CM. However, as was true for OLS NPE,

SEs calculated from the elements of the CM obtained using WLS analysis are still optimistic, since the nonlinear model is linearized in the neighborhood of the best parameter estimates.

4.4.2. The Nature of Experimental Error and the Calculation of Weights

The microbial ecologist who fits data to nonlinear (or linear) models must contend with both systematic and random errors. One type of systematic error is that produced when data are fitted to the wrong nonlinear model (Fig. 5), and it can be detected by examining the residuals ($Y_{obs} - Y_{pred}$ values) obtained for the best-fit equation (i.e., the curve calculated from the best parameter estimates). Other sources of systematic errors include (1) bias in reading instrument scales, (2) instrument calibration errors, (3) inefficiency in observing the dependent variable, and (4) variation in experimental conditions (Meyer, 1975). These types of systematic errors can also produce trends in residuals and hence may be detected, but often they go unnoticed.

In addition to systematic errors, the microbial ecologist must contend with random or "true" errors. These errors are associated with all measurements of variables, since no quantity can ever be known with certainty. In the absence of systematic errors, the residuals are themselves estimates of the measurement errors in the Y_{obs} values. These errors are as likely to be positive ($Y_{obs} > Y_{pred}$) as negative ($Y_{obs} < Y_{pred}$) (Figs. 6 and 7).

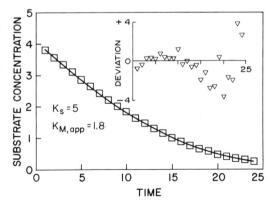

Figure 5. Systematic residuals obtained from fitting the wrong model to data. These data were obtained by solving the integrated Monod equation (43) for an S_0 below K_s and introducing slight homoscedastic errors into the solution. The data were then fitted to the integrated Michaelis–Menten equation [Eq. (31)] using the Gaussian method. The solid line is the best-fit solution of the data to the integrated Michaelis–Menten model. The residuals (inset) are not random and the apparent K_m bears no relationship to the K_s. Although these data were generated by solving the integrated Monod equation, it is not possible to turn around and fit the data to Eq. (43), because the sensitivity equations for this model are multiples of one another in the first-order region (Robinson and Tiedje, 1983).

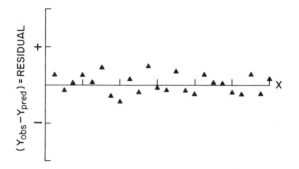

Figure 6. Homoscedastic (simple) errors.

Random errors can be divided into two types: (1) simple and (2) relative. Simple errors are defined as those having a constant standard deviation, whereas relative random errors can be defined as those for which the variance is inversely proportional to the Y_{obs} value. Simple errors are also termed homoscedastic (Zar, 1974), and yield residuals plots that are rectangular in shape (Fig. 6). In contrast, heteroscedastic or relative errors produce triangular residuals plots (Fig. 7).

The statistical consequences for violating the assumption of homoscedastic errors when OLS regression is used depend partly on the degree of heteroscedasticity in the measurement errors. Occasionally, little can be gained in using WLS when the standard deviations of the Y_{obs} values vary by less than an order of magnitude. In contrast, if the standard deviation of the measurement errors varies by more than tenfold, then WLS is usually superior to OLS (Beck and Arnold, 1977).

The splitting of random errors into simple and relative types is convenient, but in most circumstances the exact nature of the random errors

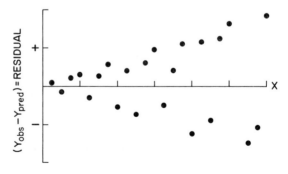

Figure 7. Heteroscedastic (relative) errors.

is unknown. Few microbial ecologists concern themselves with the exact nature of measurement errors in their experiments. To some extent this is justified, since others have shown that reliable estimates of measurement errors require at least five replicates of the dependent variable (Storer *et al.,* 1975). This degree of replication is impractical for many microbial ecologists, certainly for those whose objectives include the estimation of parameters (e.g., V_{max}) for many substrates and microbial consortia. It is not surprising, then, that microbiologists typically do not bother with WLS for parameter estimation, but instead use OLS, implicitly assuming that heteroscedastic errors little affect the parameters of biological interest.

The nature of experimental errors and the way in which they affect parameter estimates of microbiologically relevant nonlinear models have largely been studied by enzyme kineticists. Recent attention has focused on (1) whether or not errors in initial velocity measurements v are normally distributed and (2) appropriate weights to be used for estimation of V_{max} and K_m from v–S data pairs. Some investigators (Storer *et al.,* 1975; Nimmo and Mabood, 1979) found that errors in v were not normally distributed, while others (Siano *et al.,* 1975; Askelof *et al.,* 1976) observed that errors in v were normally distributed. Controversy also remains over the appropriate weighting function for WLS estimation of V_{max} and K_m from initial velocity data. Weighting functions of $1/v^{1.0}$ (Storer *et al.,* 1975), $1/v^{1.6}$ (Askelof *et al.,* 1976), and $1/v^{2.0}$ (Nimmo and Mabood, 1979), have all been recommended. Thus, there is no consensus on the nature of errors in v. Indeed, Atkins and Nimmo (1980) argue that determining the nature of the experimental errors and the appropriate weighting function should be done for *each experimental system,* a difficult recommendation to implement.

Notwithstanding the advantages of WLS NPE, it is impractical to expect all microbial ecologists and environmental biologists to undertake investigations of the nature of experimental errors in their studies. As pointed out above, the degree of replication necessary to define the measurement errors will not often be achieved. Thus, the investigator must usually assume that (1) the measurement errors are approximately Gaussian and (2) the correct weighting function is approximated by $1/Y_{obs}^2$. These assumptions can then be checked by examining the residuals. The reasonableness of this approach for estimating enzyme kinetic parameters has been shown by Cornish-Bowden and Endrenyi (1981).

4.5. Robust Regression

One major limitation to either OLS or WLS is that the worst observations have the largest residuals. Since the residuals are squared, the worst observations tend to influence the parameter estimates the greatest. Of course, the easiest way to circumvent this limitation is to ignore these "outliers." But this is a drastic procedure, because it introduces an arbitrary division between good and bad Y_{obs} values. An alternative is to use a "robust" method (Huber, 1972, 1981) in which the weight placed on a given Y_{obs} value is inversely proportional to its residual.

To date, only biochemists (Cornish-Bowden and Endrenyi, 1981; Duggleby, 1981) have applied robust methods for estimating parameters of biologically relevant models. The technique most used was originally proposed by Mosteller and Tukey (1977), where each residual is multiplied by a bisquare weight W_i. The W_i values are calculated using the following equation:

$$W_i = (1 - u_i^2)^2 \qquad (22)$$

if the absolute value of u_i is less than or equal to one. W_i equals zero when the absolute value of u_i is greater than one. The u_i values are defined as $(Y_{obs} - Y_{pred})/c$, where c is a "robustness constant." If a residual is greater than c, then this observation is ignored ($W_i = 0$). Observations with residuals less than c receive fractional weights ($0 < W_i < 1$), with the Y_{obs} values having the smallest residuals receiving bisquare weights nearest to unity.

Biweight regression may be used in conjunction with WLS (and many other parameter estimation methods) for NPE. For robust WLS, the u_i values equal $w_i(Y_{obs} - Y_{pred})$. In this case, the influence of the worst weighted residuals is minimized. Biweight WLS was used by Cornish-Bowden and Endrenyi (1981) and Duggleby (1981) to estimate V_{max} and K_m from initial velocity data. Duggleby (1981) found that unless he used bisquare weighting, the presence of outliers largely prevented him from estimating Michaelis–Menten parameters via the Gaussian method.

In order to apply Mosteller and Tukey's method, a value for c must be chosen. If c is large (e.g., 100), then W_i will equal one for all Y_{obs} values and the method becomes equivalent to the least-squares technique. Mosteller and Tukey (1977) suggested a value for c equal to six times the mean absolute residual (i.e., six times the sum of the absolute values of the $Y_{obs} - Y_{pred}$ values divided by the number of observations). This value for c is such that the vast majority of observations will be given weights greater than zero. Mosteller and Tukey (1977) show that several other values for c will work well.

Biweight regression and other robust methods must be recognized for what they are, namely, techniques for minimizing the influence of data points that do not agree with the chosen model. Robust regression is not appropriate if there is doubt regarding the nature of the mathematical model that best represents the experimental data. For example, it is inappropriate to fit data that conform to a saturation kinetic process to a one-term exponential model using biweight regression.

4.6. Alternative Objective Functions for the Estimation of Microbial Parameters

If certain conditions are met, there are objective functions that are statistically superior to either Eq. (6) or Eq. (21). Two such functions are the maximum likelihood and maximum *a posteriori* equations.

4.6.1. Maximum Likelihood Estimation

Maximum likelihood (ML) estimation is superior to OLS analysis when the covariance matrix of the measurement errors is known (Beck and Arnold, 1977). This matrix contains not only the variances of the measurement errors, but also the covariances of the measurement errors. The covariances of the measurement errors are generally assumed to be zero (an assumption required for both OLS and WLS analysis), which is not always correct. The equations needed for implementing ML analysis appear in Bard (1974) and Beck and Arnold (1977). Knowledge of the covariance matrix of the measurement errors makes it difficult to use ML because of the degree of replication necessary to define this matrix. For this reason, it is unlikely that microbial ecologists will routinely use ML NPE despite its advantages.

4.6.2. Maximum a Posteriori Estimation

Maximum *a posteriori* (MAP) estimation for both linear and nonlinear models requires (1) knowledge of the error covariance matrix and (2) provisional estimates of the parameters and their standard errors. One of the unique features of MAP estimation is that the sensitivity equations of the parameters may be proportional, a condition that rules out OLS, WLS, or ML NPE (Beck and Arnold, 1977). This is an exception to the rule that when the sensitivity equations are multiples of one another, unique parameter estimates cannot be obtained (Section 3.2.6). The equations needed for applying MAP estimation appear in Bard (1974) and Beck and Arnold (1977). As was true for ML estimation, knowledge of the covariance matrix of the measurement errors makes implementation of MAP estimation a difficult proposition.

The OLS, WLS, and ML estimation procedures are simplifications of MAP estimation (Beck and Arnold, 1977). As stated above, MAP estimation is appropriate when (1) good information is available on the measurement errors and (2) the investigator has prior knowledge of the parameters and their SEs. When the latter information is poor, MAP simplifies to ML estimation. When information about both the parameters and the error covariance matrix is poor, MAP estimation simplifies to OLS. The MAP estimation method reduces to WLS when subjective information about the parameter values is poor, but some information about the relative differences of errors in the Y_{obs} values is available.

4.7. Jackknife Estimation

4.7.1. Bias in Parameters Determined Using NPE Methods

Biases are introduced into parameter estimates obtained via least squares analysis because the measurement errors are assumed to have a particular structure (e.g., OLS assumes that all measurement errors are normally distributed with a mean of zero and a standard deviation of one). This bias can be eliminated if, instead of assuming that the measurement errors conform to a particular distribution, the actual distribution of the measurement errors is used. Several techniques for eliminating biases in parameter estimates exist, but the one most applied to models of biological relevance is the jackknife method.

Since the initial description of the jackknife by Quenouille (1956), this method has been recognized as a powerful means of eliminating the biases in parameters estimated using least squares methods (Miller, 1974; Hinkley, 1977; Fox et al., 1980). The biggest advantage of this technique, aside from its applicability to both linear and nonlinear models, is that jackknife estimation does not require a normal error distribution or homoscedastic errors for its approximate validity (Fox et al., 1980). When used in conjunction with OLS, jackknifing is a means of eliminating biases in parameter estimates that result from assuming that the measurement errors are homoscedastic.

Three different jackknife methods have been described in the statistical literature: (1) the general jackknife (Fox et al., 1980), (2) the linear jackknife (Hinkley, 1977), and (3) the weighted linear jackknife (Hinkley, 1977). I am aware of only one study in which these three methods were compared for the evaluation of parameters of a model of interest to microbial ecologists, namely the study by Oppenheimer et al. (1981). These authors used the general, linear, and weighted jackknife methods to estimate V_{max} and K_m from initial velocity data. Further, Oppenheimer et al. (1981) demonstrated the superiority of jackknifing for making interexperimental comparisons of kinetic parameters. Interexperimental comparisons of parameters are often equivocal when done using the SEs estimated from the CM and RMS, because these SEs are statistically optimistic. The SEs determined using jackknife methods are more realistic (i.e., more conservative) than those calculated using elements of the CM, and hence more appropriate for statistical comparisons of parameters.

The jackknife methods are relatively simple to implement. For the general jackknife, a given data set is divided into groups. In the extreme case, each group contains one data pair. Next, n regressions are done to obtain n "pseudovalues" (Hinkley, 1977; Fox et al., 1980) of the parameters. The pseudovalues for each parameter are defined as

$$P_i = nP_0 - (n - 1)P_{-i} \qquad (23)$$

where P_i is the ith pseudovalue, P_0 is the estimate of the parameter obtained for analysis of all n data pairs, and P_{-i} is the estimate of the parameter obtained for analysis of $n - 1$ data pairs with the ith observation missing. The jackknifed estimate of the parameter is then calculated according to

$$J(P) = \text{SUM } P_i/n \qquad (24)$$

with the variance of $J(P)$ defined as

$$\text{VAR}[J(P)] = [\text{SUM } (P_i - P_0)^2]/[n\,(n - 1)] \qquad (25)$$

Miller (1974) recommends that a geometric average (i.e., averaging the natural logarithms of the pseudovalues) be used in order to stabilize the variances of the pseudovalues.

The linear and weighted jackknife procedures require only one regression instead of $n + 1$ regressions, and thus are computationally more efficient than the general jackknife. These approximations to the standard jackknife are obtained through a Taylor series expansion to linear terms (the same technique of linearizing the RSS function for nonlinear models) of the equation for the general jackknife (Oppenheimer et al., 1981). Equations for implementation of these variants of the general jackknife can be found in Miller (1974), Hinkley (1977), and Fox et al. (1980).

Few investigators, aside from Oppenheimer et al. (1981), have used jackknifing to estimate parameters of models of interest to microbial ecologists. In one study (Cornish-Bowden and Wong, 1978), the general jackknife was applied to analysis of initial velocity data obtained for liver alcohol dehydrogenase. These authors found that the degree of ill-conditioning in their model prevented them from estimating SEs for particular parameters using the CM. Instead, these authors estimated the SEs of these parameters from the jackknife pseudovalues. Cornish-Bowden and

Wong (1978) were so taken with jackknifing that they recommend it "should be seriously considered as a general technique for analysis of enzyme kinetic results." In a much earlier study, Dammkoehler (1966) used the general jackknife in conjunction with the Levenberg–Marquardt NPE algorithm to estimate parameters of models proposed to describe the kinetics of *E. coli* pyrophosphatase. He, too, emphasized the power of the general jackknife for estimating parameters of ill-conditioned enzymatic models. These few references to jackknife estimation suggest that this technique should be given more attention by environmental biologists, particularly those interested in comparing kinetic parameters obtained for different organisms and habitats.

4.8. Bootstrapping

Like jackknifing, bootstrapping is a means of eliminating statistical bias introduced by assuming experimental errors conform to a certain distribution (e.g., the Gaussian curve). But here the similarity ends. Jackknifing is a re-sampling strategy in which data values are systematically withdrawn from the original data set and the remaining values ($n - 1$; where n equals the number of data pairs) fitted to the function of choice. In contrast, bootstrapping is a re-sampling strategy in which many subsamples (e.g., 1000 sets of n data pairs) are randomly selected from the original set and analyzed (Efron, 1982). Bootstrapping has the advantage over jackknifing that the distributions of the parameters are actually generated during analysis. This permits interval estimates of parameters to be calculated directly, rather than constructing them from parameter estimates and their respective standard errors. Theoreticians have shown that jackknifing is a simplified form of bootstrapping (Efron, 1982), and although bootstrapping is a powerful, simple statistical method it has not yet been used in microbiological investigations due to its newness.

Implementation of the bootstrap is arithmetically simpler than the jackknife, although the former method requires much more computing power than jackknifing. Assume we wish to generate interval estimates for u and A of Eq. (10) using bootstrapping, given the six N–t data pairs in Table I. The first step is to write each of the six data pairs on separate cards, which are then placed into a hat. Next, one card is picked at random from the hat and the data pair recorded. The card is returned and a second card is drawn, the data pair recorded, the card returned, and so on until the first bootstrap sample of six N–t data pairs has been obtained. This entire process is repeated many times (e.g., 1000). Once the bootstrap samples are generated they are fitted to Eq. (10), thus producing 1000 estimates of u and A. Lastly, frequency histograms of the 1000 bootstrap estimates are constructed. With this done, interval estimates of u

and A may be approximated by finding, for example, the parameter intervals giving 95% of the areas under the respective frequency histograms of u and A.

Obtaining random sub-sets of the original data set need not be done as indicated in the above paragraph. This process may be accomplished using a pseudo-random number generator, a ubiquitous function among BASIC languages implemented for microcomputers. But the time required to generate and analyze the bootstrap samples can be prohibitive. For this reason, bootstrap analysis is best done on main-frame computer systems or on microcomputers equipped with arithmetic processing ("number crunching") boards.

Bootstrapping is the method of choice if statistically conservative information about the precision of parameter estimates is desired. Of the three methods, (1) the asymptotic approach using Eq. (19), (2) jackknifing, and (3) bootstrapping, the third method is the most conservative. In other words, interval estimates produced using these three techniques are the largest for bootstrapping and the smallest for the asymptotic approach. Interval estimates constructed via jackknifing are intermediate in statistical conservatism between those generated using Eq. (19) and bootstrapping. Notwithstanding the statistical superiority of bootstrapping, the need for high-speed computational facilities may often force the microbiologist to adopt the computationally simpler, but still statistically conservative, method of jackknifing.

Bootstrapping, like jackknifing, is a method with varied applications. Its use is not limited to estimating the parameters of nonlinear (and linear) models. Bootstrapping may be used in any case where the variance of a statistic (e.g., the mean of ten replicates) is desired, and where the analyst cannot assume the data conform to a certain distribution. Indeed the real power of this method stems from its wide applicability and general validity. The only assumption required for bootstrapping is the data set must be truly representative of the sampled population (Efron, 1982).

5. Residuals Analysis

Several techniques for examining residuals ($Y_{obs} - Y_{pred}$) exist (Draper and Smith, 1981), but simply plotting the residuals against the independent variable gives a great deal of information. Residuals analysis should be used to check the validity of the assumptions made about the measurement errors. It can suggest the appropriate weighting function for WLS NPE and detect correlated measurement errors.

Correlated measurement errors arise when the error in one Y_{obs} value influences the error in a Y_{obs} value obtained for a different value of the independent variable (Beck and Arnold, 1977). Correlated measurement errors show up as trends in residuals plots and can result from fitting data to the wrong model (Fig. 5). Correlated errors also may result when experimental devices are used (Kaspar and Tiedje, 1980; Robinson and Tiedje, 1982) to take data at short intervals between measurements. Regardless of the cause, ignoring the presence of correlated measurement errors can have statistically serious consequences.

5.1. Influence of Correlated Measurement Errors on Parameters Estimated via Least Squares Analysis

If data with correlated errors are fitted to models (linear or nonlinear) via least squares analysis, then underestimates of the SEs of the parameters result. This situation worsens as the number of data points increases. Esener *et al.* (1981) demonstrated that the variance of B for the linear model $Y = BX$ is directly proportional to the number of data points fitted to this equation in the presence of correlated measurement errors. Similar conclusions hold for variances of parameters of nonlinear models estimated via OLS or WLS (Beck and Arnold, 1977). Thus, investigators should not delude themselves into thinking that more data points are better for least squares estimation of parameters, unless they know that the measurement errors are uncorrelated.

The parameters of models, and their respective SEs, can be reliably estimated using techniques other than OLS or WLS analysis when the measurement errors are correlated. If the error covariance matrix is known, then ML estimation can be used to estimate the parameters and their SEs (Beck and Arnold, 1977). Since MAP estimation requires that the error covariance matrix be known, this technique also can be used to obtain more reliable estimates of parameters from data with correlated errors.

5.2. Number of Data Points Required for Detection of Correlated Errors

Eight to ten data points are usually sufficient to define a curve for NPE, but this data density may not be sufficient to detect correlated measurement errors. It has been my experience that nearly twice this range (upwards of 20 data points) may be needed to detect the presence of correlation among measurement errors.

6. Model Discrimination

Given a data set, an investigator is often concerned with determining which of several models best represents that data set. Models of increasing complexity typically lower the RSS below that obtained for a simpler model. However, the question is whether or not the reduction in the RSS is statistically significant. Of the several methods that exist for discriminating among competing models, Beck and Arnold (1977) recommend an F-test.

The F-test to discriminate among competing models is outlined in Table II. The models proposed to represent the data are arranged in order of increasing complexity, both with respect to their functional forms and the number of parameters. Each model is then fitted to the data using the appropriate parameter estimation method. For nonlinear models, NPE methods should be used, since analysis with linearized forms will not typically yield the minimum RSS value. The changes in the RSS going from the simplest to the more complex models are then calculated. For the pairwise combinations of interest, the changes in the RSS values are divided by the RMS obtained for the more complex model of each pair (Table II). These quotients are compared with the appropriate tabular F-value (numerator degrees of freedom, one; denominator degrees of freedom, $n - p$, where p is the number of parameters of the more complex model). If the calculated F-value is less than the tabular F-value, then the data are adequately described by the simpler of the two models. This test is typically carried out at the 5% level of statistical significance. As many models as the investigator is willing to propose can be fitted to the data and compared to determine which is best.

Some microbiologists have used other criteria to discriminate among competing models. Koch (1982) used a corrected coefficient of determination (Seber, 1977) to compare the fits of several different models to growth data obtained for *E. coli*. Koch (1982) concluded that the Monod model is not generally valid for this bacterium. Similar conclusions were reached by others (Powell, 1967; Dabes *et al.*, 1973), but they must be cautiously accepted, since these authors merely ranked the fits of the var-

Table II. Discrimination among Competing Models Using the F-Test According to Beck and Arnold (1977)[a]

Model 1	RSS_1	$RSS_1/(n - p_1)$	—
Model 2	RSS_2	$RSS_2/(n - p_2)$	$(RSS_1 - RSS_2)(n - p_2)/RSS_2$
Model 3	RSS_3	$RSS_3/(n - p_3)$	$(RSS_2 - RSS_3)(n - p_3)/RSS_3$
⋮	⋮	⋮	⋮
Model i	RSS_i	$RSS_i/(n - p_i)$	$(RSS_{i-1} - RSS_i)(n - p_i/RSS_i$

[a]The last column gives the calculated F value, which is compared with the value of the F statistic for one degree of freedom in the numerator and $n - p_i$ degrees of freedom in the denominator. The subscript i indicates the number of parameters for the ith model. Note that the differences in the RSS values for the different models are expressed as absolute quantities.

ious growth models according to the minimum RMS (Powell, 1967) or RSS (Dabes *et al.,* 1973) values. These studies emphasize the need for model-building tests in the quantitative analysis of microbiological phenomena.

The practical limit to model discrimination is partly set by the level of measurement error. The *F*-test (and other techniques) can discriminate between competing models up to a point, but when the error level is high, this test no longer detects differences in the fits (Table III). Thus a simpler model may fit data as poorly as a more complex model when the data are "noisy." Although in some cases a careful examination of the residuals may suggest that the more complex model is better, the SEs of the parameter may exceed the parameter estimates themselves.

The *F*-test for model discrimination is only approximate for nonlinear models. Unfortunately, the degree to which this test is approximate is a function of how nonlinear the models are that are being compared. This should be reminiscent of the statistical limitations of calculating standard errors of parameters of nonlinear models using Eq. (19). Due to the approximate nature of the *F*-test for nonlinear models, recent conclusions drawn from its application (Robinson and Characklis, 1984; Simkins and Alexander, 1984) must be cautiously accepted.

One statistically appropriate way of comparing the goodness-of-fits of nonlinear models is by using bootstrapping (Section 4.8). In addition to generating frequency histograms for the parameters of nonlinear models, bootstrapping may also be used to construct histograms for the residual sums-of-squares obtained by fitting many bootstrap samples to different nonlinear models. Visual inspection of the RSS histograms will usually reveal which nonlinear model best represents the experimental data.

Table III. Comparison of Linear, First-Order, and Integrated Michaelis–Menten Fits to Michaelis–Menten Progress Curve Data for Various Levels of Homoscedastic (Simple) Errors[a]

Error level SD (% of S_0)	Michaelis–Menten decay RSS	First-order decay RSS	Zero-order decay RSS
0.1	3.36×10^{-5}	1.68×10^{-2}*	1.88×10^{-2}*
0.5	5.47×10^{-4}	1.46×10^{-2}*	2.32×10^{-2}*
1.0	5.13×10^{-3}	1.86×10^{-2}*	2.23×10^{-2}*
5.0	1.15×10^{-1}	1.50×10^{-1}	1.21×10^{-1}

[a]Michaelis–Menten decay is described by Eq. (31), first-order decay is described by Eq. (26), and zero-order decay is given by $Y = S_0 - Bt$. The RSS values were obtained for nonlinear fits (excluding the linear model) to error-containing data generated via Monte Carlo simulation (Harbaugh and Bonham-Carter, 1970). The asterisk indicates that the simpler models do not adequately represent the progress curve data as determined using the *F*-test (Table II) at the 5% level of statistical significance.

7. Optimal Experimental Design

Investigators should seek to design experiments that maximize the quality of information that can be extracted from data. Optimally designed experiments are desirable since they provide the highest quality of statistical information for a given expenditure of resources. The definition of an optimally designed experiment depends on the investigator's goal. A microbiologist may need to choose conditions that maximize differences among competing models. On the other hand, it may be necessary to design optimal experiments for parameter estimation. When time is an independent variable, experiments may be designed to be of optimal duration. Each one of these three experimental goals can lead to different optimal designs.

7.1. Optimal Experiments for Model Discrimination

Model discrimination is difficult if experimental conditions are chosen that do not accent differences among competing models proposed to represent the process of interest. For example, assume we are interested in deciding whether or not a product P_f from a substrate S is reversibly or irreversibly bound to an inorganic matrix (Fig. 8). The mathematical models describing this process yield similar results over time, up until $t = t_1$. Thus, in order to discriminate between these two kinetic models, it is necessary to make measurements of P_f beyond t_1, e.g., until $t = t_2$ (Fig. 8).

Several mathematical criteria for designing optimal experiments for model discrimination exist, but one of the simplest seeks to locate the

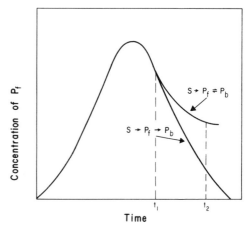

Figure 8. Curves for the concentration of a free product P_f produced by biotransformation of a substrate S. The two curves show the predicted behavior for the case where (1) P_f binds reversibly to an inorganic matrix and (2) P_f adsorbs irreversibly to an inorganic matrix. The two curves are identical until $t = t_1$.

combination of values of the independent variable(s) that maximizes the difference between the minimum RSS values for two competing models (Fedorov, 1972). The equations needed to apply this statistical criterion are supplied by Fedorov (1972) and Beck and Arnold (1977). There are other techniques, some of which are difficult to implement, and these have been surveyed by Hill (1978).

Model discrimination, like several topics in the field of NPE, is an active area of research by theoreticians. One manifestation of this activity is the controversy (Hill, 1978) over which of the many techniques proposed for designing discrimination experiments is best. Beck and Arnold (1977) illustrate this problem by demonstrating that the application of two different criteria for designing experiments to discriminate between an exponential versus a trigonometric model leads to different recommendations. Due to this controversy, a microbial ecologist interested in designing optimal experiments for model discrimination should seek the advice of an expert in regression analysis. Experimental resources may be saved if this is done before data collection is begun.

7.2. Optimal Experiments for Parameter Estimation

Once an investigator knows that a particular function adequately describes a process under study, experiments may be designed to estimate the parameters of that model with the highest possible precision (i.e., with the lowest possible variance).

7.2.1. Nonlinear Models

A great deal of theoretical work has been done on the design of optimal experiments for parameter estimation. Box and Lucas (1959) laid the foundations for much of the statistical research in this area. These authors showed that the best possible design for an experiment for parameter estimation entails choosing the values of the independent variable that minimize the determinant (DET) of the CM, which is equivalent to maximizing the DET of the SM (Section 4.3.2b). When this criterion is used, then the resulting parameter estimates have the lowest variances possible (Box and Lucas, 1959). Including additional data points away from the optimal design points will only lessen the precision of the parameter estimates (Fedorov, 1972; Duggleby, 1979; Endrenyi and Chan, 1981; Currie, 1982). Of course, determining the values of the independent variable that yield the optimal parameter estimation design requires some estimates of the parameters beforehand. Hence, an experiment designed to check the validity of a model, which should be done before designing optimal experiments for parameter estimation, does not provide parameter values with minimum variances. Such an experiment, however, yields parameter estimates needed to determine the optimal design using the Box–Lucas criterion.

An experimental design for obtaining parameter estimates with minimum variances can be radically different from a design for checking the validity of a model. For example, to check whether or not a given decay process is first order, an investigator typically monitors substrate depletion with time until the reaction is nearly complete (e.g., when the substrate concentration is 5% of its starting value). The resulting data are then fitted to the first-order decay model [Eq. (26)] and the goodness of fit assessed in some manner. However, this experimental design does not provide the best estimate of the first-order decay coefficient k. Using the Box–Lucas design criterion, it can be shown that the best estimate of k will be obtained by taking all measurements when $t = 1/k$. The precision of the k estimate obtained for this design can be an order of magnitude greater than the precision of the k estimate obtained from the more traditional suboptimal design (J. A. Robinson, unpublished results).

The application of the Box–Lucas criterion for parameter estimation has received a great deal of attention by pharmacokineticists (Cobelli *et al.*, 1983; Mori and DiStefano, 1979; DiStefano, 1980, 1981; Cobelli and DiStefano, 1980). The importance of optimal designs to pharmacologists interested in parameter estimation stems from their reliance on blood samples for data. Due to this constraint, the number of samples and the intervals between samplings must be minimized. The models of interest to these investigators usually consist of several exponential terms that describe the partitioning of metabolites among various tissues. The minimum sampling schedules derived by pharmacokineticists can be used by microbial ecologists studying the first-order partitioning of substrates among various pools in natural habitats (e.g., the fate of added nitrogen in soil), since these different processes can be described by similar exponential functions.

Enzyme kineticists and statisticians have examined the ability of various experimental designs for estimating V_{max} and K_m from initial velocity data. For measurement errors that are homoscedastic, the best estimates of V_{max} and K_m (i.e., those with minimum variances) are obtained when the initial velocity is measured at two design points. One point is the highest substrate concentration attainable and the other corresponds to the substrate concentration at which the initial velocity is one-half of the rate at the highest substrate concentration (Duggleby, 1979; Endrenyi and Chan, 1981; Currie, 1982). In contrast, when the measurement errors are relative and the coefficient of variation is constant, the optimal design points are the highest and lowest substrate concentrations attainable (Endrenyi and Chan, 1981).

The number of optimal design points for parameter estimation equals the number of parameters of the nonlinear model, and the values of the independent variable(s) that yield these design points can sometimes be determined from analysis of the sensitivity equations. Intuitively, the optimal design should be one for which the sensitivity of the dependent variable to changes in the parameter(s) is greatest. For the differential form of the Michaelis–Menten model [Eq. (1)] the sensitivity equations for V_{max} [Eq. (2)] and K_m [Eq. (3)] are maximal when S equals infinity and K_m, respectively. These substrate concentrations are the optimal design points obtained using the Box–Lucas criterion for homoscedastic errors. A similar analysis of the sensitivity equation for first-order decay [Eq. (26)], and appearance [Eq. (5)], reveals that the sensitivity equation for k is maximal when $t = 1/k$. Thus, not only can an examination of the sensitivity equations be used to predict the identifiability of a nonlinear model, these equations can also be used to obtain information on the optimal design for parameter estimation (Fedorov, 1972; Bard, 1974; Beck and Arnold, 1977; Pierce *et al.*, 1981).

7.2.2. Optimal Design for the Estimation of B and A in Y = BX + A

The linear model $Y = BX + A$ is so routinely used that giving the optimal design for estimating B and A is appropriate here. The optimal design for estimating B and A is different from the traditional linear or geometric design required to check that the data are linear. To obtain minimum variance estimates of B and A, the experimenter needs only to measure the value of Y at the highest and lowest values of X (Beck and Arnold, 1977). Of course, these optimal design points must be replicated, e.g., five replicate Y_{obs} values for each of the two values of X. The difference between the variances of B and A for the optimal versus a suboptimal design may be 300% (Beck and Arnold, 1977). This two-point design should only be applied when $Y = BX + A$ correctly describes the relationship between Y and X, and when the measurement errors are normally distributed, have constant variance, and are uncorrelated (Beck and Arnold, 1977).

7.3. Experiments of Optimal Duration

In the investigation of dynamic systems, strict application of the Box–Lucas design criterion is impractical. An investigator generally cannot set the time value precisely at the optimal design point(s) [e.g., $t = 1/k$ for a first-order process] and repeatedly measure the value of the dependent variable. Instead, the investigator usually has a choice of (1) measuring the dependent variable at evenly spaced intervals (linear design) or (2) bunching up the measurements at the beginning or end of the experiment (geometric design). For evenly spaced measurements, criteria have been developed (Fedorov, 1972) that give the optimal duration for an experiment. Such an experiment provides estimates of the parameters with minimum variances. Of course, these minimum variances will be greater than those obtained by strict application of the Box–Lucas design criterion.

Fedorov's (1972) criteria have been applied to the design of experiments for estimating parameters of some models used by microbial ecologists and environmental biologists. Given homoscedastic errors and evenly spaced measurements, the optimal duration for a first-order decay process is the time the initial substrate concentration equals 18.4% of its starting value (Beck and Arnold, 1977). This time is different from that of $t = 1/k$ given by applying the Box–Lucas design criterion to the first-order decay model. The design of experiments of optimal duration for enzymatic models has also received attention (Heineken *et al.*, 1967).

8. Some Models of Interest to Microbial Ecologists

In this section, I discuss fitting data to several models currently used by microbiologists. This section is not comprehensive; rather, it illustrates the types of nonlinear models applicable to microbiological phenomena. For some models, I give the sensitivity equations to complete the information required for parameter estimation via the Gaussian method (Section 4.3.2b). Although these equations are relatively simple to derive using basic methods of differentiation (Thomas, 1972), they can be approximated by numerical (algebraic) methods (Beck and Arnold, 1977; Duggleby, 1981). Main-frame computer packages for NPE often approximate these equations using numerical methods (Draper and Smith, 1981).

8.1. Exponential Models

8.1.1. Decay

The exponential decay equation receives a great deal of use by microbial ecologists, particularly those studying the fate of xenobiotics in natural environments (Baughman *et al.*, 1980; Paris *et al.*, 1982; Horowitz *et al.*, 1983; Suflita *et al.*, 1983). This equation has the form

$$S = S_0 e^{-kt} \tag{26}$$

where S is the substrate concentration at time t, S_0 is the substrate concentration at $t = 0$, and k is the first-order decay coefficient. Equation (26) is a two-parameter nonlinear model if S_0 is treated as another unknown. In practice, treating an initial condition (e.g., S_0) as another parameter to be estimated is more realistic than assuming that this condition is known with certainty. The sensitivity equations needed for updating S_0 and k via the Gaussian method according to Eq. (15) are

$$dS/dS_0 = e^{-kt} \qquad (27)$$

and

$$dS/dk = -S_0 t e^{-kt} \qquad (28)$$

Equation (26) can be linearized by taking the natural logarithm of both sides of this expression. This linearization can be used to obtain initial estimates of S_0 and k for subsequent improvement via an NPE method. The linearized version of Eq. (26) does have the advantage over the nonlinear version that heteroscedastic errors, if present in the untransformed dependent variable, are approximately transformed into homoscedastic errors (Zar, 1974).

If an investigator chooses to estimate k and S_0 using only the linearized form, then the untransformed data should be plotted against the solution to Eq. (26) solved for the S_0 and k estimates obtained from analysis of the linearized data [a recommendation that holds for any nonlinear model (Silvert, 1979)]. Plotting the transformed data against the best-fit linearized form of Eq. (26) can make agreement between substrate depletion data and this kinetic model look deceptively good.

8.1.2. Appearance

The first-order appearance equation [Eq. (5)] is another two-parameter nonlinear model of interest to microbial ecologists. The sensitivity equation for k (defined in this case as the product formation rate coefficient) is different only in sign from the sensitivity equation for the decay coefficient k of Eq. (26). For Eq. (5), the sensitivity of P to changes in S_0 equals one minus Eq. (27).

8.1.2a. Linearized Forms. A linearized form of Eq. (5) that permits simultaneous estimation of S_0 and k cannot be derived (Section 3.2.5). When S_0 is known with certainty, the slope of a plot of either $\ln(P/S_0)$ or $\ln P$ versus t gives a provisional estimate of k.

8.1.2b. Variations of Equation (5). Variations of the first-order appearance model have been used by microbiologists (Larson, 1980) to describe more complex product appearance behavior. Two of these equations are

$$P = S_0\{1 - \exp[-k(t - c)]\} \qquad (29)$$

and

$$P = S_0\{1 - \exp[-k(t - c)^2]\} \qquad (30)$$

Equation (29) is identical to Eq. (5), except that it allows for a positive displacement (namely, c) along the t axis. Equation (30), because of the $(t - c)^2$ term, gives an S-shaped curve and also allows for a lag period equal to c.

The parameters of Eqs. (29) and (30) have been estimated using NPE methods for CO_2 formation from alkyl benzene sulfonates (Larson, 1980), but the sensitivity equations for k and c suggest that estimates of these parameters are inaccurate. The sensitivity equations for k and c differ only by constants. This ill-conditioning results in high correlation between estimates of k and c, which in turn casts doubt on the accuracy of estimates of these parameters. This situation reinforces the benefits of plotting the sensitivity equations for the given model(s) prior to data collection.

8.2. Michaelis–Menten Model

8.2.1. Differential Form

Development of a statistically reliable method for estimating V_{max} and K_m using the differential form of the Michaelis–Menten model [Eq. (1), where Y is the initial rate of substrate consumption or product formation, $A = V_{max}$, $B = K_m$, and X is the substrate concentration] has been the focus of considerable research since Michaelis and Menten (1913) proposed this nonlinear model. Enzyme kineticists currently feel that estimating V_{max} and K_m from untransformed v–S data pairs is superior to estimating these parameters using linearized forms of Eq. (1) (Garfinkel et al., 1977; Cleland, 1979; Atkins and Nimmo, 1980), although, as pointed out by Atkins and Nimmo (1980), even nonlinear estimation of V_{max} and K_m is not entirely satisfactory. These authors recommend that the nonparametric direct linear plot (Eisenthal and Cornish-Bowden, 1974; Porter and Trager, 1977; and Cornish-Bowden et al., 1978) be used to obtain estimates of V_{max} and K_m and their respective SEs, in spite of the statistical limitations of this method (Kohberger, 1980). However, many enzyme kineticists continue to rely on linearized forms of Eq. (1) to estimate V_{max} and K_m despite the statistical admonitions of their colleagues (Wilkinson, 1961; Dowd and Riggs, 1965; Cornish-Bowden, 1979; Atkins and Nimmo, 1980).

With few exceptions (Betlach *et al.*, 1981), microbiologists, too, rely on linearized forms of Eq. (1) to estimate V_{max} and K_m from initial velocity data, particularly the Lineweaver–Burk linearization. If microbiologists continue to rely on linearized forms of Eq. (1), they at least should abandon the Lineweaver–Burk expression [Eq. (4)] and adopt the Hanes equation ($S/v = K_m/V_{max} + S/V_{max}$). This linearization of Eq. (1) is the most reliable of the linearizations for estimating V_{max} and K_m from initial velocity data (Dowd and Riggs, 1965).

Duggleby (1981) recently described a BASIC microcomputer program for estimating V_{max} and K_m from v–S data pairs, using the Gaussian method. This program is applicable to any two-parameter model [e.g., Eq. (26)], and can readily be changed to accommodate models with three or more parameters. It allows an investigator to select the weighting function if WLS NPE is appropriate. Further, biweight regression can be used to reduce the influence of outliers. The sensitivity equations are approximated using numerical differentiation, although they can be incorporated into the program itself. Finally, an examination of the code for this program should eliminate the misconception that NPE necessarily requires a main-frame computer.

8.2.2. Integral Form

The differential form of the Michaelis–Menten model [Eq. (1)] can be integrated to give the following equation:

$$V_{max}t = S_0 - S + K_m \ln(S_0/S) \tag{31}$$

This equation has the advantage over Eq. (1) that estimates of V_{max} and K_m can be obtained from one substrate depletion curve. However, estimating V_{max} and K_m from only a single progress curve is not recommended (Cornish-Bowden, 1979). A minimum of three progress curves should be run at different values of S_0 to test for product inhibition and other obfuscating factors (Yun and Suelter, 1977; Cornish-Bowden, 1979).

The sensitivity equations for the parameters of Eq. (31) are

$$\frac{dS}{dV_{max}} = \frac{t}{1 + K_m/S} \tag{32}$$

$$\frac{dS}{dK_m} = \frac{\ln(S_0/S)}{1 + K_m/S} \tag{33}$$

and

$$\frac{dS}{dS_0} = \frac{1 + K_m/S_0}{1 + K_m/S} \tag{34}$$

Only a few microbial ecologists (Betlach and Tiedje, 1981; Robinson and Tiedje, 1982, 1984) have fitted progress curve data directly to Eq. (31). A possible reason for this is the implicit nature of Eq. (31). This nonlinear model cannot be explicitly solved for S; hence, fitting data directly to it requires that the Y_{pred} (equals S_{pred}) values needed in the iterative process must be numerically approximated [see methods in Burden *et al.* (1978)].

In contrast to microbiologists, enzymologists (Atkins and Nimmo, 1973; Nimmo and Atkins, 1974; Duggleby and Morrison, 1977) have largely used the product appearance form of the integrated Michaelis–Menten model. Some researchers (Atkins and Nimmo, 1980) currently feel this equation is not reliable for estimating V_{max} and K_m, and that it should be used only for estimating initial velocities. The sensitivity equations needed for fitting progress curve data to the product form of Eq. (31) can be found elsewhere (Atkins and Nimmo, 1973; Duggleby and Morrison, 1977).

Most microbiologists (Strayer and Tiedje, 1978; Schauer *et al.,* 1982; Lovley *et al.,* 1982; Suflita *et al.,* 1983) use linearized forms of Eq. (31) to estimate V_{max} and K_m from substrate depletion data. However, the use of any linearized form of Eq. (31) violates the assumption of an essentially error-free independent variable, because S appears on both sides of these linearized forms.

Cornish-Bowden (1979) gives three different linearizations for Eq. (31). Of these three, Robinson and Characklis (1984) showed that the most reliable one for estimating V_{max} and K_m from linearized S–t data is

$$\frac{t}{S_0 - S} = \frac{K_m}{V_{max}} \frac{\ln(S_0/S)}{S_0 - S} + \frac{1}{V_{max}} \tag{35}$$

In addition, this linearization is the least sensitive to errors in S_0. Robinson and Characklis (1984) found that a variation of the linearization most used by microbiologists (Strayer and Tiedje, 1978; Schauer *et al.,* 1982; Lovely *et al.,* 1982) is the least reliable of the several possible linearizations of Eq. (31).

Since there are several linearized forms of Eq. (31), it might be tempting to fit linearized S–t data to all of these forms and choose the one for which the coefficient of determination (r^2) is highest. However, the corrected sum of squares of the dependent variable, which appears in the denominator of the function that defines r^2 (Zar, 1974; Draper and Smith, 1981), is different for each linearization. Thus, comparing the r^2 values for different linearizations of Eq. (31) (or any nonlinear model) is not a valid criterion for judging which linearization is best.

8.3. Other Uptake Models

8.3.1. Wright–Hobbie Kinetics

Since its initial description, the Wright–Hobbie kinetic model (Wright and Hobbie, 1966; Hobbie and Crawford, 1969) has been widely applied in analysis of uptake data obtained for aquatic samples. The validity of this model is currently in doubt, but nonetheless the Wright–Hobbie equation frequently describes field data (Van Es and Meyer-Reil, 1982).

The Wright–Hobbie equation is a linearized variant of Eq. (1). Only recently has the general superiority of fitting untransformed data to the nonlinear form of the Wright–Hobbie model been appreciated (Li, 1983). Li (1983) provides the equations for estimating V_{max} and T_t from untransformed uptake data, as well as showing the statistical disadvantages of estimating these parameters via linear least squares analysis.

8.3.2. An Integral Uptake Equation

The following nonlinear model has been used to estimate cellular growth rates from ^{14}C-glutamate (Taylor, 1979) and ^{35}SO$_4^{2-}$ (Cuhel *et al.,* 1982) uptake data:

$$I = [Q_0/(k \ln 2)] [\exp(k \ln 2t) - 1] \qquad (36)$$

where I is the accumulation of the radionuclide in cellular constituents as a function of time t, Q_0 is the initial rate of radionuclide incorporation, and k is the reciprocal of the doubling time. Taylor (1979) and Cuhel *et al.* (1982) used NPE methods to estimate k in Eq. (36). They specified the program (and hence the algorithm) used to fit data to this nonlinear model, a practice that should be adopted by all investigators who fit data to nonlinear models using NPE techniques.

Equation (36) is identical to the colonization equation of Caldwell *et al.* (1981) [Eq. (10)], where $I = N$, $Q_0 = A$, and $k \ln 2 = u$. Thus the sensitivity equations for updating A and u of Eq. (10) may be used to update the parameters of Eq. (36). Like Eq. (10), Eq. (36) cannot be linearized by taking the natural logarithm of both sides of this nonlinear equation.

8.4. Growth Models

8.4.1. Steady-State Forms

The Monod equation is another commonly used nonlinear microbiological model. It is identical to Eq. (1) (where Y is the growth rate, $A = u_{max}$, $B = K_s$, and X is the substrate concentration). Writing the Monod model in the form of Eq. (1) implies that the growth rate and substrate concentration are the dependent and independent variables, respectively. However, the Monod equation is typically used for analysis of chemostat data where the dependent variable is the steady-state substrate concentration and the independent variable is dilution rate (equals growth rate). Hence, the appropriate form of the Monod model for estimation of u_{max} and K_s from chemostat data is

$$S_{ss} = K_s D/(u_{max} - D) \tag{37}$$

where S_{ss} is the steady-state substrate concentration, K_s is the half-saturation constant for growth, u_{max} is the maximum specific growth rate, and D is the dilution rate.

For Eq. (37), estimates of u_{max} and K_s cannot be updated using the sensitivity equations derived from Eq. (1). The sensitivity equations needed for application of the Gaussian NPE technique must be derived from Eq. (37), and these are

$$dS_{ss}/du_{max} = -K_s D/(D - u_{max})^2 \tag{38}$$

and

$$dS_{ss}/dK_s = D/(D - u_{max}) \tag{39}$$

8.4.1a. Linearized Forms. Analogous to the differential Michaelis–Menten model [Eq. (1)], three linearizations can be derived for Eq. (37). No one has reported on the ability of these different linearizations to provide reliable estimates of u_{max} and K_s from S_{ss}–D data. If the analogy with Eq. (1) holds, the double-reciprocal form of Eq. (37) should be the worst and the Hanes form of Eq. (37) should be the best of the linearizations for this nonlinear model. The latter linearization has the following form:

$$D/S_{ss} = (u_{max}/K_s) - (D/K_s) \tag{40}$$

However, as is true for all linearizations of nonlinear models, Eq. (40) should be used primarily to obtain provisional estimates of u_{max} and K_s.

8.4.1b. Alternative Growth Models. Growth data do not always conform to the Monod equation (Powell, 1967; Dabes *et al.,* 1973; Condrey, 1982; Koch, 1982). Models such as the Blackman (1905) bilinear and Powell (1967) diffusional models better describe growth data for some organisms than Eq. (1). Indeed, Condrey (1982) showed that Monod's (1942) data are more consistent with the Blackman kinetic model than with Eq. (1). Presumably, the resistance of investigators to adopt the Blackman model is based on the fact that it is discontinuous (Condrey, 1982), which makes it difficult to use for modeling steady-state behavior. In contrast, the Powell model, like the Monod equation, expresses the growth rate as a continuous function of the substrate concentration. This steady-state growth model is

$$S_{ss} = RD + DK_s/(u_{max} - D) \qquad (41)$$

where R is the parameter that accounts for the diffusional resistance a substrate experiences in traversing the cellular membrane (Powell, 1967).

Data can be fitted to Eq. (41) using the same sensitivity equations for u_{max} and K_s derived from Eq. (37). The sensitivity equation for R (dS_{ss}/dR) equals the dilution rate D.

Equation (41) cannot be linearized in the ways used to linearize the Monod model. Thus, initial estimates of R, u_{max}, and K_s must be determined either from previous experience or by guesswork.

8.4.2. Integral Growth Models

8.4.2a. Exponential Growth. The exponential growth equation is used by many investigators to estimate batch growth rates. It is standard practice to fit optical density data to the linearized version of this expression (Koch, 1981). This equation is identical to Eq. (26), except that the exponential term is positive and the dependent variable is cellular mass or numbers, not the substrate concentration. The sensitivity equations for the growth rate u and the initial population density N_0 are $N_0 t e^{ut}$ and e^{ut}, respectively.

8.4.2b. The Integrated Monod Equation. The rate at which substrate is consumed in a batch system can be described by the following equation:

$$dS/dt = -[u_{max}S/(K_s + S)] (X/Y) \qquad (42)$$

where X is the biomass concentration at time t and Y is the cellular yield coefficient. The biomass concentration in Eq. (42) can be eliminated by using the mass balance relation that gives X as a function of S, S_0, Y, and the initial biomass concentration X_0 [namely, $X = Y(S_0 - S) + X_0$]. After this substitution and integration, the following implicit nonlinear function is obtained:

$$C_1 \ln\{[Y(S_0 - S) + X_0]/X_0\} - C_2 \ln(S/S_0) = u_{max}t \qquad (43)$$

where

$$C_1 = (YK_s + YS_0 + X_0)/(YS_0 + X_0)$$
$$C_2 = YK_s/(YS_0 + X_0)$$

An equation for the increase in cellular mass with time can be derived by solving the mass balance relation for S and substituting the result for S in Eq. (43) (Pirt, 1975; Corman and Pave, 1983).

The advantage of using Eq. (43) is the same as that for using the integrated Michaelis–Menten model [Eq. (31)]: the parameters (u_{max}, K_s, and Y, in this case) can be estimated from a single substrate depletion curve. Partly for this reason, microbiologists (Powell, 1967; Robinson and Tiedje, 1983; Corman and Pave, 1983) and mathematicians (Holmberg, 1982) have investigated the general applicability of this nonlinear model for the estimation of Monod growth parameters. In some cases (Knowles *et al.*, 1965; Graham and Canale, 1982) the integrated Monod equation has been used with little regard to its limitations, which can be severe.

The sensitivity equations for u_{max}, K_s, and Y appear in Holmberg (1982) and Robinson and Tiedje (1983). These can be used to fit sigmoidal substrate depletion data to Eq. (43) using the Gaussian method. A linearized form of Eq. (43) can be derived from this model to obtain initial estimates of the parameters (Corman and Pave, 1983), or initial u_{max}, K_s, and Y estimates can be calculated using a linearized form of Eq. (42) and its dX/dt analog (Robinson and Tiedje, 1983).

Notwithstanding the advantage of Eq. (43) for estimating Monod parameters, this nonlinear model has limited applicability. One problem with its use is that the parameter estimates are highly correlated; the absolute values of the off-diagonal elements of the PCM generally exceed 0.9 (J. A. Robinson, unpublished results). Thus, parameter estimates obtained for even very good data (Robinson and Tiedje, 1983) tend to be inaccurate. Another limitation of Eq. (43) is that it often fails to predict non-steady-state bacterial growth (Powell, 1967). Finally, Eq. (43) lacks a term for maintenance metabolism, which can be significant for some bacteria. However, inclusion of a maintenance term would only make the parameter estimation problem worse.

8.5. Models with More Than One Independent Variable

8.5.1. Diffusion Equations

All of the nonlinear models discussed up to this point have one independent variable (time, substrate concentration, or dilution rate). There are many models with two independent variables applicable to microbiological phenomena. For example, the following nonlinear model describes the time and space dependences (in one dimension) of the concentration of a solute (e.g., O_2, sulfide) obeying Fickian diffusion (Thibodeaux, 1979):

$$\partial S/\partial t = D \, \partial^2 S/\partial x^2 \tag{44}$$

where $\partial S/\partial t$ is the first partial derivative of the solute concentration with time t, D is the diffusion coefficient, and $\partial^2 S/\partial x^2$ is the second partial derivative of the solute concentration with distance x. The partial derivatives are needed here because the rate of change of S with time depends on t and x.

Equation (44) describes only diffusion, but the right-hand side of this model can be amended with terms for first-, mixed-, and zero-order biological sink and/or production terms. Similar equations have been derived to describe the fate of compounds in sedimentary systems (Berner, 1980). For such models, NPE methods must be used to simultaneously estimate the diffusivity of the solute D as well as the parameters of the biological term(s).

In recent years, microelectrodes have been used to investigate the dynamics of O_2 consumption in sediments and microbial mats on a microscale (e.g., Revsbech et al., 1980; Revsbech and Jørgensen, 1983; Revsbech and Ward, 1983). These devices offer unique opportunities for the application of NPE methods to simultaneously estimate both the physical and biological parameters in diffusion equations. Engineers have applied similar technologies for estimating the parameteres of heat flow models (Beck and Arnold, 1977), which are similar to equations that describe solute diffusion (Thibodeaux, 1979). Further, strategies have emerged from the application of optimal design criteria that predict values of (1) the independent variables, both time and distance, and (2) the number of probes (e.g., microelectrodes) needed for optimal estimation of parameters in models like Eq. (44) (Beck and Arnold, 1977). This entire area of NPE could lend increased sophistication to microelectrode studies by microbial ecologists, and deserves attention by these investigators.

9. Concluding Remarks

NPE is a branch of statistics that provides more than just techniques for estimating the parameters of nonlinear models. Typically, an investigator begins with a model(s) believed to represent a given microbial process. NPE methods may be used, first, to design model discrimination experiments, and second, to obtain minimum variance estimates of the parameters of the best model. These design criteria, when appropriately applied, will maximize the quality of statistical information gained from a given expenditure of experimental resources.

The development of algorithms for updating parameters of nonlinear models is an ongoing process. This stems from the fact that no algorithm for parameter estimation is universally applicable, although some techniques are much better than others. For this reason, the microbial ecologist or environmental biologist should specify the NPE algorithm (e.g., Levenberg–Marquardt, Gaussian, Box–Kanemasu variation of the Gaussian method) used to estimate the parameters of nonlinear methods.

Estimates of the precision with which parameters are determined should be reported with estimates of the parameters themselves. Point estimates of microbial parameters should not be considered sufficient. Variances of parameters can be calculated from the covariance matrix of the parameters, but these variance estimates can be unrealistic. When knowledge of the variances of the parameters is critical, a more conservative technique, such as jackknifing (or bootstrapping, if high-speed computational facilities are available), should be used. Regardless of the method used, specification of the technique used to estimate the variances of nonlinear parameters is essential.

Linearized forms of nonlinear models should be relied on to provide provisional parameter estimates and for illustrative purposes. The wide availability of microcomputers and the ease with which certain NPE methods (e.g., the Gaussian method) can be implemented make linear least squares analysis of transformed data difficult to defend. When an investigator chooses to use a linearized form of a nonlinear model, some attempts should be made to determine if alternative linearizations are more reliable.

The microbial ecologist should appreciate that NPE is a dynamic field in statistics. New techniques and strategies are continually emerging. This makes staying abreast of developments in this technical field difficult for the nonspecialist. Nonetheless, discovering what is possible in this area of statistical research can only enhance the development of the quantitative approach in microbial ecology. If nothing else, cultivating an awareness of NPE methods provides the environmental biologist with the rudiments of a language with which to communicate experimental needs to the statistician.

Note added in proof: In the May 1984 issue of *Byte* magazine (a monthly journal for microcomputer enthusiasts), M. Caceci and W. Caderis illustrate the application of the Simplex algorithm (Nelder and Mead, 1965) to estimating parameters of nonlinear models. This algorithm has several advantages over the method (viz., Gaussian) illustrated in Section 4.3.2b. The authors delineate these advantages and show how the Simplex algorithm may be used to estimate V_{max} and K_m from initial velocity data. The Simplex method is one of the newest algorithms proposed for fitting data to models and deserves the attention of microbial ecologists interested in NPE.

ACKNOWLEDGMENTS. I thank William Characklis (Director) and Keith Cooksey (Co-Director) of the Institute for Biological and Chemical Process Analysis (IPA), Bozeman, Montana, for financial support and encouragement. I thank Maarten Siebel and Richard Lewis for reviews of the manuscript. Thanks are given to Pat, Russell, and Nancy for continuing support.

References

Arthenari, T. S., and Dodge, Y., 1981, *Mathematical Programming in Statistics,* Wiley, New York.

Askelof, P., Korsfeldt, M., and Mannervik, B., 1976, Error structure of enzyme kinetic experiments, *Eur. J. Biochem.* **69**:61–67.

Atkins, G. L., and Nimmo, I. A., 1973, The reliability of Michaelis constants and maximum velocities estimated by using the integrated Michaelis–Menten equation, *Biochem. J.* **135**:779–784.

Atkins, G. L., and Nimmo, I. A., 1980, Current trends in the estimation of Michaelis–Menten parameters, *Anal. Biochem.* **104**:1–9.

Bard, Y., 1970, Comparison of gradient methods for the solution of nonlinear parameter estimation problems, *SIAM J. Numer. Anal.* **7**:157–186.

Bard, Y., 1974, *Nonlinear Parameter Estimation,* Academic Press, New York.

Baughman, G. L., Paris, D. F., and Steen, W. C., 1980, Quantitative expression of biotransformation rate, in: *Biotransformation and Fate of Chemicals in the Aquatic Environment* (A. S. Maki, K. L. Dickson, and J. Cairns, Jr., eds.), pp. 67–86, American Society for Microbiology, Washington, D.C.

Beck, J. V., and Arnold, K. J., 1977, *Parameter Estimation in Engineering and Science,* Wiley, New York.

Berner, R. A., 1980, *Early Diagenesis: A Theoretical Approach,* Princeton University Press, Princeton.

Betlach, M. R., and Tiedje, J. M., 1981, Kinetic explanation for accumulation of nitrite, nitric oxide, and nitrous oxide during bacterial denitrification, *Appl. Environ. Microbiol.* **42**:1074–1084.

Betlach, M. R., Tiedje, J. M., and Firestone, R. B., 1981, Assimilatory nitrate uptake in *Pseudomonas fluorescens* studied using nitrogen-13, *Appl. Environ. Microbiol.* **129**:135–140.

Blackman, F. F., 1905, Optima and limiting factors, *Ann. Bot. (Lond.)* **19**:281–295.

Box, G. E. P., and Kanemasu, H., 1972, Topics in model building, Part II, On non-linear least squares, Technical Report No. 321, Department of Statistics, University of Wisconsin, Madison, Wisconsin.

Box, G. E. P., and Lucas, H. L., 1959, Design of experiments in non-linear situations, *Biometrika* **46**:77–90.

Burden, R. L., Faires, J. D., and Reynolds, A. C., 1978, *Numerical Analysis,* Prindle, Weber and Schmidt, Boston.

Caldwell, D. E., Brannan, D. K., Morris, M. E., and Betlach, M. R., 1981, Quantitation of microbial growth on surfaces, *Microb. Ecol.* **7**:1–11.

Cleland, W. W., 1979, Statistical analysis of enzyme kinetic data, *Meth. Enzymol.* **63**:103–183.

Cobelli, C., and DiStefano III, J. J., 1980, Parameter and structural concepts and ambiguities: A critical review and analysis, *Am. J. Physiol.* **239**:R7–R24.

Cobelli, C., DiStefano III, J. J., and Ruggeri, A., 1983, Minimal sampling schedules for identification of dynamic models of metabolic systems of clinical interest: Case studies for two liver function tests, *Math. Biosci.* **63**:173–186.

Condrey, R. E., 1982, The chemostat and Blackman kinetics, *Biotechnol. Bioeng.* **24**:1705–1709.

Cooper, R. A., and Weekes, A. J., 1983, *Data, Models and Statistical Analysis,* Barnes and Noble, Totowa, New Jersey.

Corman, A., and Pave, A., 1983, On parameter estimation of Monod's bacterial growth model from batch culture data, *J. Gen. Appl. Microbiol.* **29**:91–101.

Cornish-Bowden, A., 1979, *Fundamentals of Enzyme Kinetics,* Butterworth, Boston.

Cornish-Bowden, A., and Endrenyi, L., 1981, Fitting of enzyme kinetic data without prior knowledge of weights, *Biochem. J.* **193**:1005–1008.

Cornish-Bowden, A., and Wong, J. T.-F., 1978, Evaluation of rate constants for enzyme-catalyzed reactions by the jackknife technique, *Biochem. J.* **175**:969–976.

Cornish-Bowden, A., Porter, W. R., and Trager, W. F., 1978, Evaluation of distribution-free confidence limits for enzyme kinetic parameters, *J. Theor. Biol.* **74**:163–175.

Cuhel, R. L., Taylor, C. D., and Jannasch, H. W., 1982, Assimilatory sulfur metabolism in marine microorganisms: Considerations for the application of sulfate incorporation into protein as a measurement of natural population protein synthesis, *Appl. Environ. Microbiol.* **43**:160–168.

Currie, D., 1982, Estimating Michaelis–Menten parameters: Bias, variance and experimental design, *Biometrics* **38**:907–919.

Dabes, J. N., Finn, R. K., and Wilke, C. R., 1973, Equations of substrate-linked growth: The case for Blackman kinetics, *Biotechnol. Bioeng.* **15**:1159–1177.

Dammkoehler, R. A., 1966, A computational procedure for parameter estimation applicable to certain non-linear models of enzyme kinetics, *J. Biol. Chem.* **241**:1955–1957.

Davies, M., and Whiting, I. J., 1972, A modified form of Levenberg's correction, in: *Numerical Methods for Non-linear Optimization* (F. A. Lootsma, ed.), pp. 191–201, Academic Press, London.

De Villiers, N., and Glasser, D., 1981, A continuation method for nonlinear regression, *SIAM J. Numer. Anal.* **18**:1139–1154.

DiStefano III, J. J., 1980, Design and optimization of tracer experiments in physiology and medicine, *Fed. Proc.* **39**:84–90.

DiStefano III, J. J., 1981, Optimized blood sampling protocols and sequential design of kinetic experiments, *Am. J. Physiol.* **240**:R259–R265.

Dixon, W. J., Brown, M. B., Engleman, L., Frane, J. W., Hill, M. A., Jennrich, R. I., and Toporek, J. D., 1981, *Biomedical Computer Programs,* University of California Press, Los Angeles.

Dowd, J. E., and Riggs, D. S., 1965, A comparison of estimates of Michaelis–Menten kinetic constants from various linear transformations, *J. Biol. Chem.* **240**:863–869.

Draper, N. R., and Smith, 1981, *Applied Regression Analysis,* Wiley, New York.

Duggleby, R. G., 1979, Experimental designs for estimating the kinetic parameters for enzyme-catalyzed reactions, *J. Theor. Biol.* **81**:671–684.

Duggleby, R. G., 1981, A nonlinear regression program for small computers, *Anal. Biochem.* **110**:9–18.

Duggleby, R. G., and Morrison, J. F., 1977, The analysis of progress curves for enzyme-catalyzed reactions by nonlinear regression, *Biochem. Biophys. Acta* **481**:297–312.

Efron, B., 1982, *The Jackknife, the Bootstrap and Other Resampling Plans,* Society for Industrial and Applied Mathematics, Philadelphia.

Eisenthal, R., and Cornish-Bowden, A., 1974, The direct linear plot, *Biochem. J.* **139**:715–720.

Endrenyi, L., and Chan, F.-Y., 1981, Optimal design of experiments for the estimation of precise hyperbolic kinetic and binding parameters, *J. Theor. Biol.* **90**:241–263.

Esener, A. A., Roels, I. A., and Kossen, N. W. F., 1981, On the statistical analysis of batch data, *Biotechnol. Bioeng.* **23**:2391–2396.

Fedorov, V. V., 1972, *Theory of Optimal Experiments,* Academic Press, New York.

Fox, T., Hinkley, D., and Larntz, K., 1980, Jackknifing in nonlinear regression, *Technometrics* **22**:29–33.

Garfinkel, L., Kohn, M. C., and Garfinkel, D., 1977, Systems analysis in enzyme kinetics, *CRC Crit. Rev. Bioeng.* **2**:329–361.

Goldberg, A. S., 1968, *Topics in Regression Analysis,* Macmillan, London.

Graham, J. M., and Canale, R. P., 1982, Experimental and modeling studies of a four-trophic level predator–prey system, *Microb. Ecol.* **8**:217–232.

Harbaugh, J., and Bonham-Carter, G., 1970, *Computer Simulation in Geology,* Wiley, New York.

Heineken, F. G., Tsuchiya, and Aris, R., 1967, On the accuracy of determining rate constants in enzymatic reactions, *Math. Biosci.* **1**:115–141.

Hill, P. D. H., 1978, A review of experimental design procedures for regression model discrimination, *Technometrics* **20**:15–21.

Hinkley, D. V., 1977, Jackknifing in unbalanced situations, *Technometrics* **19**:285–292.

Hobbie, J. E., and Crawford, C. C., 1969, Respiration corrections for bacterial uptake of dissolved organic compounds in natural waters, *Limnol. Oceanogr.* **14**:528–532.

Holmberg, A., 1982, On the practical identifiability of microbial growth models incorporating Michaelis–Menten type nonlinearities, *Math. Biosci.* **62**:23–43.

Horowitz, A., Suflita, J. M., and Tiedje, J. M., 1983, Reductive dehalogenations of halobenzoates by anaerobic lake sediment microorganisms, *Appl. Environ. Microbiol.* **45**:1459–1465.

Huber, P. J., 1972, Robust statistics: A review, *Ann. Math. Stat.* **43**:1041–1067.

Huber, P. J., 1981, *Robust Statistics,* Wiley, New York.

Jennrich, R. I., and Ralston, M. L., 1979, Fitting nonlinear models to data, *Annu. Rev. Bioeng.* **8**:195–238.

Kasper, H. F., and Tiedje, J. M., 1980, Response of electron capture detector to H_2, O_2, N_2O, NO and N_2, *J. Chromatogr.* **193**:142–147.

Knowles, G., Downing, A. L., and Barrett, M. J., 1965, Determination of kinetic constants for nitrifying bacteria in mixed culture, with the aid of an electronic computer, *J. Gen. Microbiol.* **38**:263–278.

Koch, A., 1981, Growth measurement, in: *Manual of Methods for General Bacteriology* (P.

Gerhardt, R. G. E. Murray, R. N. Costilow, E. N. Nester, W. A. Wood, N. R. Krieg, and G. B. Phillips, eds.), pp. 179–207, American Society for Microbiology, Washington, D.C.

Koch, A., 1982, Multistep kinetics: Choice of models for the growth of bacteria, *J. Theor. Biol.* **98**:401–417.

Koeppe, P., and Hamann, C., 1980, A program for non-linear regression analysis to be used on desk-top computers, *Comp. Programs Biomed.* **12**:121–128.

Kohberger, R. C., 1980, Statistical evaluation of the direct linear plot method for estimation of enzyme kinetic parameters, *Anal. Biochem.* **101**:1–6.

Larson, R. J., 1980, Role of biodegradation kinetics in predicting environmental fate, in: *Biotransformation and Fate of Chemicals in the Aquatic Environment* (A. W. Maki, K. L. Dickson, and J. Cairns, Jr., eds.), pp. 67–86, American Society for Microbiology, Washington, D.C.

Li, W. K. W., 1983, Consideration of errors in estimating kinetic parameters based on Michaelis–Menten formalism in microbial ecology, *Limnol. Oceanogr.* **28**:185–190.

Lovley, D. R., Dwyer, D. F., and Klug, M. K., 1982, Kinetic analysis of competition between sulfate reducers and methanogens for hydrogen in sediments, *Appl. Environ. Microbiol.* **43**:1373–1379.

Meyer, S. L., 1975, *Data Analysis for Scientists and Engineers,* Wiley, New York.

Michaelis, M., and Menten, M. L., 1913, Kinetics of invertase action, *Z. Biochem.* **49**:333.

Miller, R. G., 1974, The jackknife—A review, *Biometrika* **61**:1–15.

Monod, J., 1942, *Recherches sur la Croissance de Cultures Bacteriennes,* Herman, Paris.

Mori, F., and DiStefano III, J. J., 1979, Optimal nonuniform sampling interval and test-input design for identification of physiological systems from very limited data, *IEEE Trans. Automat. Control* **AC-26**:893–900.

Mosteller, F., and Tukey, J. W., 1977, *Data Analysis and Regression,* Addison-Wesley, Reading, Massachusetts.

Nelder, J.A., and Mead, R., 1965, A Simplex method for function minimization, *Computing Journal* **7**:308–313.

Nimmo, I. A., and Atkins, G. L., 1974, A comparison of two methods for fitting the integrated Michaelis–Menten equation, *Biochem. J.* **141**:913–914.

Nimmo, I. A., and Mabood, S. F., 1979, The nature of random experimental error encountered when acetylcholine hydrolase and alcohol dehydrogenase are assayed, *Anal. Biochem.* **94**:265–269.

Oppenheimer, L., Capizzi, T. P., and Miwa, G. T., 1981, Application of jackknife procedures to inter-experimental comparisons of parameter estimates for the Michaelis–Menten equation, *Biochem. J.* **197**:721–729.

Paris, D. F., Wolfe, N. L., and Steen, W. D., 1982, Structure activity relationships in microbial transformations of phenols, *Appl. Environ. Microbiol.* **44**:153–158.

Pierce, T. H., Cukier, R. I., and Dye, J. L., 1981, Application of nonlinear sensitivity analysis to enzyme mechanisms, *Math. Biosci.* **56**:175–208.

Pirt, S. J., 1975, *Principles of Microbe and Cell Cultivation,* Wiley, New York.

Porter, W. R., and Trager, W. F., 1977, Improved non-parametric statistical methods for the estimation of Michaelis–Menten kinetic parameters by the direct linear plot, *Biochem. J.* **61**:293–302.

Powell, E. O., 1967, The growth rate of microorganisms as a function of substrate concentration, in: *Microbial Physiology and Continuous Culture* (E. O. Powell, C. G. T. Evans, R. E. Strange, and D. W. Tempest, eds.), pp. 34–56, Her Majesty's Stationary Office, London.

Quenouille, M. H., 1956, Notes on bias in estimation, *Biometrika* **43**:353–360.

Revsbech, N. P., and Jørgensen, B. B., 1983, Photosynthesis of benthic microflora measured with high spatial resolution by the oxygen microprofile method: Capabilities and limitations of the method, *Limnol. Oceanogr.* **28**:749–756.

Revsbech, N. P., and Ward, D. M., 1983, Oxygen microelectrode that is insensitive to medium chemical composition: Use in an acid microbial mat dominated by *Cyanidium caldarium, Appl. Environ. Microbiol.* **45**:755–759.

Revsbech, N. P., Sørensen, J., Blackburn, T. H., and Lomholt, J. P., 1980, Distribution of oxygen in marine sediments measured with microelectrodes, *Limnol. Oceanogr.* **25**:403–411.

Roberts, D. V., 1977, *Enzyme Kinetics,* Cambridge University Press, London.

Robinson, J. A., and Characklis, W. G., 1984, Simultaneous estimation of V_{max}, K_m and the rate of endogenous substrate production (R) from progress curve data, *Microb. Ecol.* **10**:165–178.

Robinson, J. A., and Tiedje, J. M., 1982, Kinetics of hydrogen consumption by rumen fluid, anaerobic digestor sludge, and sediment, *Appl. Environ. Microbiol.* **44**:1374–1384.

Robinson, J. A., and Tiedje, J. M., 1983, Nonlinear estimation of Monod growth kinetic parameters from a single substrate depletion curve, *Appl. Environ. Microbiol.* **45**:1453–1458.

Robinson, J. A., and Tiedje, J. M., 1984, Competition between sulfate-reducing and methanogenic bacteria for H_2 under resting and growing conditions, *Arch. Microbiol.* **137**:26–32.

Schauer, N. L., Brown, D. P., and Ferry, J. G., 1982, Kinetics of formate metabolism in *Methanobacterium formicicum* and *Methanobacterium hungatei, Appl. Environ. Microbiol.* **44**:549–554.

Seber, G. A. F., 1977, *Linear Regression Analysis,* Wiley, New York.

Siano, D. B., Zyskind, J. W., and Fromm, H. J., 1975, A computer program for fitting and statistically analyzing initial rate data applied to bovine hexokinase type III isozyme, *Arch. Biochem. Biophys.* **170**:587–600.

Silvert, W., 1979, Practical curve fitting, *Limnol. Oceanogr.* **24**:767–773.

Simkins, S., and Alexander, M., 1984, Models for mineralization kinetics with the variables of substrate concentration and population density, *Appl. Environ. Microbiol.* **47**:1229–1306.

Storer, A. C., Darlison, M. G., and Cornish-Bowden, A., 1975, The nature of experimental error in enzyme kinetic measurements, *Biochem. J.* **151**:361–367.

Strayer, R. F., and Tiedje, J. M., 1978, Kinetic parameters of the conversion of methane precursors to methane in a hypereutrophic lake sediment, *Appl. Environ. Microbiol.* **36**:330–340.

Suflita, J. M., Robinson, J. A., and Tiedje, J. M., 1983, Kinetics of microbial dehalogenation of haloaromatic substrates in methanogenic environments, *Appl. Environ. Microbiol.* **45**:1466–1473.

Taylor, C. D., 1979, Growth of a bacterium under a high-pressure oxy-helium atmosphere, *Appl. Environ. Microbiol.* **36**:42–49.

Thibodeaux, L. J., 1979, *Chemodynamics, Environmental Movement of Chemicals in Air, Water and Soil,* Wiley, New York.

Thomas, G. B., Jr., 1972, *Calculus and Analytical Geometry,* Addison-Wesley, Reading, Massachusetts.

Van Es, F. B., and Meyer-Reil, L.-A., 1982, Biomass and metabolic activity of heterotrophic marine bacteria, in: *Advances in Microbial Ecology,* Vol. 6 (K. C. Marshall, ed.), pp. 111–170, Plenum Press, New York.

Wilkinson, G. N., 1961, Statistical estimations in enzyme kinetics, *Biochem. J.* **80**:324–332.

Wright, R. T., and Hobbie, J. E., 1966, Use of glucose and acetate by bacteria and algae in aquatic ecosystems, *Ecology* **47**:447–464.

Yun, S.-L., and Suelter, C. H., 1977, A simple method for calculating K_m and V from a single enzyme reaction progress curve, *Biochem. Biophys. Acta* **480**:1–13.

Zar, J. H., 1974, *Biostatistical Analysis,* Prentice-Hall, Englewood Cliffs, New Jersey.

3

Neglected Niches

The Microbial Ecology of the Gastrointestinal Tract

ADRIAN LEE

1. Introduction

Every day we excrete 100–200 g of feces. Given that 75% of the wet weight is composed of bacteria (Stephen and Cummings, 1980) and that each gram contains 1×10^{11} organisms belonging to up to 400 different species (Moore and Holdeman, 1974), it is clear that we are the outer casing of possibly one of the most highly evolved and complex microbial ecosystems of them all. Freter *et al.* (1983a) have commented on the apparent paradox that 100 years of intensive research has not brought us close to an understanding of what controls the indigenous microbiota* of the gastrointestinal tract. This lack of progress does not seem so surprising if we consider the intestine as a continuous culture vessel containing at the one time hundreds of organisms in steady state conditions. The study of even two or three organisms in steady state in a culture vessel is difficult enough. And yet, here the culture vessel of the gut compounds the problem by being composed of living animal cells.

The microbiology of the intestine has been the focus of many reviews

---

*Many workers in gastrointestinal microbiology still refer to the microbial flora of the tract (microflora). Taxonomically this term is inappropriate and the current convention is to use the word biota (microbiota). Familiarity with the words "normal flora" still results in "microbiota" sounding out of place. However, use of the correct term should be encouraged, and it has been used throughout this chapter.

ADRIAN LEE • School of Microbiology, University of New South Wales, Kensington, New South Wales 2033, Australia.

and it is clear that the data base on this topic has expanded dramatically in recent years. However, few of these articles consider the intestinal microbiota from a strictly ecological viewpoint; the aim of this review is to discuss the microbiology of the gut in such a framework. Thus, workers in other areas of microbial ecology will be able to compare the gut ecosystem to their own system. Researchers currently investigating the intestine may be prompted to rethink their approach. This review represents a personal viewpoint, which makes no attempt to review all the current literature, but highlights our areas of ignorance and challenges some current concepts.

The major emphasis is on the human gut ecosystem in health and disease and the contribution made by the results of studies on animals to this knowledge. The important but very different ecosystem, the rumen, is not discussed in this chapter; it has recently been very ably reviewed (Hobson and Wallace, 1982a,b). The fascinating world of the mouth and its associated organisms has also been discussed elsewhere (Bowden *et al.,* 1979).

2. Species Diversity in the Gastrointestinal Tract

As mentioned above, hundreds of different species of bacteria colonize the human intestine at any·one time. The major organisms are listed in Table I.

The intestine is an example of a community with very high species diversity with an even distribution of dominant species. A feature of ecosystems with high species diversity in the climax population is high stability. As Alexander (1971) has stated, "Though not directly proven for microbial communities it is likely that the tendency to maintain a balance among the resident organisms is enforced with increasing species diversity."

If one studies the succession and establishment of most microbial ecosystems, be it in a liter of milk, an effluent pond, or a baby's gut, the trend is from a low diversity index (Atlas, 1984) to a high index. The larger the number of species, the greater the ability to cope with minor changes, i.e., a change in one or more populations induced by modification in some environmental variable brings about responses in other populations in such a manner that the original biological fluctuation is opposed. With major environmental changes, species diversity declines and the community becomes less stable.

Certainly, the adult intestinal tract is a very stable system; alien or allochthonous organisms have great difficulty surviving. Radical changes in diet have surprisingly little influence on the composition of the normal

Table I. Predominant Bacterial Species Isolated from Human Feces[a]

Organism	Count[b]	Percent[c]
Bacteroides fragilis ss. *vulgatus*	11	12
Fusobacterium prausnitzii	10.5	7
Bifidobacterium adolescentis	10.5	6
Eubacterium acrofaciens	10.5	6
Peptostreptococcus productus II	10.5	6
Bacteroides fragilis ss. *thetaiotaomicron*	10.5	4
Eubacterium eligens	10	4
Peptostreptococcus productus I	10	3
Eubacterium biforme	10	3
E. aerofaciens III	10	2
Eubacterium rectale I	10	2
B. fragilis ss. *distasonis*	10	2
E. rectale II	10	2
B. fragilis ss. a	10	2
E. rectale IV	10	2
Bifidobacterium longum	10	2
Gemmiger formicilis nov. gen. nov. sp.	10	2
Bifidobacterium infantis	10	2
Ruminococcus bromii	10	1
Lactobacillus acidophilus	10	1
Bifidobacterium breve	10	1
Ruminococcus albus	10	1
B. fragilis ss. b	9.5	1
E. rectale III-F	9.5	1
Eubacterium ventriosum	9.5	1
Total		76

[a]From Moore and Holdeman (1974).
[b]Mean $\log^{10}$ count/g feces (dry weight).
[c]Percent of total count/g (dry weight).

microbiota. To understand the ecology of the gut, it is important to consider what contributes to this high species diversity and how it can be altered.

2.1. Factors Influencing Species Diversity

2.1.1. Nutrient Diversity

If two organisms in a continuous culture system have closely overlapping preferences for a growth-limiting substrate, they will compete severely for the common nutrient source, and thus the inferior organism will be excluded. However, if they have different preferences, i.e., the degree of overlapping is less, they may coexist, since these organisms will then occupy different ecological niches. This has been well stated by

Yoon *et al.* (1977). To explain the large numbers of different stable populations in the intestine, one has to postulate large numbers of distinct growth-limiting substrates.

This simple concept has been rarely discussed with regard to the intestinal tract, and yet it highlights the impossibility of us ever coming to a complete understanding of factors controlling the populations of intestinal bacteria. Thus, Freter *et al.* (1983b) state, "indigenous intestinal bacteria are controlled by substrate competition, i.e., that each species is more efficient than the rest in utilizing one or a few particular substrates and that the population level of that species is controlled by the concentration of these few limiting substrates." Given there are 400 different species present, we are talking about a very large number of potential substrates. Where do these substrates come from? Certainly, as Lamanna (1972) has commented, "the omnivorous gluttony" of man supplies a large number of potential foodstuffs for our bacteria. One only has to think of one's diet over the past 24 hr to appreciate this. However, the composition of the adult fecal microbiota is not easily changed by diet (Macy and Probst, 1979; Rasic and Kurman, 1983). Strict vegetarians and persons on a normal mixed diet have little difference in the major species of bacteria colonizing their intestines (Aries *et al.*, 1971).

On reflection, it is not surprising that diet has so little effect on fecal microbiota. The feces are only representative of the organisms in the large bowel, and substrates entering this section of the gastrointestinal tract are likely to be less varied than the substrates entering the small bowel, i.e., they have been extensively digested in the stomach and small intestine and much of this digest is absorbed before the chyme reaches the large intestine. Also, after intestinal cells of animal origin complete their life cycle in the villi they are shed into the lumen and a mixture of secretory and membrane glycoprotein is carried down to the lower levels of the intestinal tract (Prizont and Konigsberg, 1981). As shown by Hoskins and Boulding (1981), bacteria in the large bowel possess glycosidases, which degrade the glycoproteins, releasing sugars from the oligosaccharide chains that may be utilized as carbon sources for bacterial growth. This utilization of carbohydrates could enable the selective growth of certain bacteria in the colonic environment; therefore such populations become independent of dietary sources for nutrients.

Gut bacteria further contribute to a widening range of available substrates by their own degradative action, thus increasing the opportunity for more species to survive; innumerable microbial interrelationships have been found in the gut where one species utilizes materials excreted by other species. Hungate (1978) has classified these materials as (1) primary substrate leakage, e.g., a primary substrate such as cellulose is digested by extracellular enzymes elaborated by some bacteria; the soluble products are utilized by non-fiber-digesters; and (2) waste product utilization; a degradation waste product of one organism is utilized by another.

The range of biochemical activities of gut microorganisms is comprehensively reviewed by Prins (1977). Their complexity certainly makes it easy to appreciate how so many species evolve to coexist in the intestine. One also begins to appreciate the naivité of trying to look for individual factors controlling concentrations of bacteria in the intestine; this point will be discussed further (Section 3.1). So far only the nutritional colonizers of the intestinal lumen have been considered, i.e., organisms utilizing unique substrates and presumably multiplying with a growth rate at least as high as the dilution rate. There are other mechanisms whereby organisms can survive in high numbers in spite of high passage rates of digesta, either by attachment to the gut surface or by sequestration from the main stream.

2.1.2. Colonization of Intestinal Surfaces

The surfaces of most animal species studied are colonized by large numbers of microorganisms. Different surfaces in the gut further contribute to the high species diversity of the gut organisms.

The bacteria of the intestinal ecosystem have evolved with their mammalian hosts. This process has been continuing for some 150 million years; thus it is not surprising that extremely specialized adaptations to intestinal habitats have developed. Some of the most sophisticated mechanisms are found on the gut surface and are described in detail elsewhere (Lee, 1980). Figure 1 illustrates a range of distinct habitats on the rodent gut surface where some of the adaptations occur. Part of the stomach is colonized by populations of lactobacilli, which attach end-on to the stomach wall. A specific attachment has evolved with these bacteria, i.e., rat isolates will attach only to rat epithelia, whereas chicken isolates attach only to chicken cells (Suegara *et al.,* 1975). For organisms to survive on the surface of the small intestine they have to overcome two pressures likely to remove them. First, the relatively rapid flow rate of the intestinal content, and second, the continued turnover of intestinal epithelial cells, which migrate from the bottom of the crypts of Lieberkune and are sloughed off into the gut lumen. The organisms shown in the diagram on the ileal surface have developed a very complex attachment site to counter ingesta flow. A filamentous morphology allows them to successfully contact substrate in the lumen content and utilize it, in competition with the limited small bowel microbiota. The organisms appear to have evolved a complex life cycle to overcome the problem of host cell turnover: as the epthelial cell migrates toward the top of the villus, a dif-

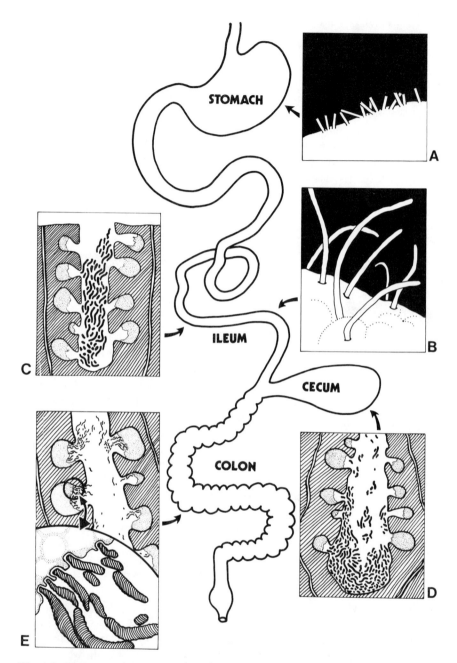

Figure 1. Diagrammatic representation of the surface-associated microbiota of the gastrointestinal tract: (A) the stomach lining is covered with lactobacilli, (B) the mucosa of the terminal ileum has a fuzzy coat of filamentous bacteria inserted into the tissue, (C) the crypts of the ileal mucosa are packed with spiral microorganisms, (D) cecal crypts are also full of spiral bacteria, but with a different morphology from the ileal organisms, and (E) the goblet cells of the colonic mucosa are lined with spiral bacteria inserted into the cell surface.

ferent form of the organism is seen to be assembled within the attached filaments. It is suggested that this "holdfast" is released as the epithelial cell is sloughed off into the ileal lumen and, being motile, can now swim down to the base of a crypt and start the cycle over again (Chase and Erlandsen, 1976).

Another group of attached organisms are found in the colon closely associated with mucus-producing goblet cells. Thus, three different surface structures of the rat intestine provide habitats for three different autochthonous* organisms.

Another heavily colonized habitat is the mucus blanket that lines the lumen of the intestine. Rozee *et al.* (1982) have shown this layer in the mouse intestine to contain a rich variety of bacteria and protozoa. The longer retention time of this material in the gut would allow organisms with different metabolic activities to survive in this mucus, whereas they would not remain established in the intestinal lumen.

2.1.3. The Intestinal Crypt. Neglected Niches

The wall of the small intestine consists of many folds of tissue and millions of fingerlike projections or villi, which provide the large surface area necessary to optimize absorption of nutrients from the lumen. At the bottom of these villi are crypts lined with mucus-secreting goblet cells. The lining of the large bowel is smooth, but is also pitted with many crypts, invaginations lined with goblet cells that secrete mucus into the gut, providing a lubricant at its surface. The mucus of these crypts provides a niche for yet another group of autochthonous organisms. These bacteria have been little studied, but provide a further example of special adaptation. Examples may be found by examination of sections of rodent bowel; their location is illustrated in Fig. 1.

Very different populations of bacteria inhabit the crypts of the ileum, the cecum, and the colon. The characteristic feature of all these bacteria is a spiral morphology, being either *Borrelia, Treponema, Spirillum* spp., or species from a previously unidentified genus (Phillips and Lee, 1983). Presumably, this morphology must give the organisms a selective advantage in these sites. The common substance in these habitats is mucus, composed of viscous polymeric mucins, which is continually excreted out of the crypts. To remain in this sequestered site protected from competition by the many lumen organisms, the crypt bacteria must be able to withstand the mucus flow. They could do this by having a growth rate

---

*This term, originally used for soil organisms, was proposed by Dubos to describe groups of bacteria that were considered to have evolved with their host and would normally be isolated from all animals. It is a useful word to distinguish these organisms from the transient residents of the intestine that may be isolated from time to time (Dubos *et al.*, 1965).

greater than the washout rate of mucus, but this is unlikely. A more feasible explanation is motility. Brock (1966) has speculated that helical movement would enable organisms to move through debris and viscous media that would impede other organisms. Many of these gut spirals have now been cultured in our laboratory (Phillips and Lee, 1983) and so this hypothesis could be tested. The velocity of the spiral bacterium from the rat ileum in solutions of methylcellulose of varying viscosities was compared with the velocity of other motile gut inhabitants. At viscosities as high as 40 cP the spiral darted along at a speed of 82 μm/sec at 21°C, whereas gut isolates of *Escherichia coli* and *Pseudomonas aeruginosa* were sluggish and could hardly move at all (M. Phillips, personal communication). This physical advantage is all that is necessary to explain the selective colonization of the crypts. The reason for different spirals in crypts from different parts of the gut presumably reflects other environmental differences at these sites. Thus, Sakata and Englehardt (1981) have shown that neutral mucin decreases and acid mucin increases in the epithelium from the cecum to the distal colon in the rodent. Also, the spirals from the small intestinal crypts are microaerophilic, compared to the anaerobic spirochetes seen in cecal crypts, reflecting different oxygen tensions at these sites.

Mucosa-associated spiral bacteria are widespread throughout the animal kingdom and are seen in large numbers in the intestinal crypts of cats (Macfie, 1916), dogs (Leach *et al.,* 1973), pigs (Harris and Kinyon, 1974), and monkeys (Takeuchi *et al.,* 1974). There is controversy over the presence of a mucosa-associated microbiota in man; however, dense colonization has been seen in a proportion of normal appendices and colons (Takeuchi *et al.,* 1974).

2.2. The Host as a Contributor to and a Beneficiary of the Microbial Ecosystem of the Intestinal Tract

An important concept when considering any microbial ecosystem is the theme of adaptation of organisms over time to fill all available niches in the environment. The previous discussion on the origins of the high species diversity in the gut illustrated this point. However, when trying to explain the microbiology of the animal gut in ecological terms, the contribution of the host to the system is often neglected. The animal gut has adapted to its microbiota so that it can cope with it or benefit from it. Thus, when the animal host uses food of animal origin that is readily digested, nutritional relationships between mammal and microorganism are basically competitive. The animal has evolved mechanisms to get "first bite at the cherry," i.e., stomach acid to reduce the numbers of competing organisms and a long, absorptive small intestine, which moves the foodstuff along by peristalsis too quickly for organisms to multiply. If the animal uses food of plant origin that is indigestible, the host has capitalized on its gut ecosystem by allowing the organisms to digest the food in a fermentation vat (the rumen) before it enters the acid stomach and motile small bowel.

Just as the host makes major contributions to the composition of the intestinal microbiota, so do these organisms contribute to host well-being. These contributions have been reviewed previously and most are not relevant to this discussion (H. A. Gordon and Pesti, 1971). Their subtlety is well illustrated by the work of Roediger (1980), which suggests that the fatty acids produced by gut anaerobes are the major source of energy for the colonic mucosa, i.e., the colonocytes have evolved ketogenic enzymes, a good example of mutualism.

The contribution of the intestinal microbiota most relevant to this chapter is the protective effect against potential human pathogens. This protective effect is dependent on ecological interactions and thus will form the major emphasis of the review.

3. Colonization Resistance

One consequence of the stability of the gut ecosystem discussed above is that introduced or allochthonous organisms have difficulty in establishing in the intestine. If the alien organism has a pathogenic potential, then the normal microbiota is acting as a defense mechanism. Thus, the infectious dose in man for *Salmonella* species other than *S. typhi* is about 10^6-10^7 (Taylor and McCoy, 1969), whereas in infants, where the climax stage of the gut ecosystem has not been reached, as few as 40 organisms can initiate infection (Lipson, 1976). The term colonization resistance coined by Van der Waaij *et al.* (1971) is the most descriptive term for this phenomenon and will be used throughout the rest of this review. It is important to realize that this is not an absolute term, but describes the state of stability of the ecosystem with regard to potential new colonizers of the gut. Thus, the original definition of colonization resistance (CR) was expressed as the number of bacteria (log) that, when given as an oral dose, resulted in colonization of 50% of the animals treated, for a prolonged period. A normal animal with a normal microbiota exerting a suppressive effect on introduced organisms would have a high CR, i.e., 10^8-10^9, which is the dose required to achieve 50% colonization. In an antibiotic-treated animal where introduced organisms faced less competition, on the other hand, the CR was low, i.e., $10-10^2$. The organisms used to measure CR were facultative anaerobes such as *Escherichia coli*. Through continued usage, the meaning of the term has become more qualitative; thus, in a more recent paper Van der Waaij described CR as the mechanism associated with the anaerobic bacteria of the gut that controls the colonization pattern of potentially pathogenic aerobic bacteria and yeasts (Van der Waaij *et al.*, 1982). Thus CR maintains a threshold to the number of these organisms that must contaminate the oropharynx of an individual before colonization of the tract is possible. The normal level of facultative anaerobes, such as *E. coli*, can be an indication of the colonization resistance of that animal, an observation that will be referred to later in the discussion on specific pathogen free animals (Section 5.3).

Given that CR is important in host defense, it is important to consider how this effect is achieved. One problem with using an all-embracing term like CR is that there is a danger of assuming there is one mechanism. Experiments by Raibaud *et al.* (1980) in gnotobiotic mice highlight this point. Human intestinal bacteria were established in germfree mice and were shown to exert a colonization barrier against introduction of the potential human pathogen *Clostridium difficile*, yet the count of *E. coli* in the feces of these mice was 10^9/g, which indicates no colonization resistance against this facultative anaerobe.

All too often researchers in intestinal microbiology have tried to identify single factors as being responsible for interaction between the normal microbiota and pathogens. This work is discussed below (Section 3.1–3.3), but it must be appreciated that given the multiple factors that result in the diversity of the ecosystem in its climax stage, multiple factors will also be responsible for pathogen exclusion. Another unproductive but oft-used approach has been to try to determine which organisms of the microbiota are responsible for colonization resistance.

An interesting series of experiments supported the view that a variety of microorganisms are responsible for CR and that groups of these organisms become predominant and function under varying circumstances (Freter and Abrams, 1972). Groups of germ-free animals were mixed with conventional animals to allow a "normal" microbiota to establish. These "conventionalized" animals showed no difference in levels of an introduced *E. coli* strain when fed two different diets. However, when similar germ-free animals were colonized with a mixture of cultures of 45 different anaerobes isolated from conventional mice, there was a hundredfold difference in the levels of introduced *E. coli*, dependent on which of the two diets was fed, i.e., with the complete microbiota of the conventionalized animals CR was exerted on both diets, but in the animals with the more restricted biota CR was greater on one of the diets. This difference could be corrected by feeding the mice a further collection of isolates from normal animals. The conclusion was that the central functions of the normal microbiota may shift from one set of intestinal bacteria to another as the diet is changed. The following are the mechanisms of colonization resistance most often referred to in the literature.

3.1. Competition for Nutrients

Since all the autochthonous intestinal bacteria are present due to an ability to utilize one or a small number of substrates better than other organisms, and over a period of evolution these organisms have adapted so that all available substrates, i.e., niches, are utilized, it is likely that competition for limiting growth factors is the major mechanism of CR. Thus, the invading pathogen is just not as good at utilizing the substrates as the autochthonous organisms, "the specialists." The competitive nature of this phenomenon would then be the ecological explanation of the phenomenon of infectious dose; i.e., assume that pathogen A can utilize substrate x but not as well as the 10^{11} autochthonous B organisms normally found in the gut. When 10^2 of pathogen A enter the gut they cannot compete with the 10^{11} of organism B long enough to multiply and establish. However, if 10^5 of pathogen A enter the gut, then there is a greater chance that they will be able to utilize some of the available substrate x present and thus multiply. Once they have started to multiply, they are then able to produce changes in the environment that alter the substrate balance, and so alternate substrates may become available and multiplication proceeds. An added feature to infectious dose that must be remembered is the host factor, stomach acidity, which may affect the final numbers of pathogen A entering the gut.

Even though the role of competition for substrate in bacterial antagonism in the gut is frequently discussed in the literature and is logically most important in ecological terms, there is very little convincing evidence for its occurrence in this habitat. A recent paper by Guiot (1982) describes a simple technique which, if imaginatively exploited, could provide important evidence as to the role of substrate competition or inhibitors in microbial interactions in the gut. Small slices of nonnutrient agar containing low numbers of the organism of interest (in this case $10^{5.5}$ *E. coli*) are preincubated for 2 hr in intestinal content or brain heart infusion broth and then transferred to saline for further incubation. Preincubation in brain heart infusion gave rise to abundant growth, whereas only poor growth occurred after preincubation in the contents of rat cecum, i.e., there was no nutrient available for the *E. coli* to diffuse into the agar slice in the intestinal content compared to the brain heart infusion. This poor growth was completely reversed by adding brain heart infusion. This simple technique should result in much potentially useful information on colonization resistance.

3.2. Toxic Metabolites

The concept of toxic products playing some role in bacterial inter-action in the gut is attractive. Once again, however, there is no clear evidence for the importance of toxic metabolites. Freter *et al.* (1983b) state that it is impossible to account for the presence of constant low numbers of a given sensitive strain, e.g., *E. coli,* in the indigenous microbiota solely on the basis of the action of growth inhibitors produced by other indigenous species (such as the predominant anaerobes). That is, constant populations could only be maintained at one precise inhibitor concentration, a situation not likely to occur in natural systems. While this is theoretically true, it is not a relevant argument against the influence of inhibitors, since they would almost certainly be acting in concert with the effect of other growth-limiting steps. The most quoted examples of inhibitory agents are the volatile fatty acids. Bergheim *et al.* (1941) initially showed that the intestinal content of rodents contained levels of short-chain fatty acids, particularly butyric acid, in concentrations that were shown *in vitro* to be inhibitory to the growth of facultative anaerobes. Lee and Gemmell (1972) showed a strong correlation between the levels of volatile fatty acids and coliform levels in antibiotic-treated animals and baby mice over the first 3 weeks of life. As the anaerobic fusiform microbiota became established, the levels of butyric acid increased and the coliform level decreased. A major end product of the *Fusobacterium* species that dominate the adult mouse gut is butyric acid. Brockett and Tannock (1982) also showed that bran-fed mice showed decreased volatile fatty acid in the cecum and increased susceptibility to *Salmonella* infection. Others (Freter and Abrams, 1972) have suggested in experiments with gnotobiotic mice that conditions exist where similar fatty acids are present and coliform levels are depressed or not, depending on other environmental parameters. These observations are not relevant to the effect of volatile fatty acids in the normal animal, since the concentration of acids in all their gnotobiotes is much lower than in normal animals and the lowest level of coliforms in these animals is relatively high, at 10^6–10^7, indicating that colonization resistance was depressed.

In a more recent paper (Freter *et al.,* 1983c), it is stated that fatty acids do not account for growth inhibition of established *E. coli* populations in the rodent gut, but the authors suggest that the acids may well protect a host by delaying the multiplication of an invading bacterium until it has been washed out.

The only other toxic metabolite proposed to influence microbial interaction in the gut is hydrogen sulfide. Once again this work comes from Freter *et al.* (1983b), who suggest an interesting linkup with population control by substrate limitation. They show that there were substrates present in their continuous culture system that supported rapid growth of implanted bacterial strains only in the absence of the normally present H_2S concentration. This system worked not only against an *E. coli* strain, but also against a mouse isolate of a strain of a *Fusobacterium* sp. and a *Eubacterium* sp. Unfortunately, it could not be determined if the levels of these two organisms achieved in the *in vitro* experiment approximated the levels normally found *in vivo*.

3.3. Bacterial Barriers

Ability to multiply in the intestine is not the only essential requirement for invading pathogens. In order to either invade the tissues or damage them, the organisms or their toxic products must get very close to the gut surface. Since these gut surfaces are colonized with complex populations of mucosa-associated organisms (Lee, 1980), it seems likely that these bacteria could interfere with the development of disease. There is some circumstantial evidence for a protective effect by the surface-associated microbiota in the ileum of rats. As illustrated in Fig. 1, the ileum is covered with a dense mat of unusual filamentous prokaryotes. *Salmonella* species invade the rat gut at this site and, in studies on growing rats, a negative correlation was found between the establishment of these populations and susceptibility to infection with *Salmonella* (Garland *et al.*, 1982). The invading salmonellae, which are firmly attached to the ileal surface, are never found in rats where the filamentous organisms have established. Rozee *et al.* (1982) have suggested that the presence of a normal, continuous, heavily colonized, and physiologically active mucus blanket may preclude tissue colonization even when potential pathogens are repeatedly introduced into the digestive tract. Often this mucus layer is not seen, due to the preparative methods used in specimen collection. More attention needs to be given to fixation methods that preserve the mucus blanket in order to gain a more realistic impression of the barriers that confront potential pathogens. This is particularly important with the use of scanning electron microscopy (Garland *et al.*, 1979).

4. Perturbation of the Gut Ecosystem

The importance of a stable intestinal microbiota is best appreciated when environmental changes result in major perturbations of the ecosystem. The host may contract a serious disease or experience nutritional disturbance (Savage, 1977a,b). Study of these perturbations, either naturally or experimentally induced, is most likely to provide information on factors influencing the colonization of pathogens and aid our understanding of how these organisms cause disease.

4.1. Consequences of the Vacated Niche

Perturbation of the gut ecosystem implies the removal of certain groups of microorganisms, thus resulting in a vacating of a previously occupied niche; this niche is now available for potentially pathogenic microorganisms. The ecosystem can also be perturbed by changes in the inflow of nutrients or growth factors such that the established biota change their metabolic activity and new niches become available.

The ultimate perturbed ecosystem would appear to be the germ-free animal. However, complete elimination of the microbiota does not occur naturally. Comparison between germ-free and conventional animals establishes only that intestinal organisms influence the morphology and histological appearance of the gut mucosa and that the virgin gut is readily colonized by pathogens with deleterious effects (Coates and Fuller, 1977). Findings on the germ-free animals may not be relevant to the functioning of the normal gut ecology, i.e., the niche occupied by a pathogen following a perturbation that allows the organism to exert its pathogenic potential might not be present in the germ-free animal, since other organisms whose metabolic byproducts create the niche are not present. Therefore a more fruitful approach is to study natural events or experimental perturbations that approximate the natural changes.

4.2. Examples of Natural Perturbation

4.2.1. Chemotherapy

Over 30 years ago, Smith (1952) reported on accumulating evidence that "the complex balance which exists among microorganisms constituting the normal flora of the body is disturbed by prolonged administration of the newer antibiotics." Indeed, he commented that the importance of the normal microbiota in man was not appreciated until it was shown to be disorganized by administration of antibiotics. The consequences of these perturbations range from mild looseness in one's motions to a serious diarrhea associated with enteritis, i.e., inflammation of the bowel epithelium. Not all of the dose of orally administered antibiotics is absorbed; thus the nonabsorbed remainder of the drug exerts its effects on the normal microbiota. Colonization resistance is lowered and the ecosystem may become colonized with potential pathogens. Two scenarios are possible:

1. Lowered infective doses of established diarrheal pathogens. If, as speculated above (Section 3.1), infective dose is related to competition for available nutrients, then antibiotics should reduce the infective threshold. This is illustrated by two case reports. In the first, an 18-year-

old male with a mild diarrhea was given ampicillin, resulting in an acute exacerbation of the infection with severe diarrhea and bloody mucus. The *Salmonella typhimurium* responsible for the infection was later shown to be resistant to ampicillin (Rosenthal, 1969). A group of children with cystic fibrosis was taking broad-spectrum antibiotics as a prophylaxis against chest infection. Unfortunately, a batch of the enzyme porcine pancreatin given orally to three children as part of their treatment was shown to be contaminated with a *Salmonella* sp. Accurate quantitation of the organisms in the enzyme preparation and knowledge of the amount administered revealed that ingestion of only 44 salmonellae resulted in infection, i.e., the infective dose was greatly reduced from the normal 10^6 (Lipson, 1976). Animal experiments have confirmed this observation; thus, the infective dose of *Salmonella typhimurium* in mice is reduced from 10^6–10^7 to 1–10 by prior oral administration of streptomycin (Bohnoff *et al.,* 1964).

2. Opportunistic infection as a consequence of antibiotic administration. A severe consequence of antibiotic therapy is a syndrome called pseudomembranous colitis (PMC). Patients have profuse, bloody diarrhea with mucus, tenesmus, abdominal cramps, and fever. On sigmoidoscopy, a pseudomembrane composed of yellowish white, gray, or green exudative plaques is seen on the large bowel mucosa. In extreme cases death may result. Early reports implicated *Staphylococcus aureus* in this enterocolitis, but it is now clear that this fascinating syndrome is due to an organism previously unassociated with disease and considered a nonpathogen, *Clostridium difficile.* The perturbation by the antibiotic results in a niche becoming available that is rapidly taken up by the *Clostridium.* The organism multiplies and produces at least two toxins, one of which exerts its effect on the large bowel wall. As early as 1943 penicillin had been shown to be lethal when given to guinea pigs. Subsequent studies have revealed that this effect is caused by *Clostridium difficile.* Work with this animal model and observation of clinical cases have revealed this to be a fairly common syndrome following administration of a wide range of antibiotics (Bartlett, 1982).

The change in intestinal milieu that results in PMC not only has to permit multiplication of *Cl. difficile,* but conditions have to allow the organism to produce toxin close to the site where it exerts its effect. This multiple effect is further discussed in Section 8.

4.2.2. Anatomical and Physiological

Much of the above discussion on species diversity in the intestine refers to the large intestine. The anatomy and the function of this organ contribute to the stability of the system, e.g., long retention time of the intestinal content. However, in the small intestine, anatomical and physiological functions act to restrict the growth of microorganisms, resulting in relatively low numbers and low species diversity. The stomach acidity prevents establishment of bacteria in this site and the peristaltic activity of the small intestine keeps the chyme moving at too fast a rate to allow organisms to colonize in large numbers. There are a number of conditions in the human where these factors are altered, resulting in an abnormal bacterial biota. They have been classified by Tabaqchali (1979) into three groups; (1) gastric abnormalities resulting in decreased acid production (e.g., achlorhydria, partial gastrectomy), (2) conditions that cause stasis (e.g., surgical enteroanastomosis), and (3) conditions in which there is free communication with the large bowel (e.g., intestinal resection).

Studies of the microbiota colonizing the small intestine under these abnormal conditions reveal some interesting features about interrelationships among gut bacteria. More importantly, they highlight the subtlety of the interrelationship between the mammalian host and its microbiota, i.e., the gut ecosystem has evolved to benefit the host; where microorganisms are likely to interfere with intestinal function, they are absent or reduced in number.

One important contribution of these studies is the observation that a more anaerobic biota occurs in areas of stagnation. When the motility of the gut content is severely restricted, the numbers of aerobes and anaerobes greatly increase and the biota resembles that of the large bowel. Similar substrates must be available in both areas, but in one the flow rates are so fast the anaerobes cannot establish. This also means that the composition of the small bowel chyme compared to the large bowel content is not a major contributing factor to anaerobiosis. Presumably, the facultative aerobes scavenge the available oxygen, allowing the anaerobes to establish. It would be of interest to determine the effect of selective decontamination of facultative anaerobes on the growth of anaerobes in patients with small bowel overgrowth.

In the large bowel, *Bacteriodes* spp. have adapted to utilize bile acids. Since these compounds have no function in this site, it is of no consequence to the host, excluding the possibility of carcinogenic by-products. However, when conditions are changed so that these organisms grow in the small bowel, the results are disastrous (Savage, 1977a). Conjugated bile acids contribute to fat digestion and absorption in the normal small bowel. Thus, patients with stagnant small bowel syndrome suffer from diarrhea and malabsorption of fats.

Amino acids are usually absorbed before they reach the large bowel; however, when abnormal conditions allow the overgrowth of bacteria in the small bowel, amino acids, particularly L-tryptophan, are metabolized. This may deprive the host of essential growth factors and lead to protein malnutrition, in adults a syndrome resembling kwashiorkor and in children, failure to grow.

A significant example of interdependence of the microbiota and normal host cell function in the large bowel has recently been proposed by Roediger (1980). The fatty acids formed as a by-product of large bowel anaerobes may be a major source of energy for the colonic mucosa, particularly of the distal colon. If this is so, a consequence of perturbations of the colonic ecosystem that results in decreased butyrate production could be tissue starvation and cellular change. This change of tissue could provide a new niche for a potential pathogen, and a microbial-induced colitis could result (Cummings, 1981).

4.2.3. Stress

The problem facing researchers trying to unravel the complexity of the gut ecosystem is no better illustrated than by considering the effect of stress on intestinal populations. This topic has not been discussed in recent reviews and has been little investigated. Schaedler and Dubos (1962) noted that the composition of the bacterial biota of rodents could be rapidly and profoundly altered by a variety of unrelated disturbances, such as changes in environmental temperature, crowding in cages, fighting among the animals, and handling. These changes could be measured by an increase in the number of facultatively anaerobic Gram-negative bacilli, i.e., a decrease in colonization resistance. Holdeman *et al.* (1976), in a limited anecdotal study, suggested that levels of certain gut bacteria varied depending on the emotional status of the person being studied, i.e., the percentage of isolates that were *Bacteroides fragilis* subsp. *thetaiotaomicron* increased after an individual had an argument and in a 19-year-old woman in emotional anguish.

The influence of stress on the gut ecosystem should be further investigated, since this is a good model system, particularly in rodents, where constant subtle changes in the microbiota can be reproduced. However, a decision on which parameters to investigate is difficult; this is a constant problem in studies with the gut and highlights our ignorance in this area.

4.2.4. Nutrient Deprivation

Food differences have no marked effect on the large bowel microbiota, since small bowel epithelial cells, mucus, and other secretions are the main substrates. Starvation, on the other hand, can lead to major changes. Tannock and Savage (1974) demonstrated that starved rodents become more susceptible to *Salmonella* infection. Significantly, the filamentous organisms referred to above (Section 3.3) when discussing bacterial barriers to infection were absent in these animals. Holding of cattle in yards before slaughter predisposed them to *Salmonella* and *E. coli* infections (Brownlie and Grau, 1967; Grau *et al.*, 1969), and starvation was probably responsible. However, starvation is a form of stress and these changes might be related to the unknown factors referred to in Section 4.2.3. The animal presumably would have to be near death before the nutrient contributions of tissue cells and secretions to the large bowel were reduced to a level to affect bacterial growth. Yet, even mild starvation will lead to changes.

4.3. Experimental Perturbations of the Gut Ecosystem

One of the working hypotheses of our research group has been that research on experimental perturbations of the ecosystem is likely to provide information on factors influencing colonization of potential pathogens. Studies on two perturbations not previously investigated have revealed interesting observations relevant to the microbial ecology of the gut.

4.3.1. Physical Perturbations

One method of temporarily removing a significant number of the normal gut microbiota is to induce a fairly violent diarrhea with magnesium sulfate. This compound is given by mouth and has a direct action on the intestinal mucosa, releasing the hormone cholecystokininpancreozymin, which affects intestinal secretions and induces a very watery diarrhea. Lumen organisms are flushed out of the intestine and a significant number of the mucosa-associated bacteria also disappear (Phillips *et al.*, 1978). A further study of more ecological importance was an investigation of the recolonization of the gut following this physical perturbation (Phillips and Lee, 1984). The lumen populations appeared to reestablish rapidly; also, the populations within the crypts of the ileum and cecum were comparable to those of normal animals within 2–4 days. However, a layer of spiral bacteria colonized sites on the surface of the colon where a surface-associated biota is not normally seen. These organisms had one end inserted into the microvillus border and were still present in animals 6 months after cessation of $MgSO_4$ treatment. A possible explanation for this phenomenon is that a barrier to colonization exists in the normal colon that excludes microorganisms, including the spiral organisms, from the mucosal surface. Removal of such a barrier during perturbation would enable organisms to reach the surface and proliferate. The mucus layers and their complement of associated organisms were absent after the severe episodes of diarrhea.

In some of these experiments two other populations were found in the recolonized animals in areas not normally colonized. Thick layers of a rod-shaped bacterium with a Gram-negative type ultrastructure were seen in ileum and cecum. The ileal mucosa was colonized by myco- plasma-like organisms firmly attached between the microvilli. These observations reveal an important phenomenon that may have relevance to gastrointestinal disease; i.e., a perturbation of the ecosystem by a phys- ical event (purging) leads to alterations in gut surfaces such that they have an altered reactivity to microorganisms. In a different system, Ramphal *et al.* (1980) showed that cells injured in the trachea by mechanical intu- bation were more easily colonized by attached *Pseudomonas aeruginosa.* Also, the *Pseudomonas* attached more readily to cells damaged by influ- enza virus infection. The name "opportunistic adherence" (strictly, adhe- sion is the correct term) was given to this phenomenon, and they pro- posed that alterations of the cell surfaces or cell injury facilitates the adhesion and that adhesion to injured cells may be a key to the patho- genesis of opportunistic *Pseudomonas* infections. In a study of cases of protracted diarrhea in infants caused by enterocyte-attached *Escherichia coli,* it was suggested that previous infection of the intestine by rotavi- ruses contributed to the pathogenesis of the *E. coli* infection (Rothbaum *et al.,* 1982).

In experiments with a continuous culture system and gnotobiotic mice, it was found that if strains of *E. coli* were implanted before estab- lishment of an antagonistic microbiota, they remained in the system, but cultures of the same strains added after the establishment of the normal microbiota were rapidly washed out of both systems (Freter *et al.,* 1983c). The authors speculate that in the first instance, the *E. coli* get to adhesion sites first, although no evidence is presented for this.

There are a number of diseases where perturbation or differences in the intestinal ecosystem are a prerequisite to infection, e.g., pseudomem- branous colitis and necrotizing enteritis. In others, altered reactivity to antigens in the intestinal tract may be the result, e.g., inflammatory bowel disease. Unrelated physical disturbances in the intestine may be an as yet unsuspected predisposing factor in some of these diseases. More work should be done on recolonization of disturbed gut ecosystems, looking at both lumen and surface organisms.

4.3.2. Anatomical Modification

Experimental surgical modifications of the intestinal tract can pro- duce major perturbations in the ecosystem and thus provide a useful model for investigation of factors influencing colonization. Cecectomy of rodents is one good example. Surgical removal of the cecum of both mice and rats results in a gut ecosystem with a grossly reduced CR, using the useful criterion of levels of coliform bacteria in the large bowel. Numbers in cecectomized rodents remain at 10^9 per gram of feces or colonic content indefinitely, compared to 10^3-10^4 in sham-operated controls, although the stress of surgery will initially lift the coliform levels in these controls (A. Lee, unpublished data). Microscopic examination of the intestinal content of cecectomized animals reveals that many of the fusiform bacteria that predominate in the normal cecum, colon, and feces are absent. Levels of volatile fatty acids are very low in the treated rodents. Here, then, is another perturbed system for inclusion in experiments investigating the effect of the normal microbiota on potential pathogens.

Researchers who want to study the behavior of any organisms in the rodent intestine have at their disposal a wide range of experimental models with reduced colonization resistance, e.g., (1) penicillin treatment (1 g/liter in drinking water), (2) $MgSO_4$ treatment, (3) stress (mild starvation, handling, etc.), and (4) cecectomy.

5. Consequences of an Immature Microbiota

So far, only climax communities of microorganisms in the intestinal tract of adult animals have been considered in this chapter. As is the case with any ecosystem, the establishment of the climax community follows a series of fairly constant stages. The microbial succession from birth to weaning has been well documented for many animals; however, the factors influencing this succession are still poorly understood (Smith, 1965). Given that diarrheal disease in childhood is one of the major killers in the world today and loss of newborn livestock is a major economic factor in animal husbandry, study of the development of the immature microbiota in humans and domestic animals should be a major research priority.

5.1. The Neonatal Animal

The overall patterns of colonization of a wide range of animal species were determined many years ago by Smith (1965). Of these animals, the one studied most has been the mouse (Schaedler *et al.*, 1965; Lee *et al.*, 1971; Lee and Gemmell, 1972). Given that the alimentary tract is sterile at birth, it is not surprising that the establishment of the intestinal microbiota is a fairly constant succession with different groups of bacteria establishing at approximately the same times. Inoculation of the tract is presumably via the mother's fecal material contaminating the teat. Initially the intestinal milieu favors the establishment of aerobes and facultative anaerobes. The exact timing depends on the level of organisms present in the mother. Thus, with a mother having a high coliform level the baby may be colonized by these organisms at a concentration of 10^9/g gut content by 3 days of age, whereas if the mother has very low levels of coliforms the offspring may never be colonized with these organisms. Enterococci and lactobacilli colonize all young mice within 2–3 days, reflecting the more uniform numbers in the mothers. Even though the *Bacteroides* sp. and a mixed group of anaerobic fusiform bacteria, which are the dominant members of the adult mouse biota, are presumably inoculated in as large numbers in the infant mouse, they do not colonize the gut in the early stages. These organisms only start to appear around 11–14 days, just after it can be shown that the animal first ingests solid food. This time was accurately measured by incorporating 10% activated charcoal in the mouse's pelleted diet. Removal of the stomachs of animals from day 1 onward revealed that the pristine white stomachs of the baby first showed specks of carbon at day 10. This is illustrated pictorially in the paper of Lee and Gemmell (1972). The solid food most likely is necessary for establishment of anaerobic conditions. It is often claimed that facultative anaerobes help create anaerobiosis by scavenging residual oxygen, i.e., a true succession is occurring, but this is unlikely with the extremely oxygen-sensitive fusiforms. Thus, when mice are kept on a pelleted cow's milk diet instead of conventional pellets past the weaning period and into adulthood, coliforms remain at very high levels (10^9/g), as do the strictly anaerobic *Bacteroides* sp. (10^{10}/g), whereas the fusiforms do not establish. The normal diet therefore contributes to the anaerobic milieu of the young mouse gut required for the climax community to establish. This happens by the completion of weaning at about 3 weeks of age. This time corresponds to the development of colonization resistance in the mouse and correlates almost exactly with the establishment of the strictly anaerobic fusiforms, i.e., at this time the concentration of the facultatively anaerobic coliforms decreases dramatically. Predictably, the fusiforms did not establish in the young mice described above that were fed pelleted cow's milk, i.e., colonization resistance does not develop (Lee and Gemmell, 1972).

The major consequence of the immature gut ecosystem in the mouse is that the infant mouse, not having colonization resistance, is more likely to be infected by potential pathogens. Thus, *Candida albicans* can colonize mice if fed to infants but not when given to adults (Hector and Domer, 1982). The rat follows a colonization pattern similar to that of the mouse: 100% will die when given *Salmonella enteritidis* at day 7 of age, 85% at 3 weeks, and 10% at 10 weeks. The susceptibility of young calves and pigs to gastrointestinal infection is a major animal husbandry problem. It should be pointed out, however, that host factors unrelated to microbial ecology can reduce this problem, i.e., the colostrum or first milk produced by the mothers of these animals can protect the young by neutralizing the pathogenic potential of the infecting organisms (Porter *et al.,* 1977).

In summary, achievement of the climax populations within the gut ecosystems of animals is not strictly a succession in the ecological sense; rather, it is dependent on anatomical development of the tract and the dietary habits of the animal, particularly relating to weaning and consumption of solid food. With birds, the situation is simpler, with the time of establishment of the climax community depending on chance, since chicks eat solid food from the day of hatching. Thus, the period of vulnerability of this ecosystem can be drastically shortened by eliminating this chance acquisition of autochthonous microorganisms, i.e., if chicks are fed cecal content from adult birds on the first day of life, colonization resistance is established immediately, since the adult biota establishes immediately (Rantala and Nurmi, 1973). The levels of contamination by *Salmonella* sp. in these birds are significantly lower than in untreated flocks, a result that has obvious commercial application (see Section 10.1).

5.2. The Infant Bowel

There have been many studies on the development of the microbiota of the human intestine. Most of these studies investigated groups of different children at different ages on different diets. The first long-term prospective study of the microbial ecology of the large bowel of healthy breast-fed and formula-fed infants living in an industrialized community was carried out recently in our laboratory (Stark and Lee, 1982b). The succession of bacterial populations of seven breast-fed and seven formula-fed infants was examined during the first year of life. Whereas the numbers of babies were small, results were consistent and raise several interesting questions. In the first week, the gastrointestinal tract of the neonate is seeded with a wide variety of organisms from the birth canal and the infant's surroundings. Organisms best suited to the intestinal environment become established by a process of natural selection. A diet of breast milk creates an environment favoring the development of a simple biota of bifidobacteria, a few other anaerobes, and small numbers of facultatively anaerobic bacteria. A formula diet also allows bifidobacteria to reach high population densities equal to those occurring in the climax community in the large bowel of adults while permitting *Bacteroides* spp., clostridia, and anaerobic streptococci to colonize more frequently, and facultatively anaerobic bacteria to reach higher levels. The introduction of solid food to the breast-fed infant causes a major perturbation in the gut ecosystem, with a rapid rise in the number of enterobacteria and enterococci, followed by progressive colonization by *Bacteroides* spp., clostridia, and anaerobic streptococci. The addition of solid food to the diet of the formula-fed infant does not have such an impact on the gastrointestinal biota. Facultative anaerobes remain numerous, while colonization with anaerobes other than bifidobacteria continues; as the amount of solid food in the diet increases, the fecal bacterial biota of breast-fed and formula-fed infants approaches that of adults, i.e., the climax community is achieved at approximately 1 year of age.

Thus, unlike the situation with adults, there is clear evidence that diet does dramatically influence the composition of the bowel microbiota in infants. One can speculate that this is a consequence of evolution; in the adult it is important to have a highly diverse population of microorganisms to achieve the homeostatic mechanisms described above (Section 2). Before an adult biota is achieved, however, it would be of benefit to have some inhibitory mechanisms present. The breast milk apparently creates an environment with an acetate buffer at pH 5–6 in the intestine of the breast-fed child, which inhibits the growth of Gram-negative facultative anaerobes (Bullen and Tearle, 1976). Lactoferrin, an iron-binding protein present in breast milk, may also prevent colonization by potential pathogens by competing for the iron necessary for growth in the intestinal lumen (Bullen *et al.,* 1972). The unnatural substitution of formula for breast milk eliminates this controlling factor, thus allowing facultative anaerobes to colonize. This is presumably one of the factors responsible for a higher incidence of gastrointestinal infection in formula-fed compared with exclusively breast-fed children (Kanaaneh, 1972; France *et al.,* 1980).

Whereas this inhibitory mechanism is of benefit to the infant, an interesting possibility arises at weaning based on ecological logic (ecologic!!). At weaning, the intestinal milieu is greatly disturbed, with rapid changes in bacterial populations. This is in contrast to the homeostasis of the intestinal environment of the formula-fed infants, in which aerobic populations have been established in large numbers for several months and would be better adapted to their niche and therefore better able to compete against enteric pathogens than the more recently established aerobic organisms in the breast-fed babies. Thus, it appears that when solid foods are added to the diet the breast-fed child could be more susceptible to gastrointestinal infection than the formula-fed infant. There is little information available on the incidence of such infections at weaning in breast-fed compared to bottle-fed children in developed countries. However, the perturbations occurring in the gut ecosystem of the breast-fed infant at weaning may be an important factor in the pathogenesis of weanling diarrhea, a common disease in underdeveloped countries, where a prolonged and inefficient weaning process is accompanied by malnutrition and intestinal infection (J. E. Gordon, 1971).

5.3. Specific Pathogen Free (SPF) Animals

The use of SPF animals is a very good example showing how lack of appreciation of ecological principles has created a problem and how the application of knowledge on the microbiology of the gastrointestinal tract may provide a solution.

In the 1960s a large amount of effort was put into the production of specific pathogen free animals (SPF), particularly rodents, with the expectation they would eventually replace conventional animals in most research programs. Many projects had failed due to the carriage of pathogens in these conventional stock. The animals were originally derived from germ-free stock or by caesarian delivery under aseptic conditions. They were then maintained by skilled technical staff under barrier conditions with sterile diets, bedding, and water. The problem was how to establish an intestinal microbiota. At that time, Schaedler *et al.* (1965) and Dubos *et al.* (1965) had published their important papers on the composition of the rodent microbiota. Laudibly, cultures of the organisms grown by these workers were used to colonize these SPF mice (the so-called Schaedler's flora). Unfortunately, the limited anaerobic techniques available to this research team meant that the dominant members of the rodent intestine were not cultured. Schaedler's flora contributed no CR to these animals, yet these SPF animals had extremely low coliform levels when the colony had reached commercial proportions and animals were sold. Examination of cecal smears from all animals at all stages of production showed that the typical fusiform microflora of the normal rodent developed randomly. This correlated with coliform levels, which were 10^9/g of gut content at the start of the process and 10^3/g of content when the fusiforms established (A. Lee, unpublished data). These mouse organisms colonized the animals accidentally in spite of the strict barrier conditions imposed. The widespread nature of these bacteria is remarkable. A box of germ-free mice was placed in an enclosed room in a country property far removed from any animal colonies and supposedly vermin proof. The fusiform flora established within 3 weeks (A. Lee, unpublished data).

Even though our knowledge of the mouse intestine has improved over the last 10 years, the production of SPF rodents does not take this information into account. The starter cultures used by various authors over the last 15 years to establish SPF animals from germ-free stock are shown in Table II. None of these combinations of starter cultures would

Table II. Starter Cultures Used in the Production of SPF Rodent Colonies

Microorganisms	Reference
Lactobacillus sp., *Bacillus subtilis,* and *Streptococcus glycerinaeus*	Heine and Thunert (1968)
Clostridium difficile and *Lactobacillus acidophilus*	Christie *et al.* (1968)
Lactobacillus sp., *B. subtilis,* and *C. difficile*	Pesti *et al.* (1969)
Lactobacillus sp., *E. coli, Streptococcus* sp., *Clostridium* sp., and *Bacteroides* sp.	Sasaki *et al.* (1970)
Lactobacillus sp., *B. subtilis, E. coli, Streptococcus faecalis,* and *Enterobacter* sp.	Perrot (1976)
Nonpathogenic *E. coli, Staphylococcus* sp., and *Streptococcus* sp.	Burek (1978)

confer CR on these animals. The final climax communities are likely to be very different combinations of bacteria from the original cultures and vary from colony to colony, which is inconsistent with the aim of producing standardized experimental animals. A recent survey of some of the leading mouse producers in Australia revealed great differences in the number and type of facultative anaerobes present. Thus, one colony was free of Enterobacteriaceae, another was very heavily colonized with *Proteus mirabilis* and *Escherichia coli,* while a third had 30% of animals colonized with *Enterobacter cloacae* (A. Lee and J. O'Rourke, unpublished data). Under the stress of experimentation these organisms could potentially cause very different effects, thus influencing experimental results. For example, large groups of experiments on total body irradiation of mice using animals colonized with *Enterobacter* had to be abandoned due to deaths caused by this organism.

Inevitably, even well-maintained rodent colonies become colonized with Enterobacteriaceae. There is an urgent need to understand the ecological mechanisms whereby these organisms suddenly become able to establish. It is not due to sudden exposure to the bacteria. Given that these animal colonies will continually be required to be culled out and new SPF colonies established, there is a need to have more efficient and consistent methods of colonizing germ-free stock with a biota that rapidly confers colonization resistance. The experiments of Freter on establishment of such a biota with pure cultures show this approach to be impractical (Freter and Abrams, 1972). Between 50 and 90 cultures were required and these appeared to behave differently, depending on the diet of the animals. The more logical approach would be to keep stocks of starter animals known to have a good CR under protected conditions, i.e., in isolators. When germ-free animals need to be colonized, either they can be mixed with some of these animals or cecal content can be fed to them. This is being done at the Animal Production Unit of the Australian Atomic Energy Research Establishment in Sydney. Balb/c mice proven to be free of Enterobacteriaceae are maintained under germ-free conditions in isolators. This "E-free" colony has shown good colonization resistance for more than 2 years (A. Lee *et al.*, unpublished data).

6. Immune Mechanisms

The immune system of animals has a capacity to respond to the antigens of the normal microbiota of the gastrointestinal tract and to those of potential gut pathogens. Thus, low levels of "natural antibody" can be found in the serum to many intestinal organisms (Foo *et al.*, 1974) and immunity can develop to typhoid fever, cholera, and shigellosis. Often it has been assumed that immune mechanisms have a controlling influence on populations in the gut, i.e., an example of host contribution to the ecosystem. In their early papers on colonization resistance, Van der Waaij and Heidt (1977) explored this possibility. Resident colonizing intestinal bacteria reaching high concentrations in the digestive tract were found to be coated with the immunoglobulin IgA.

Dubos (1965) postulated that autochthonous microbes would not be immunogenic in their native hosts. This situation would have arisen through a long adaptive process. This hypothesis appeared to be confirmed by the work of Foo and Lee (1972, 1974). A strain of *Bacteroides* isolated from mice was found to be nonimmunogenic in mice but highly immunogenic in other animal species, e.g., sheep. The bacteria were found to have a cross-reacting antigen with mouse intestinal tissue. Animals do not form antibodies against their own tissue substances. This is analogous to the "molecular mimicry" observed by Damian (1964). Other workers also observed that some indigenous microbes may not induce antibody formation (Berg and Savage, 1972, 1975).

These examples fit nicely with the hypothesis and are logically pleasing; thus they have been quoted often and have become part of established dogma (Mims, 1982). However, these findings need to be placed in perspective. How would this property of molecular mimicry be of benefit to both host and organism? One could argue that the organisms would be better able to colonize the gut surfaces. The host would be protected from the possible deleterious effects of antigen–antibody reactions close to or within the gut tissues. The *Bacteroides* strain used in the mouse experiments described above was not a surface colonizer, but lives in the gut lumen. Some fusiform bacteria known to colonize the intestinal surfaces were found to be highly immunogenic (A. Lee, unpublished data). The limited numbers of the autochthonous microbiota that show this phenomenon of immunological responsiveness suggest that this property is of only minor consequence in the microbial ecology of the gut.

Close examination of the literature would suggest that immune mechanisms in general have very little (if any) influence on the composition of the gut microbial ecosystem. Germ-free animals can be mono-contaminated with pure cultures of a wide range of normal or pathogenic gut microorganisms and intestinal populations continually remain high. Immunity is important in protecting the host from invasion by potential pathogens. Inhibition of bacterial adhesion has been proposed to be a major biological function of secretory IgA (McNabb and Tomasi, 1981). Even this conclusion is open to debate. Thus, the work of Van der Waaij and Heidt (1977) shows that nearly all Enterobacteriaceae in the gut in concentrations greater than 10^5/g are coated with IgA, and Berg (1980) has shown that these organisms readily translocate into tissues at this concentration. It has been suggested that there may be a synergism between ecological and immunological control mechanisms of intestinal biota. Shedlofsky and Freter (1974) state that *Vibrio cholerae* colonized in greater numbers in immunized germ-free animals compared to the immunized germ-free animals inoculated with a biota of antagonistic bacteria. A summary of some of this work is shown in Table III. These data seem too inconsistent and the differences between groups too small to warrant the acceptance in the literature of this hypothesis. This is particularly so since the antagonistic microbiota used was one that would

Table III. Suggested Synergism between Immunity and Bacterial Antagonism in Limiting the Growth of *Vibrio cholerae* in Ceca of Gnotobiotic Mice[a]

Microbiota	Immunization	Geometric mean count per cecum	Range log number
Experiment 1			
None	Previous oral exposure to *V. cholerae*	3.1×10^9	9.4–9.6
None	None	6.7×10^9	9.6–10
Experiment 2			
Five facultative anaerobes	Previous oral exposure to *V. cholerae*	7.1×10^6	6.3–7.7
Five facultative anaerobes	None	3.4×10^7	6.8–8.3
Experiment 3			
Five facultative anaerobes	Previous oral exposure to *V. cholerae*	5.4×10^5	4.0–7.7
Five facultative anaerobes	None	3.2×10^7	6.2–8.4

[a]Summarized from Shedlofsky and Freter (1974).

not confer normal CR on these animals, i.e., it was a mixture of five facultative anaerobes (*Streptococcus faecalis, Enterobacter aerogenes, Proteus
vulgaris,* and two strains of *Escherichia coli*). Results of more recent
experiments from this group (Freter *et al.,* 1983c) appear in conflict with
the earlier results. Strains of *E. coli* remained at higher levels if they were
allowed to establish in germ-free animals first before the antagonistic
biota was administered. According to the hypothesis on synergism
between ecological and immunological control mechanisms, these animals should be "immunized" to the *E. coli* and numbers should be lower
when in the presence of the antagonistic microbiota. While comparison
between these very different experiments is not strictly valid, this situation highlights the need for more information on the role of the immune
system. On present evidence, fluctuations in microbial populations can
be explained in purely ecological terms; there appears to be no need to
include a role for the immune system.

One series of observations that could challenge my disregard for the
immune system and its influence on the gut ecosystem are the reports of
differences in the gut microbial populations in immunodeficiency states,
e.g., increased numbers of aerobic and anaerobic organisms in the prox-
imal jejunum of patients with agammaglobulinemia (Ament *et al.,* 1973).
However, these hosts are likely to be stressed and this could be causing
the differences. There may also be other controlling factors, i.e., agam-
maglobulinemics are achlorhydric. Thus, the small intestine is not protected by the low pH from seeding of bacteria from the stomach.

7. Adaptation to the Ecosystem: Important Determinants of Microbial Pathogenicity

The stability of the gut microbial ecosystem acting as an important
defense mechanism against intestinal pathogens has been discussed in
Section 4. Perturbations in this system may allow pathogens to establish.
Some successful pathogens, however, are able to compete with the microbiota in the normal host or in situations where only minor changes have
occurred. The property that allows a pathogen to compete for a niche or
occupy a vacated niche may therefore be considered as a determinant of
microbial pathogenicity. Understanding of these virulence factors for colonization may be as important as an understanding of how a pathogen
causes disease symptoms. If an organism cannot successfully colonize its
target niche, then it will not be able to exert its pathogenic potential.
These ecological determinants of microbial pathogenicity are discussed
below. Adhesion is the only determinant to have received adequate attention in the research literature.

7.1. Adhesion to Intestinal Surfaces

The importance of adhesion to intestinal surfaces as a first stage of many of the major diarrheal diseases is well established and has been well reviewed (Savage, 1980; Candy, 1980). Some of these diseases have been well researched and the specificity of the mechanisms involved highlights the subtlety of the evolution of the relationship between host and parasite; e.g., strains of *Escherichia coli* causing diarrhea in young pigs have acquired a plasmid-mediated fimbrial antigen designated K88 (Shipley *et al.,* 1978). This "adhesin" allows the bacterium to attach to the brush border of cells lining the porcine small bowel. Once attached to the tissue, the organism produces a toxin that causes the diarrheal symptoms. Toxin-producing strains are avirulent if they lack the plasmid and cannot attach to the surface. Certain strains of pig lack the specific receptors for the *E. coli* adhesin and are then resistant to diarrhea after challenge with this organism (Sellwood *et al.,* 1975). Human strains of disease-causing *E. coli* have similar antigens of different attachment specificities. These have been designated as colonization factors (Evans *et al.,* 1978).

Once an intestinal surface is colonized by normal microorganisms, an introduced pathogen even with specific tissue adhesins would have difficulty in establishing. Thus, it is not surprising that the pathogens with the most sophisticated attachment mechanisms have specificity for tissue surfaces not normally colonized by other bacteria, as in, e.g., *E. coli* infection.

7.2. Mucus Colonization

As discussed in Section 2.1.3. the mucus layer that coats the intestinal surface and fills intestinal crypts provides a niche for a variety of spiral shaped organisms in most animals studied. Adaptation to this environment is therefore possible by intestinal pathogens, and would provide a means of close contact between the organism and the tissue surface, as an alternative to adhesion. Indeed, the assumption of the importance of adhesion in the pathogenesis of a number of intestinal diseases may be ill-founded. Savage (1982) has made the point that the experimental evidence for adhesion is most commonly obtained from studies in which bacterial and epithelial membranes are exposed to each other in test tubes. Furthermore, Costerton *et al.* (1981) emphasize that "the clear demonstration that a particular bacterium adheres specifically to the surfaces of cells of a washed intestinal epithelium may have little relevance to bacterial adhesion to tissues in the complex mixture of animal and bacterial polymers that covers the epithelial surface of live animals." This is well illustrated in studies on the pathogenesis of the recently recognized human intestinal pathogen *Campylobacter jejuni*. This organism has the ability to induce an inflammatory response in intestinal tissue and also produce a form of enterotoxin. Many workers assumed that adhesion would have to be the mechanism of association with tissues and looked for specific adhesins (Cinco *et al.*, 1983). Claims for adhesion were based on the occasional observation of single cells appearing to be close to intestinal cells. Thus, Naess *et al.* (1983) reported that *C. jejuni* adhered to isolated porcine brushborders *in vitro* but only when the ratio of bacteria to brushborders was greater than 100:1. The best way to visualize adhesion was stated to be by scanning electron microscopy, but this technique does not reveal the mode of firm adhesion. There was no evidence of the extensive attachment found with the pathogenic *E. coli* strains. At the Second International Workshop on *Campylobacter* Infection at Brussels in 1983, an alternative hypothesis relating to mucus colonization was put forward to explain how this organism associated with intestinal tissue (Lee *et al.*, 1983).

Campylobacter jejuni has a spiral morphology similar to that of organisms previously described as inhabitants of intestinal crypts, and has been isolated from normal rodents, dogs, and pigs. Recent observations with the scanning electron microscope of mice colonized by *C. jejuni* have revealed sheets of *C. jejuni* on the colonic surface (Merrell *et al.*, 1981; Field *et al.*, 1981). Thus, it was considered possible that *C. jejuni* is a mucosa-associated bacterium specifically adapted to the mucus environment of the intestinal surface. This hypothesis was tested in gnotobiotic Balb/c mice maintained in plastic isolators. Mice were inoculated by the orogastric route with 0.1 ml of a thick suspension (10^{10}/ml) of a human isolate of *C. jejuni* for three successive days. Animals were killed after 1 week and tissue specimens were examined.

The cecal crypts of control mice were seen to be empty, whereas the majority of the cecal crypts in the inoculated mice were colonized by large numbers of campylobacters. Only a few of the crypts, in the ileum or colon, were colonized and by smaller numbers. This suggested a preferential colonization of the cecal mucosa by these bacteria.

Germ-free mice and SPF mice with no normal mucosa-associated biota were also inoculated with *Campylobacter* cultures and the cecal mucosa examined for bacteria. The colonization of crypts was not as heavy as in the gnotobiotic animals. Variable colonization of intestinal crypts depending on the microbial status of the animal has been observed with another microaerophilic spiral organism, isolated from the rodent gut, which appears to colonize similar locations to *Campylobacter* spp. Due to its characteristic morphology, this organism can be recognized in conventional animals, where it preferentially colonizes the small intestine, compared to the gnotobiotic mouse or rat, where the colon and the cecum are heavily colonized. This observation may be relevant to the finding that *C. jejuni* causes crypt lesions in the small bowel of conventional dogs (Macartney *et al.*, 1982), whereas cecitis and colitis are found in gnotobiotic dogs (Prescott *et al.*, 1981, 1982).

The reasons for this association with intestinal surfaces and the apparent affinity for mucus have been investigated. Spiral bacteria in other systems are well suited to a viscous environment due to their characteristic motility. This property would be an ecological advantage in the mucus-filled crypts. Therefore, videotapes made using cultures of campylobacters and other gut organisms in solutions of varying viscosity were observed and the motility measured using a planimeter. Cultures of a human *Campylobacter* isolate and the spiral inhabitant of intestinal crypts moved more effectively than other gut organisms in solutions with a viscosity approximating intestinal mucus (A. Lee *et al.*, unpublished data.)

These findings confirmed the hypothesis that *C. jejuni* is a surface-associated organism in the mouse intestine. Histological studies in normal dogs suggest that the same will be found with these animals.

The ability of *Campylobacter* to associate closely with intestinal surfaces may have important consequences with respect to the pathogenicity of human infection, which is poorly understood. Histological sections of biopsies from patients with severe infection show an acute colitis with inflammatory infiltration of the lamina propria, with crypt abscesses being an important feature (Blaser *et al.*, 1980; Colgan *et al.*, 1980; Duffy *et al.*, 1980; McKendrick *et al.*, 1982). Recently, inflammatory bowel disease has been studied in a colony of marmosets; destruction of crypt abscesses was found to be associated with colonization by *C. jejuni* (Cisneros *et al.*, 1981).

Motility and capacity to colonize mucus rather than to fix firmly on the epithelial surface have also been suggested as being one of the most important determinants of pathogenesis in cholera (Savage, 1982). Nonmotile *Vibrio cholerae* mutants are less efficient in penetrating mucus gel than motile strains (Freter *et al.*, 1981).

In keeping with the concept of crypt association proposed above, Freter *et al.* (1981) state that "one can imagine that motile bacteria like Alice must keep running fast just to stay in place." Motility is energetically expensive and must therefore be very important for the survival of these organisms in the intestine.

Vibrio cholerae produces enzymes that catalyze hydrolysis of mucinous glycoproteins and it is claimed that the organism is well equipped metabolically to colonize mucus gel (Savage, 1982). Ability to degrade the mucin matrix may not be an advantage to the crypt colonizers like *Campylobacter* and the normal spiral organisms of the crypt. If the organisms in a packed crypt digested the mucins, then they would lose the selective advantage of movement in a viscous environment. No catabolism occurred in experiments with purified intestinal mucins and pure cultures of these bacteria. *Campylobacter* and gut spiral organisms generally do not utilize carbohydrates and presumably they obtain their energy source from the proteinaceous material that is present in the intestinal mucus (A. Robertson and A. Lee, unpublished data).

Associated with motility and ability to survive in mucus is the phenomenon of chemotaxis. Many strains of *Vibrio cholerae* are attracted to the mucus gel. This can be shown *in vitro* (Freter and O'Brien, 1981; Freter *et al.*, 1981). The importance of chemotaxis in colonization of these organisms is not known and needs further study. Certainly an ability to swim toward the intestinal cell surface would enable an organism to avoid washout due to the peristaltic movement of bowel contents. Jones and Isaacson (1983) suggest that chemotaxis alone may be sufficient to hold a bacterium at a mucilagenous surface. Attractants released by surfaces of living cells may form gradients of attraction, and organisms swimming up such gradients would remain predominantly in the running mode and have less chance of moving away from the surface.

7.3. Nutrient Avidity

The most satisfying hypothesis relating to the protective effect of the normal intestinal biota is the concept of adaptation to available substrates. Any pathogen that can establish without major perturbation to the ecosystem must be able to successfully compete for one or more of these substrates or utilize a substrate that is not attacked by the normal organisms. There is no information available on the preferred substrates of pathogens in the intestinal tract. One reason is because it is difficult to know where to start experimentally. This phenomenon has recently been examined using a continuous culture system (Freter *et al.*, 1983a,b,c). Instinctively, it is easy to reject the conclusions from any such *in vitro* system because it is extremely hard to accept that conditions can ever approximate the *in vivo* environment with its complex living epithelial surface. This research team, however, shows that the microbial composition of their continuous culture system (CF cultures) is in the same ecological balance as found in the mouse cecum. They speculate as to the success of their system by stating that the indigenous bacteria of the large intestine occur in thick layers on the gut epithelium, i.e., most of the bacteria adhere to each other rather than the mucosa itself. The CF cultures were found to have thick layers of bacteria adhering to the wall of the culture vessel and if the formation of these layers was prevented, the culture no longer resembled the intestinal biota in its function. Unfortunately, no data were presented on this attached layer nor on its similarity to the well-defined layers found in the mouse cecum.

The relevance of the above work to this discussion on adaptation to the ecosystem as a determinant of pathogenicity relates to the study on implantation of *E. coli* in the intestinal tract (Freter *et al.,* 1983c). Taking *E. coli* as a facultatively anaerobic prototype of an intestinal pathogen, the ways that organisms may compete for the same limiting nutrient were investigated. A mathematical model was described that predicts that two or more bacterial strains that compete in the gut for the same limiting nutrient can coexist if the metabolically less efficient strains have specific adhesion sites available. This could well be the situation with a potential intestinal pathogen. Freter's studies are important since they are the most rigorous attempt yet to apply ecological principles to the investigation and interpretation of observations on intestinal microorganisms. However, the conclusions reached go much further than the presented data allow. Thus, there is no evidence that *E. coli* attaches to the surface of the rodent gut, and the proposed model is premature. Nonetheless, the concepts presented are very interesting and should serve as a basis for further investigation.

8. Behaviorism in the Intestinal Tract: The Importance of What Organisms Do Rather Than What They Are

As cultural techniques improved, particularly the development of anaerobic chambers that allowed the isolation of the extremely oxygen-sensitive bacteria that dominate the intestinal biota, the interest of workers in intestinal microbiota centered on the identity of gut organisms and their quantitation. Too much emphasis has been placed on what the intestinal organisms are; a much greater need is to investigate what they do. Thus, Macy and Probst (1979) make the comment that "diet which one might intuitively think is the factor most likely to influence the kind and numbers of microbes has very little effect on the human fecal flora." This topic has been discussed in Section 2.1.1 and it is suggested that populations tend to be similar in people on different diets because the nutrients mainly available to organisms are of host origin, e.g., dead tissue cells and glycoproteins, and thus are similar. While this may be true in part, these studies tend to ignore one very important point. Exactly the same population of organisms in identical concentrations may behave very differently in situations where the intestinal milieu is altered, i.e., where there are slight differences in available substrate due to dietary differences. Moore and Holdeman (1974) and Finegold *et al.* (1974) carried out definitive studies on the fecal microbiota of high- and low-risk colon cancer groups. All this work revealed very little difference and the conclusion could be made that the gut bacteria have little influence on colon cancer. However, $10^{11}/g$ *Bacteroides* sp. in one group could be behaving very differently from $10^{11}/g$ *Bacteroides* sp. of the identical serotype in the other group. Thus, Reddy *et al.* (1975) showed that people who initially ate a high-beef diet and subsequently shifted to a nonmeat diet were found to have a decrease in fecal β-glucuronidase even though major differences in the microbiota were not noted. This enzyme can reactivate various carcinogens and thus increase the risk of bowel cancer.

The need for greater emphasis on the behavior of the gut microbiota was stressed by Miller and Hoskins (1981), who suggested the following approach to a study of the intestinal ecosystem: (1) study the gut microflora in terms of subpopulations with functional activities that are postulated to affect health; (2) develop methods for estimating the population densities of these subpopulations based on their functional activities; and (3) determine if their population densities in subjects at risk for a given illness differ from healthy subjects.

Using this approach, these workers developed a semiquantitative, most probable number method that measured the mucin-degrading potential of dilutions of human feces. They found that bacteria that do degrade mucin oligosaccharides can be regarded as a functionally distinct subset of normal fecal microbiota comprising 1% of the total bacterial populations. One criticism of this method is that by diluting out the samples, the balance of populations in the ecosystem is being changed and it could be possible that dilution out of organisms in lower concentrations might be removing essential growth factors for the dominant organisms; it might be that the 1% are not the mucin-degraders, but that the mucin-degraders are unable to perform due to the lack of growth factors produced by the 1%, that is, at a dilution of ¹⁄₁₀₀ many mucin-degraders could still be present but unable to grow because other organisms on which they are dependent have been diluted out. Nevertheless, the concept of Miller and Hoskins (1981), of considering the enzymic potential of the total microbiota rather than being concerned with isolating single strains, is important.

Another problem with the enumerate and speciate approach to intestinal microbiology is that only organisms present in the highest concentrations tend to be investigated. Brock (1966), in his general treatise on microbial ecosystems, commented, "it is impossible to evaluate, from numbers alone, the ecological significance of an organism in a given habitat. One must have data on its mass, its metabolic activities, and its possible ability to produce and react to substances which have unique biological activities. Only then can numbers be meaningful." He speculates that 10^6 organisms/ml is the lower population limit for an effect in an ecosystem.

A good model to illustrate the complexity of the gut microbiota with regard to concentration versus behavior is the human pathogen *Clostridium difficile* in the human or rodent large bowel. Reference to the data in Table IV shows that four different ecostates may occur with this organism in the intestine: (1) The normal microbiota exerts colonization resistance such that the clostridia do not establish; (2) the normal microbiota is absent; the organism colonizes and produces toxin but no symptoms; (3) the normal microbiota is absent; the organism colonizes and produces toxin, and classical pseudomembranous colitis (PMC) results; and (4) the clostridia colonize in significant numbers but produce no toxin.

Differences between human and mouse susceptibility may be due to toxin specificity; also, absence of symptoms in the human in the presence of cytotoxin may be evidence that this is not the same toxin responsible for PMC. However, other explanations are possible and indicate the breadth of investigations required before we can come to a thorough understanding of the interplay between the normal microbiota and potential pathogens. Figure 2 indicates these different scenarios.

In summary, even though a microbiota may permit establishment and colonization of a potential pathogen, the behavior of this pathogen may be influenced by this microbiota in many different indirect ways. Studies on the ecology of intestinal pathogens must be directed more toward these behavioral interactions.

Another model that illustrates this point and requires more investigation with an ecological bias is infant botulism (Arnon *et al.,* 1982). This tragic disease appears to be a consequence of the intestinal milieu of the infant allowing *Clostridium botulinum* to produce active toxin.

Table IV. *Clostridium difficile* in the Intestinal Tract of Human and Mouse in Different Ecological States

	Human			Mouse	
	Infant	Adult	Perturbed adult	Infant	Adult
Level of *Clostridium difficile*[a]	10^3–10^{7b}	$<10^c$	10^2–10^{9c}	10^3–10^{5d}	$<10^{3d}$
Toxin production	+	−	+	+	−
Symptoms	−	−	+	−	−

[a]Concentration of bacteria per gram of intestinal content.
[b]Stark *et al.* (1982).
[c]Viscidi *et al.* (1981).
[d]A. Lee (unpublished data).

1 <u>Clostridium difficile</u> in the perturbed adult intestine

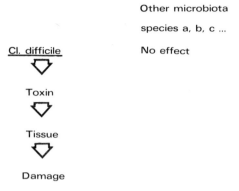

Other microbiota

species a, b, c ...

<u>Cl. difficile</u>   No effect

Toxin

Tissue

Damage

2 <u>Clostridium difficile</u> in the infant intestine

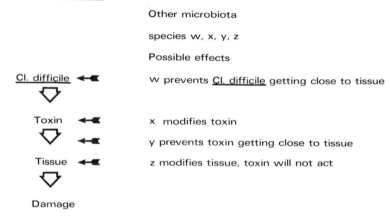

Other microbiota

species w, x, y, z

Possible effects

<u>Cl. difficile</u>   w prevents <u>Cl. difficile</u> getting close to tissue

Toxin   x modifies toxin

          y prevents toxin getting close to tissue

Tissue   z modifies tissue, toxin will not act

Damage

Figure 2. Pathogenic potential of *Clostridium difficile* in the human intestine in different ecological states. Possible roles of the other microbiota.

9. Man versus Mouse: The Relevance of Current Concepts of Intestinal Ecology

Rene Dubos and Russel Schaedler were the first to systematically study the organisms of the intestinal tract and to speculate on their influence on the host (Dubos and Schaedler, 1964). Out of this laboratory also came the first detailed descriptions of surface-associated organisms (Savage *et al.*, 1968) and one of the first attempts to apply strictly anaerobic methodologies to the study of nonruminant animals (Lee *et al.*, 1968).

The basis of much current understanding and comment of the microbiology of the intestinal tract originates from this work. However, it must be remembered that the experimental model in these studies was the rodent gut, a situation far removed from the human digestive tract, the size of the lumen relative to the tissue is so different that gaseous and redox balances are likely to be totally dissimilar. The cecum is a dominant influence on the composition of the rodent microbiota; in man it carries little significance. Simple examination of a Gram stain of feces from mouse and man reveals a completely different balance of microbial populations.

The conclusions drawn from Dubos' group and those of other workers studying rodents are referred to in discussions of human gut microbiology often with not enough comment as to their relevance. Volatile fatty acids (VFA) are given as an example of a protective function of the gut microbiota (Rasic and Kurman, 1983). A negative correlation exists between levels of facultative anaerobes such as coliforms in the rodent gut and, in particular, levels of butyric acid; when butyric acid is present the levels of $E.$ $coli$ will drop to 10^3-10^5/g feces or less. The level of $E.$ $coli$ in human feces is often as high as 10^6-10^7/g, levels usually associated with decreased VFA levels in the rodent. Yet, in her study on the development of the human microbiota, Stark (1983) found many instances where the fatty acid levels had reached the adult level but the coliform levels were still raised.

Given that we cannot experiment adequately on the human intestine, conclusions need to be drawn from animal experimentation. However, greater care needs to be taken when extrapolating results, due to the great differences between the human and small-animal gut ecosystems. Possibly alternate models need to be found. Results of recent experiments by Raibaud et $al.$ (1980) appear promising and worthy of considerable attention. These workers have attempted to establish a human microbiota in germ-free animals by feeding them human feces. On theoretical grounds, due to differences in anatomy it would appear unlikely that this model could mimic the human gut. However, close examination of the data reveals that the balance of the human populations appears to be maintained, although, interestingly, the coliform levels do appear slightly reduced in the mouse system. Colonization resistance as in the human is maintained. Provided these animals are studied relatively soon after colonization, they may be more relevant for experimental studies on the interaction with potential human gut pathogens than normal mice.

The development of small, easily handled, germ-free isolators that fit one cage of mice should make this system very attractive to microbial ecologists.

A recently reported observation using this system shows not only how the mouse ecosystem may be irrelevant to the human situation, but also that this model may catalyze completely different avenues of investigation in gut ecology. The human pathogen *Campylobacter jejuni* colonizes normal mice intestine easily to a level of 10^7/g feces. This remains constant for months with no disease symptoms evident. However, *Campylobacter jejuni* would not colonize these "heteroxenic" mice established with a normal human microbiota even for a few days (Adremont *et al.*, 1983). If this means that the human microbiota exerts a strong colonization resistance against *C. jejuni*, how do humans apparently acquire the disease so readily?

The other approach to this problem has already been referred to in Section 7.3 and is being extensively promoted by Freter *et al.* (1983a,b,c): *in vitro* continuous culture systems of human feces. Once again on theoretical grounds, the ecologist instinctively must distrust this system; however, normal balances of populations appear to be maintained. Possibly a more satisfactory solution would be to carry out experiments in both the *in vivo* heteroxenic mouse system and the *in vitro* continuous culture system. Similar results with both systems might give more confidence in their relevance to the actual human gut ecosystem.

10. Manipulation of the Intestinal Ecosystem

To satisfy the granting bodies in the increasingly competitive world of fund-seeking, we all tend to define long-term goals for our research programs. For workers in the microbiology of the intestine, a goal that hopefully sounds convincing to referees is to be able (in the long term) to manipulate the ecosystem of the gut and thus prevent the establishment of intestinal pathogens or shorten the course of intestinal disease. This has obvious humanitarian appeal when applied to animal husbandry.

Whereas most research carried out under this banner has been necessarily pure research laying a foundation of knowledge in understanding ecological principles of the gut, successful manipulations have been reported recently that show progress is being made toward the ultimate goal.

10.1. Replacement or Preseeding of the Premature or Perturbed Gut Ecosystem

Colonization of birds by species of *Salmonella* that cause intestinal disease is a major problem to the poultry industry. Young chicks are much more susceptible to colonization, presumably due to lack of colonization resistance. Nurmi and Rantala (1973) fed chicks a suspension of gut contents from healthy adult birds and so protected the chicks from challenge by *Salmonella* sp. This work has been taken forward by others who have achieved the same result by implanting a mixture of 48 different bacterial strains cultured *in vitro* (Impey *et al.,* 1982).

Implantation in humans has been spoken of since the beginnings of microbiology, when Metchnikoff (1908) postulated that the longevity of Bulgarian peasants was linked with their addiction to a goats' milk yoghurt cultured with lactobacilli. Early experiments feeding lactobacilli were variable due to the use of nonhuman bacterial isolates. Recently there seems to be a more scientific approach to the use of bacterial cultures as a dietary supplement.

Shahani and Ayebo (1980) defined the basic qualities required for bacterial cultures used for implantation: (1) they should be normal inhabitants of the host intestine or be capable of adapting to the host intestinal environment; (2) they must survive passage into the intestine and be capable of establishing; and (3) they must perform functions that are advantageous to the host.

The main organisms considered to fulfill these criteria are species of *Lactobacillus* or *Bifidobacterium*. The composition of the various preparations commercially available have been recently listed (Rasic and Kurman, 1983); examples include: (1) Infloran Berna, a freeze-dried preparation containing 10^9 *Lactobacillus acidophilus* and 10^9 *Bifidobacterium infantis* (Swiss Serum and Vaccine Institute, Berne, Switzerland, and Instituto Sierotrapico Berne, Como, Italy) and (2) Eugalon Topler Forte, a preparation of bifidobacteria (Topler Gumbolt, Dietmannsnied Allgan, West Germany). Indications for both preparations are stated to be a disturbed balance of the intestinal biota, e.g., by antibiotics. Limited acceptance of this type of formulation is indication that more needs to be done, presumably to obtain isolates that better fulfill the criteria listed above. Gorbach's group, who have some evidence that feeding lactobacilli to animals can influence the incidence of experimental colon cancer (Gorbach, 1982), have been trying to isolate human cultures with better colonizing potential (P. Conway, personal communication).

Presumably, lactobacilli and bifidobacteria are still used for historical reasons and their ease of culture. These organisms are not the logical choice for this work, since they are not the most dominant organisms of the human microbiota. An exception might be implantation in infant preweaning, where bifidobacteria are the dominant bacteria. The problem is analogous to the establishment of a satisfactory microbiota in SPF mice discussed in Section 5.3. Here the conclusion was that limited numbers of pure cultures are unsatisfactory. The answer is either large numbers of isolates or, probably most satisfactorily, the use of intestinal content of animals known to be clear of potential pathogens. The same could apply in the human, although it would be a brave investigator who would institute administration of diluted intestinal content from screened donors, even though it might be more likely of success than pure cultures!

One area where limited pure cultures might prove successful is in the prophylaxis of necrotizing enterocolitis, a disease that affects small babies in intensive care wards during the first weeks of life. This life-threatening condition occurs in about 3% of admissions. The fecal microbiota of these preterm neonates is different from either breast- or bottle-fed infants (Stark and Lee, 1982a). Lawrence *et al.* (1982) have proposed that this is due to physical isolation, the cleanliness of nursing procedures, and, in some cases, antibiotic treatment in neonatal intensive care units, which resemble germ-free laboratories. Colonization of these babies with a limited number of organisms isolated from normal healthy babies, thus providing a nonpathogenic competing microbial biota, would appear feasible and, on ecological grounds, desirable. However, once again, who would be willing to experiment with these tiny and extremely vulnerable infants?

10.2. Selective Decontamination of the Digestive Tract

There are certain groups of patients who are at risk when facultatively anaerobic Gram-negative bacilli enter the tissue, i.e., patients who have depressed or ablated immune function due to total-body irradiation or cytotoxic drugs. As the portal of entry of these organisms is almost certainly the gastrointestinal tract, it is not surprising that much effort has gone into manipulation of the gut ecosystem to prevent large numbers of these bacteria accumulating. One approach would be to totally eliminate all gut bacteria with antibiotics, but this would be self-defeating, since the grossly perturbed ecosystem resulting from this treatment would provide a haven for the antibiotic-resistant strains that inhabit hospital environments. The concept of colonization resistance discussed in Section 3 comes from the work of Van der Waaij *et al.* (1982), who recognized that the desirable goal was to eliminate existing facultative anaerobes in the host but preserve the normal anaerobic microbiota, which could then exert colonization resistance and thus prevent establishment of the small inocula of drug-resistance bacteria the host is likely to encounter. This has been put into practice as selective decontamination of the digestive tract (SDD) with very encouraging results. Such work is a major example of where a sound ecological approach to investigation of the microbiology of the intestinal tract has produced beneficial applied results. Thus, in a controlled prospective randomized trial, the effect of SDD in granulocytopenic patients (patients with depressed white cell counts and thus lacking normal cellular defense mechanisms) was studied (Sleijfer *et al.*,

1980). SDD was accomplished by oral administration of nalidixic acid, co-trimoxazole, or polymyxin E to suppress growth of aerobic Gram-negative bacteria and amphotericin B to inhibit growth of yeasts. Eighteen instances of Gram-negative or yeast infection were seen in the 52 control patients, whereas only two infections were recorded in patients who had been selectively decontaminated. Nine of the control group died of an acquired infection, but none of the SDD-treated group died.

The success of this approach seems to be a numbers game. Berg (1980) has shown that the translocation of enteric bacteria from the gastrointestinal tract into the tissues, i.e., the draining lymph nodes, liver, and spleen of mice, is proportional to the numbers in the intestinal lumen. We have also shown that if small bowel numbers become greater than $10^5/g$ of gut content, translocation of facultative anaerobes into mesenteric lymph nodes, liver, and spleen will occur (A. Lee, unpublished data). Thus, any antibiotic therapy that is likely to remove normal microbiota and influence colonization resistance could be harmful, particularly to debilitated patients.

10.3. Environmental Manipulation of the Gut Ecosystem

The more we learn about factors that influence the behavior of organisms in the intestine, then the more we will be able to successfully manipulate the gut environment to prevent organisms exerting a harmful effect. A terrible accident indicates the dramatic consequence of poorly understood manipulations. Oral mannitol solutions have been given to patients as a mechanical bowel preparation for colonoscopy, elective colorectal surgery, and for barium enemas, since the mannitol through an osmotic effect leads to a watery diarrhea, which flushes out gut content. The *E. coli* present in these patients was able to ferment the mannitol, producing the gases hydrogen and methane, which exploded in one patient as diathermy was being done, resulting in death (Keighley *et al.,* 1981).

11. Conclusion

The microbiology of the gastrointestinal tract has been well documented, but until recently there has been little effort to try to understand the factors that control these populations. Fortunately, more studies with an ecological bias are now appearing and hopefully our understanding of this complex system will increase. This chapter has attempted to discuss what is now known in ecological terms and to show the directions in which research is preceeding. On reflection, the knowledge base is still remarkably thin. Researchers with interests in the ecology of other better understood systems may well have contributions to make in this area. The gut is one of the classic microbial ecosystems, so much so that I have suggested that the study of the rodent gut would be an excellent teaching exercise to introduce school students to microbial ecology (Lee, 1984). At this superficial level the concepts appear straightforward and satisfying. However, in reality we are little further advanced than Metchnikoff and his Bulgarian peasants.

References

Alexander, M., 1971, *Microbial Ecology,* Wiley, New York.

Ament, M. E., Ochs, H. D., and Davis, S. D., 1973, Structure and function of the gastrointestinal tract in primary immunodeficiency syndromes. A study of 39 patients, *Medicine* **52**:227–248.

Andremont, A., Raibaud, P., Tancrede, C., Duval-Iflah, Y. and Ducluzeau, R., 1983, The use of germ-free mice associated with human fecal flora as an animal model to study enteric bacterial interactions, in: *Recent Advances in Bacterial Diarrheal Diseases: An International Symposium,* K. T. K. Scientific Publishers, Tokyo.

Aries, V. C., Crowther, J. S., Draser, B. S., Hill, M. J., and Ellis, F. R. 1971, The effect of a strict vegetarian diet on the faecal flora and faecal steroid concentrations, *J. Pathol.* **103**:54–56.

Arnon, S. S., Damus, K., Thompson, B., Midura, T. F. and Chin, J., 1982, Protective role of human milk against sudden death from infant botulism, *J. Pediatr.* **100**:568–573.

Atlas, R. M., 1984, Diversity of microbial communities, in: *Advances in Microbial Ecology,* Vol. 7 (K. C. Marshall, ed.), pp. 1–47, Plenum Press, New York.

Bartlett, J. G., 1982, Virulence factors of anaerobic bacteria, *Johns Hopkins Med. J.* **151**:1–9.

Berg, R. D., 1980, Mechanisms confining indigenous bacteria to the gastrointestinal tract, *Am. J. Clin. Nutr.* **33**:2472–2484.

Berg, R. D., and Savage, D. C., 1972, Immunological responses and microorganisms indigenous to the gastrointestinal tract, *Am. J. Clin. Nutr.* **25**:1364–1371.

Berg, R. D., and Savage, D. C., 1975, Immune responses of specific pathogen-free and gnotobiotic mice to antigens of indigenous and non-indigenous microorganisms, *Infect. Immunol.* **11**:320–329.

Bergheim, O., Hansjen, A. H., Pincussen, L., and Weiss, E., 1941, Relation of volatile fatty acids and hydrogen sulphide to the intestinal flora, *J. Infect. Dis.* **69**:155–166.

Blaser, M. J., Parsons, R. B., and Wang, W.-L., 1980, Acute colitis caused by *Campylobacter fetus ss. jejuni, Gastroenterology* **78**:448–453.

Bohnoff, M., Miller, C. P., and Martin, W. R., 1964, Resistance of the mouse's intestinal tract to experimental *Salmonella* infection. II. Factors responsible for its loss following streptomycin treatment, *J. Exp. Med.* **120**:817–824.

Bowden, G. H. W., Ellwood, D. C., and Hamilton, I. R., 1979, Microbial ecology of the oral cavity, in: *Advances in Microbial Ecology,* Vol. 3 (M. Alexander, ed.), pp. 135–217, Plenum Press, New York.

Brock, T. D., 1966, *Principles of Microbial Ecology,* Prentice-Hall, Englewood Cliffs, New Jersey.

Brockett, M., and Tannock, G. W., 1982, Dietary influence on microbial activities in the caecum of mice, *Can. J. Microbiol.* **28**:493–499.

Brownlie, L. E., and Grau, F. H., 1967, Effect of food intake on growth and survival of salmonellas and *Escherichia coli* in the bovine rumen, *J. Gen. Microbiol.* **46**:125–134.

Bullen, C. L., and Tearle, P. V., 1976, Bifidobacteria in the intestinal tract of infants: An *in vitro* study, *J. Med. Microbiol.* **9**:335–344.

Bullen, J. M., Rogers, H. J., and Leigh, L., 1972, Iron-binding proteins in milk and resistance to *Escherichia coli* infection in infants, *Br. Med. J.* **1**:69–75.

Burek, J. D., 1978, *Pathology of Aging Rats,* CRC Press, Boca Raton, Florida.

Candy, D. C. A., 1980, Adhesion of bacteria to mucosal surfaces—An area of increasing importance in diarrhoeal disease, *Eur. J. Pediatr.* **134**:3–8.

Chase, D. G., and Erlandsen, S. L., 1976, Evidence for a complex life cycle and endospore formation, in attached, filamentous, segmented bacterium from murine ileum, *J. Bacteriol.* **127**:572–583.

Christie, R. J., Williams, F. P., Whitney, Jr., R. A., and Johnson, D. J., 1968, Techniques used in the establishment and maintenance of a barrier mouse breeding colony, *Lab. Anim. Care* **18**:543–549.

Cinco, M., Banfi, E., and Crotti, D., 1983, Studies on the adhesive properties of campylobacters, in: *Campylobacter II* (A. D. Pearson, M. B. Skirrow, B. Rowe, J. R. Davies, and D. M. Jones, eds.), p. 112, PHLS, London.

Cisneros, R. L., Onderdonk, A. B., Bronson, R., and Shegal, R., 1981, Association of inflammatory bowel disease in a colony of cotton-top marmosets with the presence of *Campylobacter fetus* subsp. *jejuni*, in: *Abstracts of the Annual Meeting of the American Society for Microbiology 81*, p. 24 (abstract).

Coates, M. E., and Fuller, R., 1977, The gnotobiotic animal in the study of gut microbiology, in: *Microbial Ecology of the Gut* (R. T. J. Clarke and T. Bauchop, eds.), pp. 311–346, Academic Press, London.

Colgan, T., Lambert, J. R., Newman, A., and Luk, S. C., 1980, *Campylobacter jejuni* enterocolitis. A clinicopathologic study, *Arch. Pathol. Lab. Med.* **104**:571–574.

Costerton, J. W., Irvin, R. T., and Cheng, K. J., 1981, The role of bacterial surface structures in pathogenesis, *CRC Crit. Rev. Microbiol.* **8**:303–338.

Cummings, J. H., 1981, Short chain fatty acids in the human colon, *Gut* **22**:763–779.

Damian, R. T., 1964, Molecular mimicry: Antigen sharing by parasite and host and its consequences, *Am. Nat.* **XCVIII**:129–149.

Dubos, R., 1965, *Man Adapting,* Yale University Press, New Haven.

Dubos, R., and Schaedler, R., 1964, The digestive tract as an ecosystem, *Am. J. Med. Sci.* **248**:267–272.

Dubos, R., Schaedler, R. W., Costello, R., and Hoet, P., 1965, Indigenous, normal and autochthonous flora of the gastrointestinal tract, *J. Exp. Med.* **122**:67–76.

Duffy, M. C., Benson, J. B., and Rubon, S. J., 1980, Mucosal invasion in *Campylobacter* enteritis, *Am. J. Clin. Pathol.* **73**:706–708.

Evans, D. G., Evans, D. J., Tjoa, W. S., and Dupont, H. L., 1978, Detection and characterization of colonization factor of enterotoxigenic *Escherichia coli* isolated from adults with diarrhea, *Infect. Immunol.* **19**:727–736.

Field, L. H., Underwood, J. L., Pope, L. M., and Berry, L. J., 1981, Intestinal colonization of neonatal animals by *Campylobacter fetus* subsp. *jejuni, Infect. Immunol.* **33**:884–892.

Finegold, S. M., Atteberg, H. R., and Sutter, V. L. 1974, Effect of diet on human fecal flora: Comparison of Japanese and American diets, *Am. J. Clin. Nutr.* **27**:1456–1469.

Foo, M. C., and Lee, A., 1972, Immunological response of mice to members of the autochthonous intestinal microflora, *Infect. Immunol.* **6**:525–532.

Foo, M. C., and Lee, A., 1974, Antigenic cross-reaction between mouse intestine and a member of the autochthonous microflora, *Infect. Immunol.* **9:**1066–1069.

Foo, M. C., Lee, A., and Cooper, G. N., 1974, Natural antibodies and the intestinal flora of rodents, *Aust. J. Exp. Biol. Med. Sci.* **52:**321–330.

France, G. L., Marmer, D. J., and Steele, R. W., 1980, Breast feeding and *Salmonella* infection, *Am. J. Dis. Child.* **134:**147–152.

Freter, R., and Abrams, G. D., 1972, Function of various intestinal bacteria in converting germfree mice to the normal state, *Infect. Immunol.* **6:**119–126.

Freter, R., and O'Brien, P. C. M., 1981, Role of chemotaxis in the association of motile bacteria with intestinal mucosa: Fitness and virulence of nonchemotactic *Vibrio cholerae* mutants in infant mice, *Infect. Immunol.* **34:**222–233.

Freter, R., Allweiss, B., O'Brien, P. C. M., Halstead, S. A., and Macsai, M. S., 1981, Role of chemotaxis in the association of motile bacteria with intestinal mucosa: *In vitro* studies, *Infect. Immunol.* **34:**241–249.

Freter, R., Stauffer, E., Cleven, D., Holdeman, L. V., and Moore, E. C., 1983a, Continuous-flow cultures as *in vitro* models of the ecology of large intestinal flora, *Infect. Immunol.* **39:**666–675.

Freter, R., Brickner, H., Botney, M., Cleven, D., and Aranki, A., 1983b, Mechanisms that control bacterial populations in continuous flow culture models of mouse large intestinal flora, *Infect. Immunol.* **39:**676–685.

Freter, R., Brickner, H., Fekete, J., Vickerman, M. M., and Carey, K. E., 1983c, Survival and implantation of *Escherichia coli* in the intestinal tract, *Infect. Immunol.* **39:**686–703.

Garland, C. D., Lee, A., and Dickson, M. R., 1979, The preservation of surface-associated microorganisms prepared for scanning electron microscopy, *J. Microsc.* **116:**227–242.

Garland, C. D., Lee, A., and Dickson, M. R., 1982, Segmented filamentous bacteria in the rodent small intestine: Their colonization of growing animals and possible role in host resistance to *Salmonella, Microb. Ecol.* **8:**181–190.

Gorbach, S. L., 1982, The intestinal microflora and its colon cancer connection, *Infection* **10:**379–384.

Gordon, H. A., and Pesti, L., 1971, The gnotobiotic animal as a tool in the study of host microbial relationships, *Bacteriol. Rev.* **35:**390–429.

Gordon, J. E., 1971, Diarrheal disease of early childhood—World wide scope of the problem, *Ann. N.Y. Acad. Sci.* **176:**9–15.

Grau, F. H., Brownlie, L. E., and Smith, M. G., 1969, Effects of food intake on number of salmonellae and *Escherichia coli* in rumen and faeces of sheep, *J. Appl. Bacteriol.* **32:**112–117.

Guiot, H. F. L., 1982, Role of competition for substrate in bacterial antagonism in the gut, *Infect. Immunol.* **38:**887–892.

Harris, D. L., and Kinyon, J. M., 1974, Significance of anaerobic spirochetes in the intestines of animals, *Am. J. Clin. Nutr.* **27:**1297–1304.

Hector, R. F., and Domer, J. E., 1982, Mammary gland contamination as a means of establishing long-term gastrointestinal colonization of infant mice with *Candida albicans, Infect. Immunol.* **38:**788–790.

Heine, W., and Thunert, A., 1968, The establishment of a division of gnotobiology and conceptions of special barrier type animal houses, in: *Advances in Germfree Research and Gnotobiology* (M. Miyakawa and T. D. Luckey, eds.), pp. 9–15, Iliffe Books, London.

Hobson, P. N., and Wallace, R. J., 1982a, Microbial ecology and activities in the rumen: Part I, *CRC Crit. Rev. Microbiol.* **9:**165–225.

Hobson, P. N., and Wallace, R. J., 1982b, Microbial ecology and activities in the rumen: Part II, *CRC Crit. Rev. Microbiol.* **9:**253–320.

Holdeman, L. V., Good, I. J., and Moore, W. E. C., 1976, Human fecal flora: Variation in bacterial composition within individuals and a possible effect of emotional stress, *Appl. Environ. Microbiol.* **31**:359–375.

Hoskins, L. C., and Boulding, E. T., 1981, Mucin degradation in human colon ecosystems. Evidence for the existence and role of bacterial subpopulations producing glycosidases as extracellular enzymes, *J. Clin. Invest.* **67**:163–172.

Hungate, R. E., 1978, Gut microbiology, in: *Microbial Ecology* (M. W. Loutit and J. A. R. Miles, eds.), pp. 258–264, Springer-Verlag, Berlin.

Impey, C. S., Mead, G. C., and George, S. M., 1982, Competitive exclusion of salmonellas from the chick caecum using a defined mixture of bacterial isolates from the caecal microflora of an adult bird, *J. Hyg. Camb.* **89**:479–490.

Jones, G. W., and Isaacson, R. E., 1983, Proteinaceous bacterial adhesis and their receptors, *CRC Crit. Rev. Microbiol.* **10**:229–260.

Kanaaneh, H., 1972, The relationship of bottle feeding to malnutrition and gastroenteritis in a pre-industrial setting. *J. Trop. Pediatr.* **18**:302–306.

Keighley, M. R. B., Taylor, E. W., Hares, M. M., Arabi, Y., Youngs, D., Bentley, S., and Burdon, D. W., 1981, Influence of oral mannitol bowel preparation on colonic microflora and the risk of explosion during endoscopic diathermy, *Br. J. Surg.* **68**:554–556.

Lamanna, C., 1972, Needs for illuminating the microbiology of the lumen, *Am. J. Clin. Nutr.* **25**:1488–1494.

Lawrence, G., Bates, J., and Gaul, A., 1982, Pathogenesis of neonatal necrotising enterocolitis, *Lancet* **1**:137–139.

Leach, W. D., Lee, A., and Stubbs, R. P., 1973, Localization of bacteria in the gastrointestinal tract: A possible explanation of intestinal spirochaetosis, *Infect. Immunol.* **7**:961–972.

Lee, A., 1980, Normal flora of animal intestinal surfaces, in: *Adsorption of Microorganisms to Surfaces* (G. Bitton and K. C. Marshall, eds.), pp. 145–173, Wiley, New York.

Lee, A., 1984, The formalinized rat: A convenient microbial ecosystem, *Am. J. Biol. Teach.* **46**:48–52.

Lee, A., and Gemmell, E., 1972, Changes in the mouse intestinal microflora during weaning: Role of volatile fatty acids, *Infect. Immunol.* **5**:1–7.

Lee, A., Gordon, J., and Dubos, R., 1968, Enumeration of the oxygen sensitive bacteria usually present in the intestine of healthy mice, *Nature* **220**:1137–1139.

Lee, A., Gordon, J., Lee, C. J., and Dubos, R., 1971, The mouse intestinal flora with emphasis on the strict anaerobes, *J. Exp. Med.* **133**:339–352.

Lee, A., O'Rourke, J., Phillips, M. W., and Barrington, P., 1983, *Campylobacter jejuni* as a mucosa-associated organism: An ecological study, in: *Campylobacter II* (A. D. Pearson, M. B. Skirrow, B. Rowe, J. R. Davies, and D. M. Jones, eds.), pp. 112–114, PHLS, London.

Lipson, A., 1976, Infecting dose of *Salmonella, Lancet* **1**:969.

Macartney, L., McCandlish, I. A. P., Al-Mashat, R. R., and Taylor, D. J., 1982, Natural and experimental enteric infections with *Campylobacter jejuni* in dogs, in: *Campylobacter Epidemiology, Pathogenesis and Biochemistry* (D. G. Newell, ed.), p. 172, MTP Press, Lancaster, England.

Macfie, J. W. S., 1916, The morphology of certain spirochetes of man and other animals, *Ann. Trop. Med. Parasitol.* **10**:305–343.

Macy, J. M., and Probst, I., 1979, The biology of gastrointestinal *Bacteroides, Annu. Rev. Microbiol.* **33**:561–594.

McKendrick, M. W., Geddes, A. M., and Gearty, J., 1982, *Campylobacter* enteritis: A study of clinical features and rectal mucosal changes, *Scand. J. Infect. Dis.* **14**:35–38.

McNabb, P. C., and Tomasi, T. B., 1981, Host defense mechanisms at mucosal surfaces, *Annu. Rev. Microbiol.* **35**:477–496.

Merrell, B. R., Walker, R. I., and Coolbaugh, J. C., 1981, *Campylobacter fetus* ss *jejuni,* a newly recognised enteric pathogen: Morphology and intestinal colonization, *Scanning Electron Microsc.* **4**:125–131.

Metchnikoff, E., 1908, *The Prolongation of Life; Optimistic Studies,* G. P. Putnam's and Sons, London.

Miller, R. S., and Hoskins, L. C., 1981, Mucin degradation in human colon ecosystems. Fecal population densities of mucin-degrading bacteria estimated by a 'most probable number' method, *Gastroenterology* **81**:759–765.

Mims, C. A., 1982, *The Pathogenesis of Infectious Disease,* Academic Press, London.

Moore, W. E. C., and Holdeman, L. V., 1974, Human fecal flora: The normal flora of 20 Japanese-Hawaiians, *Appl. Microbiol.* **27**:961–979.

Naess, V., Johannessen, A. C., and Hofstad, T., 1983, Adherence of *Campylobacter jejuni* to porcine brushborders, in: *Campylobacter II* (A. D. Pearson, M. B. Skirrow, B. Rowe, J. R. Davies, and D. M. Jones, eds.), pp. 111–112, PHLS, London.

Nurmi, E., and Rantalaa, M., 1973, New aspects of *Salmonella* infection in broiler production, *Nature* **241**:210–211.

Perrot, A., 1976, Evolution of the digestive microflora in a unit of specified pathogen-free mice: Efficiency of the barrier, *Lab. Anim.* **10**:143–156.

Pesti, L., Kokas, E., and Gordon, H. A., 1969, Effects of *Clostridium difficile, Lactobacillus casei, Bacillus subtilis,* and *Lactobacillus* sp. as mono- and di-contaminants on the caecum of germfree mice, in: *Germfree biology* (E. A. Mirand and N. Back, eds.), pp. 179–180, Plenum Press, New York.

Phillips, M. W., and Lee, A., 1983, Isolation and characterization of a spiral bacterium from the crypts of rodent gastrointestinal tracts, *Appl. Environ. Microbiol.* **45**:675–683.

Phillips, M. W., and Lee, A., 1984, Microbial colonization of rat colonic mucosa following intestinal perturbation, *Microb. Ecol.* **10**:79–88.

Phillips, M., Lee, A., and Leach, W. D., 1978, The mucosa-associated microflora of the rat intestine: A study of normal distribution and magnesium sulphate induced diarrhoea, *Aust. J. Exp. Biol. Med. Sci.* **56**:649–662.

Porter, P., Parry, S. H., and Allen, W. D., 1977, Significance of immune mechanisms in relation to enteric infections of the gastrointestinal tract in animals, *Ciba Found. Symp.* **46**:55–75.

Prescott, J. F., Barker, I. K., Manninen, K. I., and Miniats, O. P., 1981, *Campylobacter jejuni* enteritis studies in gnotobiotic dogs, *Can. J. Comp. Med.* **45**:377–383.

Prescott, J. F., Manninen, K. I., and Barker, I. K., 1982, Experimental pathogenesis of *Campylobacter jejuni* enteritis studies in gnotobiotic dogs, pigs, and chickens, in: *Campylobacter Epidemiology, Pathogenesis, and Biochemistry* (D. G. Newell, ed.), pp. 170–171, MTP Press, Lancaster, England.

Prins, R. A., 1977, Biochemical activities of gut micro-organisms, in: *Microbial Ecology of the Gut* (R. T. J. Clarke and T. Bauchop, eds.), pp. 74–183, Academic Press, London.

Prizont, R., and Konigsberg, N., 1981, Identification of bacterial glycosidases in rat cecal contents, *Digest. Dis. Sci.* **26**:773–777.

Raibaud, P., Ducluzeau, R., Dubos, F., Hudault, S., Bewa, H., and Muller, M. C., 1980, Implantation of bacteria from the digestive tract of man and various animals into gnotobiotic mice, *Am. J. Clin. Nutr.* **33**:2440–2447.

Ramphal, R., Small, P. M., Shands, J. W., Fischlschweiger, W., and Small, P. A., 1980, Adherence of *Pseudomonas aeruginosa* to tracheal cells injured by influenza infection or by endotracheal intubation, *Infect. Immunol.* **27**:614–619.

Rantala, M., and Nurmi, E., 1973, Prevention of the growth of *Salmonella infantis* in chicks by the flora of the alimentary tract of chickens, *Br. Poult. Sci.* **14**:627–630.

Rasic, J. L., and Kurman, J. A., 1983, *Bifidobacteria and Their Role,* Brikhauser, Basel.

Reddy, B. S., Weisburger, J. H., and Wynder, E. L., 1975, Effect on high risk and low risk diets for colon carcinogenesis on fecal microflora and steroids in man, *J. Nutr.* **105**:878–884.

Roediger, W. E. W., 1980, Role of anaerobic bacteria in the metabolic welfare of the colonic mucosa in man, *Gut* **21**:793–798.

Rosenthal, S. L., 1969, Exacerbation of *Salmonella* enteritis due to ampicillin, *N. Engl. J. Med.* **280**:147–148.

Rothbaum, R., McAdams, A. J., Giannella, R., and Partin, J. C., 1982, A clinicopathologic study of enterocyte-adherent *Escherichia coli:* A cause of protracted diarrhea in infants, *Gastroenterology* **83**:441–454.

Rozee, K. R., Cooper, P., Lam, K., and Costerton, J. W., 1982, Microbial flora of the mouse ileum mucous layer and epithelial surface, *Appl. Environ. Microbiol* **43**:1451–1463.

Sakata, T., and Englehardt, W. V., 1981, Lumenal mucin in the large intestine of mice, rats and guinea pigs, *Cell. Tiss. Res.* **219**:629–635.

Sasaki, S., Onishi, N., Nishikawa, T., Suzuki, R., Maeda, R., Takahashi, T., Usuda, M., Nomura, T., and Saito, M., 1970, Monoassociation with bacteria in the intestines of germfree mice, *Keio J. Med.* **19**:87–101.

Savage, D. C., 1977a, Microbial ecology of the gastrointestinal tract, *Annu. Rev. Microbiol.* **31**:107–133.

Savage, D. C., 1977b, Interactions between the host and its microbes, in: *Microbial Ecology of the Gut* (R. T. J. Clarke and T. Bauchop, eds.) pp. 277–310, Academic Press, London.

Savage, D. C., 1980, Colonisation by and survival of pathogenic bacteria on intestinal mucosal surfaces, in: *Adsorption of Microorganisms to Surfaces* (G. Bitton and K. C. Marshall, eds.), pp. 176–206, Wiley, New York.

Savage, D. C., 1982, Association of pathogenic bacteria with mucosal surfaces in humans, *Clin. Microbiol. News* **4**:105–108.

Savage, D. C., Dubos, R., and Schaedler, R. W., 1968, The gastrointestinal epithelium and its autochthonous bacterial flora, *J. Exp. Med.* **128**:97–110.

Schaedler, R. W., and Dubos, R. J., 1962, The fecal flora of various strains of mice. Its bearing on their susceptibility to endotoxin, *J. Exp. Med.* **115**:1149–1159.

Schaedler, R. W., Dubos, R., and Costello, R., 1965, The development of the bacterial flora in the gastrointestinal tract of mice, *J. Exp. Med.* **122**:59–66.

Sellwood, R., Gibbons, R. A., Jones, G. W., and Rutter, J. M., 1975, Adhesion of enteropathogenic *Escherichia coli* to pig intestinal brush-borders: The existence of two pig phenotypes, *J. Med. Microbiol.* **8**:405–411.

Shahani, K. M., and Ayebo, A. D., 1980, Role of dietary lactobacilli in gastrointestinal microecology, *Am. J. Clin. Nutr.* **33**:2448–2457.

Shedlofsky, S., and Freter, R., 1974, Synergism between ecologic and immunologic control mechanisms of intestinal flora, *J. Inf. Dis.* **129**:296–303.

Shipley, P. L., Gyles, C. L., and Falkow, S., 1978, Characterization of plasmids that encode for the K88 colonization antigen, *Infect. Immunol.* **20**:559–566.

Sleijfer, D. T., Mulder, N. H., de Vries-Hospers, H. G., Fidler, V., Nieweg, H. O., van der Waaij, D., and van Saene, H. K. F., 1980, Infection prevention in granulocytopenic patients by selective decontamination of the digestive tract, *Eur. J Cancer* **16**:859–869.

Smith, D. T., 1952, The disturbance of the normal bacterial ecology by the administration of antibiotics with the development of new clinical syndromes, *Ann. Int. Med.* **37**:1135–1143.

Smith, H. W., 1965, The development of the flora of the alimentary tract in young animals, *J. Pathol. Bacteriol.* **90**:495–513.

Stark, P. L., 1983, The microbial ecology of the large bowel of breast and formula-fed infants during the first year of life, Ph.D. thesis, University of New South Wales, Sydney.

Stark, P. L., and Lee, A., 1982a, The bacterial colonization of the large bowel of pre-term low birth weight neonates, *J. Hyg. Camb.* **89:**59–67.

Stark, P. L., and Lee, A., 1982b, The microbial ecology of the large bowel of breast fed and formula fed infants during the first year of life, *J. Med. Microbiol.* **15:**189–203.

Stark, P. L., Lee, A., and Parsonage, B. D., 1982, Colonization of the large bowel by *Clostridium difficile* in healthy infants: Quantitative study, *Infect. Immunol.* **35:**895–899.

Stephen, A., and Cummings, J. H., 1980, The microbial contribution to human faecal mass, *J. Med. Microbiol.* **13:**45–56.

Suegara, N., Morotomi, M., Watanabe, T., Kawai, Y., and Matai, M., 1975, Behaviour of microflora in the rat stomach: Adhesion of lactobacilli to the keratinized epithelial cells of the rat stomach *in vitro, Infect. Immunol.* **12:**173–179.

Tabaqchali, S., 1979, Abnormal intestinal flora, *Ann. Ist. Super. Sanita* **15:**29–42.

Takeuchi, A., Jervis, H. R., Nakagawa, H., and Robinson, D. M., 1974, Spiral-shaped organisms on the surface colonic epithelium of the monkey and man, *Am. J. Clin. Nutr.* **27:**1287–1296.

Tannock, G. W., and Savage, D. C., 1974, Influence of dietary and environmental stress on microbial populations in the murine gastrointestinal tract, *Infect. Immunol.* **9:**591–598.

Taylor, J., and Mccoy, J. H., 1969, *Salmonella* and *Arizona* infections, in: *Food-Borne Infections and Intoxications* (H. Riemann, ed.), pp. 3–72, Academic Press, New York.

Van der Waaij, D., and Heidt, P. J., 1977, Intestinal bacterial ecology in relation to immunological factors and other defense mechanisms, in: *Food and Immunology* (L. Hambraens, C. A. Hanson, and H. McFarlane, eds.), pp. 133–141, Almquist and Wiksell, Stockholm.

Van der Waaij, D., Berghuis-de Vries, J. M., and Lekkerkerk-van der Wees, J. E. C., 1971, Colonization resistance of the digestive tract in conventional and antibiotic-treated mice, *J. Hyg.* **69:**405–413.

Van der Waaij, D., Aberson, J., Thijm, H. A., and Welling, G. W., 1982, The screening of four aminoglycosides in the selective decontamination of the digestive tract in mice, *Infection* **10:**35–40.

Viscidi, R., Willey, S., and Bartlett, J. G., 1981, Isolation rates and toxigenic potential of *Clostridium difficile* isolates from various patient populations, *Gastroenterology* **81:**5–9.

Yoon, H., Klinzing, G., and Blanch, H. W., 1977, Competition for mixed substrates by microbial populations, *Biotechnol. Bioeng.* **19:**1193–1210.

4

Ecological Constraints on Nitrogen Fixation in Agricultural Ecosystems

MARTIN ALEXANDER

1. Introduction

Recent years have witnessed striking discoveries in the genetics and bio-chemistry of nitrogen fixation, and it is likely that new and exciting find-ings will be forthcoming in the next decade. Nevertheless, it is my con-tention that, notwithstanding the far greater understanding that we now have about certain traits of nitrogen-fixing bacteria and blue-green algae (cyanobacteria), research is moving in a direction that will have only a modest impact on increasing the amount of nitrogen that is fixed in agri-cultural land. These studies are largely concerned with the genetics and biochemistry of nitrogen fixation and those ancillary physiological pro-cesses that have an effect on the fixation, such as energetics and enzyme regulation. However, nitrogen fixation in agricultural ecosystems is rarely limited because of the absence of highly active nitrogen-fixing microor-ganisms. This point has often been stressed by individuals concerned with the practical aspects of nitrogen fixation, but it seems to have been forgotten under the weight of the voluminous literature on the biochem-istry and genetics of nitrogen-fixing organisms.

The limiting factor is often not the absence of organisms with the genetic or biochemical potential to bring about appreciable nitrogen gains; rather, it is one or more ecological constraints that hold these organisms in check. The enormous progress made in recent years in

MARTIN ALEXANDER ● Laboratory of Soil Microbiology, Department of Agronomy, Cornell University, Ithaca, New York 14853.

understanding the biochemistry and genetics of nitrogen-fixing microorganisms frequently is only leading to improved organisms when no improvement is needed in the near future. Conversely, little attention is being given to overcoming the environmental stresses that prevent the presently available, active nitrogen-fixing microorganisms from doing what the laboratory and greenhouse studies suggest they ought to do. Identifying these stresses should facilitate the development of practical means for overcoming the stresses and help agricultural scientists obtain free-living or symbiotic nitrogen-fixers that not only are active in bringing about the desired reaction *in vitro,* but can also survive, grow, and be beneficial in agricultural environments.

This review will focus on *Rhizobium* and, to a lesser extent, nitrogen-fixers growing in fields of flooded rice, because, based on present evidence at least, these are the organisms that have, or may have in the near future, the greatest impact on the nitrogen economy of farmed land. Attention will be given to those environmental stresses that are probably most important in restricting the activity of either indigenous nitrogen-fixers or those that are introduced in nature as a result of inoculation.

2. The *Rhizobium* Inoculum

Although the inoculum carrier and the seed are not what the microbial ecologist considers as natural environments, they will be considered because stresses imposed on rhizobia when in the carrier or on the seed frequently determine whether or not nitrogen fixation will be appreciable.

Excellent strains of *Rhizobium* are now available for use as inocula to be distributed to farmers, yet the bacteria frequently fail to survive in sufficient number in the carriers. For example, the moisture level may be too low or too high during storage, and both insufficient and excessive moisture deleteriously affect the survival of rhizobia. Moreover, their survival is markedly affected by a restriction in aeration in the inoculum carrier (Roughley, 1968). Similarly, inocula in the temperate zone often are exposed to low or high temperatures, and many of the inocula prepared for use in the tropics are stored or are inadvertently placed in sites that reach excessively high temperatures; survival of the organism is markedly affected by such stresses. Thus, the numbers of rhizobia fall rapidly when the temperature reaches 45°C, and a three-logarithmic-order decline in abundance can occur in a short period at 55°C (D. O. Wilson and Trang, 1980). American, European, and Australian agronomists and microbiologists who have worked in the tropics frequently are amazed when they observe inoculum preparations stored in sheds that reach excessively high temperatures, and it is common knowledge among these scientists that frequently the excellent inoculum that is prepared in the United States, Europe, or Australia is worthless because of the poor storage conditions in the tropics. Nevertheless, little work has been performed to isolate strains that are tolerant of such stresses or to create new genotypes having traits that prevent the destruction in inoculum preparations of those bacteria that are extremely effective in symbiotic nitrogen fixation.

The death of root-nodule bacteria applied to and dried on seed can be dramatic, and even under ideal conditions and with only a few hours elapsing from the time the seeds are inoculated until they are sown, the declines in abundance may be appreciable. In some instances, as much as a 1000-fold fall in rhizobial numbers occurs during the first 24 hr, and even a two-logarithmic-order decline, for example, of *Rhizobium trifolii,* may take place within 1 hr after the application of the bacteria to the seed (Salema *et al.,* 1982). Appreciable declines in numbers of *R. japonicum* take place in 12 hr following their application to soybean seeds (Iswaran, 1971). Exposing the inoculated seed to 3 or more hours of sunlight often leads to poor results associated with the inoculation (Alexander and Chamblee, 1965); nevertheless, light-tolerant mutants of other species have been isolated and light tolerance of bacteria can be modified by growth conditions (Kunisawa and Stanier, 1958), so that genetic or cultural manipulation can be used and exploited to protect bacteria against solar inactivation.

With no scientific understanding of the causes of the decline and with only a modest basis for developing their approaches, a number of individuals have attempted to devise materials that can be applied to seeds to reduce the loss of viability. For example, pelleting of seeds of subterranean clover or barrel medic results in longer survival of the rhizobia prior to their introduction into dry soil (Goss and Shipton, 1965). However, pelleting does not always improve the survival of the root-nodule bacteria (Herridge and Roughley, 1974). In some instances, the extent of the decline in viable cells is so great that, independent of the seed-coating material, the sole way of having sufficient rhizobia on seeds for adequate nodulation involves the use of extremely large numbers of these bacteria (Davidson and Reuszer, 1978). Some novel approaches have been attempted to overcome the problem of poor survival on seeds; for example, the entrapment of the bacteria in a polyacrylamide gel for the inoculation of soybeans (Dommergues *et al.,* 1979).

Differences exist among *Rhizobium* strains in the capacity to survive the stresses on inoculated seeds, and these differences could be exploited in future research designed to overcome or minimize the deleterious effects that prevent already excellent nitrogen-fixing rhizobia from doing what is expected of them based on greenhouse or laboratory tests. Thus, strains of *R. trifolii* differ in their rates of decline on inoculated seed, some strains showing good and others poor survival (Philpotts, 1977). Strains of *Rhizobium* fixing nitrogen in symbiosis with cowpeas vary enormously in their susceptibility to drying, the cell densities of some falling to levels less than 1% of the original population, others enduring drying for 11 days with 50% of the original inoculum remaining viable (Osa-Afiana and Alexander, 1982a).

Seeds of several economically important legumes contain inhibitors that suppress the bacteria nodulating the developing root system (Hale and Mathers, 1977; Jain and Rewari, 1976; Thompson, 1960). Part of the toxicity can be removed by washing or soaking the seeds prior to sowing, but given the difficulties in convincing farmers merely to inoculate, it is not likely that they will be convinced to wash seeds free of, from their viewpoint, an unknown inhibitor of a possibly unimportant organism. It is not difficult to isolate mutants of *Rhizobium* that are resistant to inhibitors, a procedure that has been used to obtain strains that are not affected by antibacterial chemicals applied for seed protection (Odeyemi and Alexander, 1977a), and antibiotic-resistant mutants have also been obtained for ecological and genetic research. Hence, the development of toxin-resistant rhizobia may be a feasible means of overcoming the effect of seed toxins.

Although existing, highly effective strains of *Rhizobium* often fail to improve growth of legumes in the field because of poor bacterial persistence in inoculant preparations or on seeds, this issue unfortunately has received little study.

3. Establishment of *Rhizobium*

The ideal strain for inoculation should have, in addition to the capacity for nitrogen fixation, the ability to persist in soil and to grow in soil or around plant roots. These abilities to survive and grow are especially important for rhizobia to be used with legumes that must be replanted each year. Differences exist among species of *Rhizobium* in the ability to colonize plant roots; for example, *R. lupini* grows faster and reaches larger populations than *R. trifolii* (Chatel and Parker, 1973a). Furthermore, strains of a single species differ in their ability to colonize the roots of higher plants (Chatel and Greenwood, 1973a).

The available evidence, which admittedly is scant, indicates that the highly effective strains currently important in agriculture fail to grow under most natural soil conditions. Brief periods of proliferation may follow the moistening of dry soil, but the extent of such growth is slight and its duration is short. Moreover, the natural organic substrates in soil and at least some of those that enter in the form of residues of above-ground portions of plants fail to support appreciable replication. Under natural conditions, these organisms only increase appreciably in abundance when provided with excretions from plant roots (Pena-Cabriales and Alexander, 1983a). Hence, because existing strains of these bacteria are not well adapted for growth in the underground habitat, a significant limitation may exist on their capacity to grow in soil to reach the high population densities needed for extensive nodulation.

Many *Rhizobium* strains that are active in nitrogen fixation are unable to become established when in "competition" with indigenous strains of *Rhizobium* having low effectiveness. Numerous studies have been conducted of "competition" among *Rhizobium* strains for the nodulation of leguminous plants, and the results are remarkable in their general agreement that this is a major practical problem and frequently results in the highly active nitrogen-fixer being almost worthless in practice. In some instances, as with *R. japonicum,* some of the inoculum strains fail to form an appreciable percentage of nodules, regardless of the rate or technique of inoculation, because of the poor "competitiveness" of the organism (Boonkerd *et al.,* 1978). Sometimes, only an average of about 5% of the nodules are formed by *R. japonicum* introduced with the inoculum (Johnson *et al.,* 1965). Similarly, field studies with *R. leguminosarum* showed that the frequency of nodulation by test strains ranged from about 5 to 13% (Strivastava *et al.,* 1980). Poor "competitiveness" sometimes can be overcome by using very large numbers of the rhizobia in the inoculant, thereby increasing the number of nodules derived from the inoculum organism (Amarger, 1974). However, frequently it is not possible to produce inoculants with sufficiently large numbers of bacteria to ensure that the population on the seed overwhelms the more competitive rhizobia derived from the soil.

The physiological traits associated with "competitiveness" are wholly unknown. The sole means of obtaining the desired strains is to screen them under greenhouse conditions. Even with these presumed "competitive" strains, field conditions frequently result in circumstances that completely alter the ranking of potential competitiveness of the bacteria. Thus, some strains are good competitors with other strains of *Rhizobium* at low temperatures but fare poorly at high temperatures (Roughley *et al.,* 1980). Furthermore, a *Rhizobium* that is successful in competition with other strains in the presence of one variety of legume may be wholly unsuccessful or fare poorly in the root zone of other varieties of the same plant species. Not only does the host cultivar modify the presumed competitive abilities, but so does the particular soil into which the plant is introduced (Roughley *et al.,* 1976). In view of the frequent inability of highly effective rhizobia to displace the partially or wholly ineffective strains indigenous to soil, constraints on nitrogen fixation are frequently not the nitrogen-fixing activities of the bacterium or the symbiotic association, but rather those factors that prevent the bacteria from succeeding in their interactions with other rhizobia able to infect the same host.

Successful establishment and extensive nodulation by introduced *Rhizobium* are also greatly influenced by its very poor movement through soil. Very few cells of *R. trifolii,* for example, are able to move for distances even as short as 3 cm from the point of their introduction into soil (Hamdi, 1971). Insignificant movement occurs even in the presence of the developing root system, the boring activities of earthworms, and high rates of water infiltration (Madsen and Alexander, 1982). The nitrogen-fixing rhizobia applied to the seed are the microorganisms that one wishes to form nodules on roots in the field, but a *Rhizobium* that has poor mobility through the soil will not be able to generate nodules at any distance from the point of its first introduction into the soil with the seed, although the roots will move considerable distances laterally and vertically through soil, there to become infected by indigenous rhizobia. These indigenous bacteria often have low activity in nitrogen fixation.

4. Survival of *Rhizobium* in Soil

Highly effective rhizobia, following their introduction into soil with seeds, often do not persist in significant numbers. Nevertheless, little attention has been given to the use of strains that are able to persist in sufficient number to ensure nodulation of succeeding generations of legumes. The problem of survival is clearly illustrated by Australian studies of the nodulation of subterranean clover by *R. trifolii.* In a typical study, all of the nodules in the first year were formed by the inoculum strain, but nearly all of the nodules in the second year were formed by strains not in the original inoculum (Roughley *et al.,* 1976). In other studies of the same bacterium–legume association, 80% of the nodules in the first year were derived from the introduced bacteria. However, none of the nodules appearing at the end of the succeeding season was derived from the added bacterium, and all of the plants bore nodules generated by strains that were not as effective in the symbiosis as the originally introduced microorganism (Bergersen, 1970).

Poor survival is also evident in laboratory studies of certain soils, even those at pH values near neutrality; in one investigation, for example, the four test strains of *R. japonicum* fared poorly, and three of the four declined to fewer than 10^4 per gram in 6 weeks or less (Vidor and Miller, 1980). Nevertheless, some rhizobia maintain their viability for appreciable periods of time in other soils that are moist and are not under temperature or pH stress (Danso and Alexander, 1974).

Considerable attention was directed to the poor survival of many rhizobia as a result of a problem in Western Australia known as "second-year clover mortality." In the problem area, clover sown in certain soils became established and nodulated in the first year, but nodulation failed to occur and the plants died in the second year. The plant mortality in the second year was attributed to the inability of the root-nodule bacterium to survive from the first season to the second, an apparent consequence of the high soil temperatures and extreme desiccation prevailing in these sandy soils (Marshall *et al.*, 1963).

The problem of poor survival is not intractable, because strains vary considerably in their susceptibility to decline under natural conditions. Without considering for the moment the identity of the particular stresses that result in the loss of viability, it is evident that some strains of *R. trifolii* persist poorly in soils of Australia, whereas other strains endure in large numbers (Gibson *et al.*, 1976). Similarly, the steady-state populations of strains of the same species may be quite different at varying periods after the sowing of inoculated seeds (Chatel and Greenwood, 1973b).

As a rule, the ability to endure in large numbers is deemed to be a beneficial attribute of *Rhizobium* used for seed inoculation. On the other hand, persistence in soil may be undesirable among indigenous strains because the capacity of indigenous rhizobia to endure the various stresses and be maintained in abundance may make it difficult to introduce new and effective strains to displace the partially effective indigenous organisms.

5. Biological Stresses on *Rhizobium*

Rhizobium strains active in nitrogen fixation appear to be poor competitors with members of the indigenous community of microorganisms in soil and the rhizosphere. Evidence for an effect of indigenous microorganisms on the activity of the root-nodule bacteria in soil comes from several sources. In an early report, the failure of inoculation of subterranean clover in certain soils of Australia was attributed to the indigenous microbial community preventing colonization of the rhizosphere by the added, active nitrogen-fixing inoculum strain (Hely *et al.*, 1957). Similarly, in areas of New Zealand, it was observed that establishment of white clover was poor, even when lime, fertilizers, or micronutrients were added to the soil. Establishment was successful, however, if the size of the indigenous microbial community was reduced by the application of antimicrobial agents. The antimicrobial compounds presumably reduced the activity of components of the indigenous community, thereby permitting the rhizobium to colonize the rhizosphere and bring about nodulation (Beggs, 1961, 1964). Under more defined conditions in tests of subterranean clover (Harris, 1953) and white clover (Anderson, 1957), it was observed that the nodulation by effective *R. trifolii* was reduced, delayed, or prevented by fungi or other bacteria. The microorganisms responsible for these deleterious changes usually did not produce antibiotics acting against the rhizobia in culture, and hence the harmful effects were attributed to competition.

In recent studies, it was observed that rhizobia fail to grow in soil even if a carbon source they can use is added in reasonable amounts. If enormous concentrations of that carbon source are added, proliferation occurs; evidently, the indigenous populations are better able than the rhizobia to make use of low concentrations of the organic compound, but at the high concentrations, some is left for the slow-growing and poorly competitive root-nodule bacteria. Moreover, if antibiotics toxic to many native soil bacteria but not to rhizobia are added to soil, thereby suppressing the potential competitors, rhizobia are able to proliferate and increase markedly in numbers (Pena-Cabriales and Alexander, 1983b).

Competition for limiting supplies of nutrients probably is an extremely important interaction among natural microbial communities. Soils are known to be almost invariably deficient in readily available carbon compounds. Members of the genus *Rhizobium* grow slowly and presumably are not effective competitors in soil. Probably more than is presently realized, competition restricts the activity of nitrogen-fixing organisms, both free-living and those that, like *Rhizobium,* must compete in the rhizosphere prior to invading the host plant. Means of overcoming or minimizing the impact of competition, notwithstanding its presumed importance, have not attracted attention. One possible approach is to apply inhibitors to the seed (or possibly, for chemicals that are translocated downward, to the foliage) in the hope that these chemicals will suppress the competing organisms but not the nitrogen-fixing species. To obtain such chemicals for seed treatment is not difficult, because many of the commercially important fungicides, including those designed for treatment of legume seeds, inhibit many bacteria, and the rhizobia can be made resistant to these fungicides. Such an approach has been used with species of *Rhizobium* and with several leguminous species. The chemicals presumably act, at least in part, by inhibiting the indigenous bacteria, thus allowing the introduced root-nodule bacterium to proliferate with a minimum of competition (Odeyemi and Alexander, 1977b).

Biologically formed toxins also may influence the establishment of rhizobia on roots. For example, extracts of soils supporting clovers that nodulated poorly were toxic to *R. trifolii* growing in laboratory media, and it has been postulated that such inhibitors restrict the development of rhizobia in nature (Chatel and Parker, 1972). If, in fact, soil toxins affect the colonizing ability of these microorganisms, it should not be difficult to add the capacity to resist soil toxins to the traits of effective rhizobia.

Predation seems to be a significant factor in reducing the populations of rhizobia and other bacteria that at times are present in abundance in soil. When artificially high numbers of *Rhizobium* are added to soil, the population size is drastically diminished. This decline is paralleled by the proliferation of indigenous protozoa. The protozoa apparently feed on the large numbers of rhizobia and appreciably reduce their density without eliminating the bacteria (Danso *et al.*, 1975). Although the high rhizobial densities needed to trigger protozoan predation are not common in natural soils, the root-nodule bacteria proliferate as inoculated seeds imbibe water and the roots begin to develop. *Rhizobium* then begins to grow and may attain large population sizes; as the potential prey for protozoa become more numerous, the predators begin to feed and thereby prevent the *Rhizobium* population from becoming as large as it would be in the absence of predation. In support of the hypothesis that predation is a significant factor in the rhizosphere is the finding that suppressing protozoa with thiram resulted in enhanced colonization of beans by *R. phaseoli* (Ramirez and Alexander, 1980). In subsequent studies, it was noted that the abrupt fall in population of *R. phaseoli* around germinating seeds and the developing root system of beans was delayed if the seed was treated with chemicals that inhibited the protozoa. The inhibition of protozoa by thiram was accompanied by a high initial frequency of nodules formed by the test *Rhizobium,* a yield increase, and a greater amount of nitrogen fixed by beans (Lennox and Alexander, 1981). In this instance also, therefore, nitrogen gains by effective rhizobia are not realized because of a stress on the bacteria, the stress being predation. It is also possible that protozoa may be significant in nonrhizosphere soil, as suggested by data showing that additions of inhibitors of eukaryotes allowed for an increase in size of *Rhizobium* populations (Pena-Cabriales and Alexander, 1983b).

6. Abiotic Stresses on *Rhizobium*

6.1. Acidity

Much of the arable land of the world is acidic, and widespread agreement exists among agronomists that acidity is a major constraint for legume cultivation in these regions. Moreover, ample evidence exists that the survival and growth of many of the most active strains of *Rhizobium* that have been tested and the nodulation that these organisms effect are deleteriously influenced at low pH. On the other hand, although many legumes are acid-sensitive, even when growing on fixed nitrogen, many grow well in acid conditions. Thus, serradella grows at pH 4.0, and white clover develops readily at pH 5.0 (Mulder *et al.,* 1966). Even in a single legume species, for example, alfalfa (Jo *et al.,* 1980), tolerances to soil acidity vary among varieties, and plant breeders are currently endeavoring to develop varieties able to cope with this significant stress. However, nodulation is more markedly affected at low pH than is root development or plant growth; this has been observed for alfalfa (Jo *et al.,* 1980) and peas (Mulder *et al.,* 1966). Apparently, some phase of the infection process induced by the bacteria is inhibited at low pH, although root development and bacterial growth do not show this high sensitivity (Evans *et al.,* 1980). In other instances, as with *Medicago truncatula,* the absence of nodulation in acid soils may result from the inability of *R. meliloti* to survive or grow (Robson and Loneragan, 1970). Nodulation failures due to poor survival of rhizobia in acid soils are particularly likely when inoculation is not practiced every year, a common occurrence even in developed countries. Indeed, frequent inoculation of legumes growing in acid soils was recommended some 60 years ago (Bryan, 1923). If nodules are not formed at these low pH values, the plant may develop using fixed nitrogen, but nitrogen gains will not occur. Serradella and alfalfa, for example, continue to grow at the expense of fixed nitrogen, even when the low pH prevents dinitrogen fixation (Mulder *et al.,* 1966).

Some species of *Rhizobium* do not multiply in culture even at moderate acidities, e.g., *R. meliloti* often does not grow below pH 5.3. Other species have greater tolerances, however, and some strains of *R. phaseoli* are able to multiply at pH 3.8 (Lowendorf and Alexander, 1983). The pH range for survival is expected to be wider than that for replication, but many of the active N_2-fixers even fail to survive in sterile soil (where suppression is not a result of some harmful microorganism) at pH 5.2 (Lowendorf *et al.,* 1981). Nevertheless, strains of a single species of *Rhizobium* vary in their pH sensitivity, whether sensitivity is assessed by growth in culture, survival in sterile soil, or nodulation of host plants (Lowendorf and Alexander, 1983; Lowendorf *et al.,* 1981; Thornton and Davey, 1983). Such strains may be useful as legume inoculants because they nodulate at pH values at which other strains do not (Mulder *et al.,* 1966; Munns *et al.,* 1979) or because they can survive longer in acid soils (Lowendorf *et al.,* 1981).

Acid soils frequently contain levels of Al, Mn, or Fe that may be injurious to nodulation or growth of rhizobia. From the viewpoint of developing resistance in the N_2-fixing symbiosis, attention has only been given to Al. *Rhizobium* strains can be selected that vary in their ability to tolerate Al in culture (Keyser and Munns, 1979; Hartel and Alexander, 1983), and differences in sensitivity are also evident in the behavior of such strains in soil (Hartel *et al.,* 1983). Nevertheless, the Al stress in nature appears to affect the host plants and not the rhizobia (Hartel *et al.,* 1983).

6.2. Desiccation

Nearly all areas of the world are subject to rain-free periods of sufficient duration to result in extensive drying of soil, and the rhizobia that must survive in these soils in order to bring about nitrogen fixation in the succeeding crop may thereby become too few in number to cause extensive nodulation. With many strains of all species of *Rhizobium,* a single exposure of the soil to drying reduces the viable population by 99% or more, and several cycles of soil wetting followed by drying, a not uncommon occurrence in nature, reduce the population still further (Pena-Cabriales and Alexander, 1979). With some strains of the bacteria that infect *Lotus corniculatus,* the decline in soil kept dry for several months—again, a common occurrence in many tropical regions—may be four logarithmic orders of magnitude (Foulds, 1971). The conditions under which water is lost from the soil affects the extent of the decline (Bushby and Marshall, 1977). Without question, prolonged drought or even a short period of drying is a major stress on *Rhizobium,* yet little information exists on the soil properties affecting the reduction in population size (Chao and Alexander, 1982) and on microbial and other factors that are related to differences among strains and soils (Bushby and Marshall, 1977), although the type and amount of clay are of great importance (Osa-Afiana and Alexander, 1982b). From observations that nearly half of the viable cells of some strains of cowpea rhizobia are not killed in one drying cycle whereas more than 99% of the cells of other strains die under identical circumstances (Osa-Afiana and Alexander, 1982a), it appears that means can be devised to obtain cultures not seriously affected by the drying of soil.

6.3. Temperature

High soil temperatures may occur at seeding time, during the period of plant growth, and following harvest; in each instance, the high temperatures may be deleterious. The period following harvest should not be ignored, because survival of the bacteria at that time is important. In tropical and subtropical regions, the temperature near the soil surface is often above 40°C and sometimes may reach 60°C. Rhizobia are subject to these temperatures following harvest and before sowing, after planting into hot, dry soil (as sometimes occurs in Australia), or both. Some strains survive very poorly under such conditions; e.g., *R. trifolii* and strains nodulating *Lotus pedunculatus* and cowpeas (Brockwell and Phillips, 1970; Chatel and Parker, 1973b; Osa-Afiana and Alexander, 1982b). It is the death of *R. trifolii* in fields of Western Australia that is responsible for the poor growth of subterranean clover in the second year after planting: the rhizobia fail to survive in appreciable numbers in the hot, dry soil at the end of the first season (Chatel and Parker, 1973b). The great decline is not a problem in soils of heavy texture and is prevented to some degree by certain clays (Marshall, 1964). The sensitivity to heat is greater in moist than in dry soil (Wilkins, 1967), and marked declines are evident in moist soil even at 36°C (Danso and Alexander, 1974).

For rhizobia to colonize the rhizosphere and cause nodulation, they must grow. During this time, too, their sensitivity to temperature is of practical significance. The optimum temperatures for growth in culture vary among strains and species, and values of 27–39°C have been noted. The maximum temperatures are generally 35–39°C, but proliferation may take place up to 42°C (Allison and Minor, 1940; Bowen and Kennedy, 1959; Munévar and Wollum, 1981a). Differences in growth and colonizing abilities probably explain why some strains are more active in nodulating soybeans at low temperatures and others are more active at high temperatures (Weber and Miller, 1972). Studies of *Vicia atropurpurea* and *Medicago tribuloides* confirm that certain rhizobia form more nodules at low temperatures and others produce more of the nodules at high temperature (Pate, 1961). Hence, super N_2-fixers could easily be displaced in nature by strains of lesser effectiveness simply because of their temperature responses. Furthermore, some strains of *R. leguminosarum* fail to induce nodulation at 30°C even though they and their host, peas, grow at that temperature (Frings, 1976). In addition, the relative activity of the rhizobia is altered by temperature, so that a bacterium that is highly effective at one temperature is less active at different temperatures (Munévar and Wollum, 1981b). For these reasons, greater nitrogen gains probably can be achieved by improvements in the heat resistance of the symbiosis.

Rhizobium strains or species vary greatly in their susceptibility to high temperatures. Thus, *R. meliloti* survives better than *R. trifolii* on seed lying for long periods in hot, dry soil (Brockwell and Phillips, 1970), *R. lupini* and *R. japonicum* are less susceptible than *R. trifolii* when in dry soil at elevated temperatures (Chatel and Parker, 1973b; Marshall, 1964), and *Rhizobium* strains that nodulate and fix N_2 similarly on *Cicer arietinum* at low temperatures exhibit differences in nodulation and N_2-fixing activity at 30°C (Dart *et al.,* 1976). Hence, it should be possible to obtain or devise rhizobia able to survive, grow, or fix N_2 under conditions where elevated temperatures are of importance.

6.4. Salinity and Alkalinity

Many soils are rich in salt and have high pH values, and these areas are often considered to be undesirable for legumes. For example, berseem clover, guar, cowpeas, and lentils nodulate poorly in highly saline-alkali soils of India, and peas may be wholly devoid of nodules (Bhardwaj, 1974). Because of these sensitivities, many studies have been performed to establish the effect of salts on the growth of rhizobia in culture. These investigations have demonstrated that inhibition of growth of the bacteria usually requires high salt levels (Ethiraj *et al.,* 1972; Mendez-Castro and Alexander, 1976; Steinborn and Roughley, 1975; Rai, 1983). Comparisons of the sensitivities of microorganisms and plants, moreover, show that the bacteria are able to proliferate at salt levels that do not permit growth of the host (Bhardwaj, 1975). However, it is not correct to assume that because replication of the bacteria is not seriously affected, no problem related to the symbiosis exists. First, the rhizobia do not survive in some of these soils (Bhardwaj, 1975; Pant and Iswaran, 1970). Whether it is possible to exploit the differences among *Rhizobium* strains in salt tolerance for growth or the ability of some rhizobia to "acclimate" or mutate to even higher levels of salt tolerance, as has been shown several times (Ethiraj *et al.,* 1972; Mendez-Castro and Alexander, 1976; Steinborn and Roughley, 1975), is not now certain. Second, nodulation is more sensitive than root development, at least for soybeans (Bernstein and Ogata, 1966), and may be affected by salinity in its early phases (Singleton and Bohlool, 1984), and alkali- and salt-induced delays in nodulation of berseem clover and lentils may result in reduced yields (Bhardwaj, 1975). Singleton *et al.* (1982) suggested that because strains of *Rhizobium* exist that are able to survive at high salinities, attention should be given to the influence of salinity on aspects of the symbiosis other than the survival of the bacteria.

6.5. Fungicides

The seeds of many legumes are attacked by a variety of plant pathogens, especially fungi. The destruction may be so extensive that yields are greatly reduced. In countries in which farmers earn enough to pay for fungicides, the chemicals have been widely used, even after it was realized that they are harmful to *Rhizobium*. Seed-applied fungicides are not specific for plant pathogens or even fungi, and many kill rhizobia and render useless the added inoculant. Thus, the number of *R. phaseoli* added to bean seeds declines as a result of the application of thiram, captan, or PCNB to seeds (Graham *et al.*, 1980). Furthermore, some of the best seed-protecting chemicals also reduce nodulation by *R. phaseoli* (Graham *et al.*, 1980) and *R. japonicum* and *R. leguminosarum* (Stovold and Evans, 1980), and such decreases in nodulation of peanut lead to reduced nitrogen fixation and pod yield (Chendrayan and Prasad, 1976).

Several methods have been proposed or are used to minimize the impact of these fungicides. Because the chemicals vary greatly in their effect on rhizobia (Curley and Burton, 1975; Diatloff, 1970a), a simple approach is to use those seed protectants that are least deleterious; unfortunately, these are not always the most desirable for control of the pathogens. A second approach involves covering the fungicide-treated seeds with a coating material; poly(vinyl acetate) has been used for this purpose (Diatloff, 1970b). A third approach is based on obtaining fungicide-resistant mutants derived from effective rhizobia and applying these to the seeds together with the chemicals (Odeyemi and Alexander, 1977b). Each procedure is useful in permitting nodulation or increasing yield as compared to farmers' conventional practices. Nevertheless, because of dissimilar farming practices in many regions of the world and the continual introduction of new fungicides and different rhizobia, fungicide toxicity remains an issue.

7. Limitations on Nitrogen Fixation by Cyanobacteria

Cyanobacteria (blue-green algae) sometimes bring about significant nitrogen fixation in paddy fields. Indeed, the ability of farmers in Asia to grow lowland rice for millenia with no input of fixed nitrogen has been attributed to the activity of these microorganisms. Yet, although the amount they fix in the field at present is insufficient for high yields of rice and although species able to fix N_2 in large amounts are frequently present in paddy fields, little attention has been given to defining why so little N_2 is fixed in paddy fields under natural conditions.

Studies of the ecology of these algae have revealed that several factors limit their development or N_2-fixing activity in paddy fields. Phosphorus is often in insufficient supply in the water phase above the soil, and iron may limit algal development in waters that are at high pH or become alkaline during photosynthesis. In the absence of adequate levels of these elements, excellent N_2-fixers have little activity, but additions of these nutrients stimulate growth and N_2 fixation. Most of the cyanobacteria grow poorly below pH 6.0, and they are not active and/or are displaced by green algae in waters of even slight acidity; for these reasons, N_2 fixation and development of cyanobacteria are correlated with soil pH (Roger and Reynaud, 1979; J. T. Wilson and Alexander, 1979). Competition with algae that fix little or no N_2 may also reduce the nitrogen gain potentially brought about by the photosynthetic prokaryotes (J. T. Wilson et al., 1979). These organisms are also quite susceptible to grazing by invertebrates that sometimes flourish in the water, and ostracods (Grant et al., 1983; Grant and Alexander, 1981; J. T. Wilson et al., 1980) and daphnids (Roger and Reynaud, 1979) may proliferate at their expense and rapidly suppress the actions of excellent N_2-fixers.

Each one of these limitations probably can be overcome. Thus, the P (Roger and Reynaud, 1979) and Fe (Ryther and Kramer, 1961) demand for optimal algal growth varies appreciably, so that inoculant strains could be developed with a low nutrient demand. The acid limitation may be overcome by using or developing inoculants of acid-tolerant cyanobacteria. The occurrence of cyanobacteria in some moderately acid soils suggests that this is not a far-fetched option. Competition from algae may be reduced by using herbicides, many of which are algicidal, together with an inoculum consisting of a herbicide-resistant cyanobacterium. This has been done in laboratory studies of paddy soils treated with simetryne and a variant of *Aulosira* sp. that was resistant to this herbicide (J. T. Wilson et al., 1979). The invertebrate-grazers, but not the N_2-fixers, are suppressed by insecticides (Osa-Afiana and Alexander, 1981), so insecticides represent a feasible means to increase cyanobacterial activity. However, the cost of herbicides and insecticides may be too high for most rice farmers of the developing countries, so a more attractive approach in many instances might involve finding cyanobacteria that are better competitors or less susceptible to predation. The literature of limnology and oceanography indicates that predation-resistant cyanobacteria are reasonably common in aquatic environments.

Some of the ecological factors that govern N_2 fixation by cyanobacteria also limit growth and activity of *Azolla*. Phosphorus is a major limiting nutrient in diverse areas of the world, and its addition to paddy water enhances *Azolla* development and N_2 fixation (Talley et al., 1981; Watanabe et al., 1980). Moreover, some *Azolla* populations may be more efficient than others in using P (Talley et al., 1981). Iron may also sometimes restrict the growth and activity of the fern symbiosis (Talley et al., 1981). *Azolla*, like the cyanobacteria, fails to develop if the pH of paddy soils is low (Singh, 1977). In addition, the fern, like the free-living cyanobacteria, is subject to attack by pests, and the latter may be controlled by insecticides (Singh, 1979). Temperature, moreover, greatly affects growth, some species flourishing at warmer temperatures, others growing well at cooler temperatures (Rains and Talley, 1979; Tung and Watanabe, 1983). Work on overcoming these limitations, which reduce the magnitude of fixation by strains already having good activity, has scarcely begun.

8. Conclusion

From the standpoint of the microbial contribution to the symbiosis involving legumes and to nitrogen gains in paddy fields, the factors limiting nitrogen fixation are rarely the absence of highly effective strains of *Rhizobium* or the lack of availability of cyanobacteria potentially active in fixation. The limiting factors are nearly always the absence of strains adapted to ecologically significant stresses in soil or in paddy fields. Therefore, basic and applied research that is designed to increase food production, whether in the long or the short run, should seek to define these stresses and help to find means to overcome them.

ACKNOWLEDGMENTS. This chapter was prepared for a workshop on Priorities in Biotechnology Research for International Development sponsored by the U.S. National Academy of Sciences/National Research Council and supported by the U.S. Agency for International Development. The recommendations and background papers for the workshop have been published (National Research Council, 1982).

References

Alexander, C. W., and Chamblee, D. S., 1965, Effect of sunlight and drying on the inoculation of legumes with *Rhizobium* species, *Agron. J.* **57**:550–553.

Allison, F. E., and Minor, F. W., 1940, The effect of temperature on the growth rates of rhizobia, *J. Bacteriol.* **39**:365–371.

Amarger, N., 1974, Compétition pour la formation des nodosités sur la féverole entre souches de *Rhizobium leguminosarum* apportées par inoculation et souche indigènes, *C. R. Acad. Sci. Paris D* **279**:527–530.

Anderson, K. J., 1957, The effect of soil micro-organisms on the plant–rhizobia association, *Phyton* **8**:59–73.

Beggs, J. P., 1961, Soil sterilant aid to clover establishment indicates anti-nodulation factor in soils, *N. Z. J. Agric.* **103**:325–334.

Beggs, J. P., 1964, Growth inhibitor in the soil, *N. Z. J. Agric.* **108**:529–535.

Bergersen, F. J., 1970, Some Australian studies relating to the long-term effects of the inoculation of legume seeds, *Plant Soil* **32**:727–736.

Bernstein, L., and Ogata, G., 1966, Effects of salinity on nodulation, nitrogen fixation, and growth of soybeans and alfalfa, *Agron. J.* **58**:201–203.

Bhardwaj, K. K. R., 1974, Growth and symbiotic effectiveness of indigenous *Rhizobium* species of a saline-alkaline soil, *Proc. Ind. Natl. Sci. Acad.* **40**:540–543.

Bhardwaj, K. K. R., 1975, Survival and symbiotic characteristics of *Rhizobium* in saline-alkali soils, *Plant Soil* **43**:377–385.

Boonkerd, N., Weber, D. F., and Bezdicek, D. F., 1978, Influence of *Rhizobium japonicum* strains and inoculation methods on soybeans grown in rhizobia-populated soil, *Agron. J.* **70**:547–549.

Bowen, G. D., and Kennedy, M. M., 1959. Effect of high soil temperatures on *Rhizobium* spp., *Qd. J. Agric. Sci.* **16**:177–197.

Brockwell, J., and Phillips, L. J., 1970, Studies on seed pelleting as an aid to legume seed inoculation. 3. Survival of *Rhizobium* applied to seed sown in hot, dry soil, *Aust. J. Exp. Agric. Anim. Husb.* **10**:739–744.

Bryan, O. C., 1923, Effect of acid soils on nodule-forming bacteria, *Soil Sci.* **15**:37–40.

Bushby, H. V. A., and Marshall, K. C., 1977, Some factors affecting the survival of root-nodule bacteria on desiccation, *Soil Biol. Biochem.* **9**:143–147.

Chao, W. L., and Alexander, M., 1982, Influence of soil characteristics on the survival of *Rhizobium* in soils undergoing drying, *Soil Sci. Soc. Am. J.* **46**:949–952.

Chatel, D. L., and Greenwood, R. M., 1973a, Differences between strains of *Rhizobium trifolii* in ability to colonize soil and plant roots in the absence of their specific host plants, *Soil Biol. Biochem.* **5**:809–813.

Chatel, D. L., and Greenwood, R. M., 1973b, The colonization of host-root and soil by rhizobia. II. Strain differences in the species *Rhizobium trifolii, Soil Biol. Biochem.* **5**:433–440.

Chatel, D. L., and Parker, C. A., 1972, Inhibition of rhizobia by toxic soil-water extracts, *Soil Biol. Biochem.* **4**:289–294.

Chatel, D. L., and Parker, C. A., 1973a, The colonization of host-root and soil by rhizobia. I. Species and strain differences in the field, *Soil Biol. Biochem.* **5**:425–432.

Chatel, D. L., and Parker, C. A., 1973b, Survival of field-grown rhizobia over the dry summer period in Western Australia, *Soil Biol. Biochem.* **5**:415–423.

Chendrayan, K., and Prasad, N. N., 1976, Effect of seed treatment with Tillex and Ceresan on *Rhizobium* groundnut symbiosis, *Madras Agric. J.* **63**:520–522.

Curley, R. L., and Burton, J. C., 1975, Compatibility of *Rhizobium japonicum* with chemical seed protectants, *Agron. J.* **67**:807–808.

Danso, S. K. A., and Alexander, M., 1974, Survival of two strains of *Rhizobium* in soil, *Soil Sci. Soc. Am. Proc.* **38**:86–89.

Danso, S. K. A., Keya, S. O., and Alexander, M., 1975, Protozoa and the decline of *Rhizobium* populations added to soil, *Can. J. Microbiol.* **21**:884–895.

Dart, P., Day, J., Islam, R., and Dobereiner, J., 1976, Symbiosis in tropical grain legumes: Some effects of temperature and the composition of the rooting medium, in: *Symbiotic Nitrogen Fixation in Plants* (P. S. Nutman, ed.), pp. 361–384, Cambridge University Press, Cambridge.

Davidson, F., and Reuszer, H. W., 1978, Persistence of *Rhizobium japonicum* on the soybean seed coat under controlled temperature and humidity, *Appl. Environ. Microbiol.* **35**:94–96.

Diatloff, A., 1970a, The effects of some pesticides on root nodule bacteria and subsequent nodulation, *Aust. J. Exp. Agric. Anim. Husb.* **10**:562–567.

Diatloff, A., 1970b, Overcoming fungicide toxicity to rhizobia by insulating with a polyvinyl acetate layer, *J. Aust. Inst. Agric. Sci.* **36**:293–294.

Dommergues, Y. R., Diem, H. G., and Divies, C., 1979, Polyacrylamide-entrapped *Rhizobium* as an inoculant for legumes, *Appl. Environ. Microbiol.* **37**:779–781.

Ethiraj, S., Sharma, H. R., and Vyas, S. R., 1972, Studies on salt tolerance of rhizobia, *Ind. J. Microbiol.* **12**:87–92.

Evans, L. S., Lewin, K. F., and Vella, F. A., 1980, Effect of nutrient medium pH on symbiotic nitrogen fixation by *Rhizobium leguminosarum* and *Pisum sativum, Plant Soil* **56:**71–80.

Foulds, W., 1971, Effect of drought on three species of *Rhizobium, Plant Soil* **35:**665–667.

Frings, J. F. J., 1976, The rhizobium–pea symbiosis as affected by high temperatures, *Meded. Landbouwhogesch. Wageningen* **76/77:**1–76.

Gibson, A. H., Date, R. A., Ireland, J. A., and Brockwell, J., 1976, A comparison of competitiveness and persistence amongst five strains of *Rhizobium trifolii, Soil Biol. Biochem.* **8:**395–401.

Goss, O. M., and Shipton, W. A., 1965, Nodulation of legumes on new light land. I. Survival of rhizobia on inoculated pelleted seed held for varying periods before sowing in dry soil, *J. Agric. W. Aust.* **6:**616–621.

Graham, P. H., Ocampo, G., Ruiz, L. D., and Duque, A., 1980, Survival of *Rhizobium phaseoli* in contact with chemical seed protectants, *Agron. J.* **72:**625–627.

Grant, I. F., and Alexander, M., 1981, Grazing of blue-green algae (cyanobacteria) in flooded soils by *Cypris* sp. (Ostracoda), *Soil Sci. Soc. Am. J.* **45:**773–777.

Grant, I. F., Tirol, A. C., Aziz, T., and Watanabe, I., 1983, Regulation of invertebrate grazers as a means to enhance biomass and nitrogen fixation of cyanophyceae in wetland rice fields, *Soil Sci. Soc. Am. J.* **47:**669–675.

Hale, C. N., and Mathers, D. J., 1977, Toxicity of white clover seed diffusate and its effect on the survival of *Rhizobium trifolii, N. Z. J. Agric. Res.* **20:**69–73.

Hamdi, Y. A., 1971, Soil-water tension and the movement of rhizobia, *Soil Biol. Biochem.* **3:**121–126.

Harris, J. R., 1953, Influence of rhizosphere micro-organisms on the virulence of *Rhizobium trifolii, Nature* **172:**507–508.

Hartel, P. G., and Alexander, M., 1983, Growth and survival of cowpea rhizobia in acid, aluminum-rich soils, *Soil Sci. Soc. Am. J.* **47:**502–506.

Hartel, P. G., Whelan, A. M., and Alexander, M., 1983, Nodulation of cowpeas and survival of cowpea rhizobia in acid, aluminum-rich soils, *Soil Sci. Soc. Am. J.* **47:**514–517.

Hely, F. W., Bergersen, F. J., and Brockwell, J., 1957, Microbial antagonism in the rhizosphere as a factor in the failure of inoculation of subterranean clover, *Aust. J. Agric. Res.* **8:**24–44.

Herridge, D. F., and Roughley, R. J., 1974, Survival of some slow-growing *Rhizobium* on inoculated legume seed, *Plant Soil* **40:**441–444.

Iswaran, V., 1971, Survival of *Rhizobium japonicum* on soybean seeds, *Isr. J. Agric. Res.* **21:**79–80.

Jain, M. K., and Rewari, R. B., 1976, Studies on seed coat toxicity to rhizobia of urid *(Phaseolus mungo),* mung *(Phaseolus aureus),* and soybean *(Glycine max), Zentralbl. Bakteriol. Parasitenkd. Infektionskr. Hyg. Abt. II* **131:**163–169.

Jo, J., Yoshida, S., and Kayama, R., 1980, Acidity tolerance and symbiotic nitrogen fixation capacity of some varieties of alfalfa (in Japanese), *Nippon Sochi Gakkaishi* **26:**174–178.

Johnson, H. W., Means, U. M., and Weber, C. R., 1965, Competition for nodule sites between strains of *Rhizobium japonicum* applied as inoculum and strains in the soil, *Agron. J.* **57:**179–185.

Keyser, H. H., and Munns, D. N., 1979, Tolerance of rhizobia to acidity, aluminum, and phosphate, *Soil Sci. Soc. Am. J.* **43:**519–523.

Kunisawa, R., and Stanier, R. Y., 1958. Studies on the role of carotenoid pigments in a chemoheterotrophic bacterium, *Corynebacterium poinsettiae, Arch. Mikrobiol.* **31:**146–156.

Lennox, L. B., and Alexander, M., 1981, Fungicide enhancement of nitrogen fixation and colonization of *Phaseolus vulgaris* by *Rhizobium phaseoli, Appl Environ. Microbiol.* **41:**404–411.

Lowendorf, H. S., and Alexander, M., 1983, Identification of *Rhizobium phaseoli* strains that are tolerant or sensitive to soil acidity, *Appl. Environ. Microbiol.* **45**:737–742.

Lowendorf, H. S., Baya, A. M., and Alexander, M., 1981, Survival of *Rhizobium* in acid soils, *Appl. Environ. Microbiol.* **42**:951–957.

Madsen, E. L., and Alexander, M., 1982, Transport of *Rhizobium* and *Pseudomonas* through soil, *Soil Sci. Soc. Am. J.* **46**:557–560.

Marshall, K. C., 1964, Survival of root-nodule bacteria in dry soils exposed to high temperatures, *Aust. J. Agric. Res.* **15**:273–281.

Marshall, K. C., Mulcahy, M. J., and Chowdhury, M. S., 1963, Second-year clover mortality in Western Australia—A microbiological problem, *J. Aust. Inst. Agric. Sci.* **29**:160–164.

Mendez-Castro, F. A., and Alexander, M., 1976, Acclimation of *Rhizobium* to salts, increasing temperature and acidity, *Rev. Latinoam. Microbiol.* **18**:155–158.

Mulder, E. G., Lie, T. A., Dilz, K., and Houwers, A., 1966, Effect of pH on symbiotic nitrogen fixation of some leguminous plants, in: *Symposium of the 9th International Congress of Microbiology,* pp. 133–151, Ivanovski Institute of Virology, Moscow.

Munévar, F., and Wollum II, A. G., 1981a, Growth of *Rhizobium japonicum* strains at temperatures above 27°C, *Appl. Environ. Microbiol.* **42**:272–276.

Munévar, F., and Wollum II, A. G., 1981b, Effect of high root temperature and *Rhizobium* strain on nodulation, nitrogen fixation, and growth of soybeans, *Soil Sci. Soc. Am. J.* **45**:1113–1120.

Munns, D. N., Keyser, H. H., Fogle, V. W., Hohenberg, J. S., Righetti, T. L., Lauter, D. L., Zaroug, M. G., Clarkin, K. L., and Whitacre, K. W., 1979, Tolerance of soil acidity in symbioses of mung bean with rhizobia, *Agron. J.* **71**:256–260.

Munns, D. N., Hohenberg, J. S., Righetti, T. L., and Lauter, D. J., 1981, Soil acidity tolerance of symbiotic and nitrogen-fertilized soybeans, *Agron. J.* **73**:407–410.

National Research Council, 1982, *Priorities in Biotechnology Research for International Development,* National Academy Press, Washington, D.C.

Odeyemi, O., and Alexander, M., 1977a, Resistance of *Rhizobium* strains to phygon, spergon, and thiram, *Appl. Environ. Microbiol.* **33**:784–790.

Odeyemi, O., and Alexander, M., 1977b, Use of fungicide-resistant rhizobia for legume inoculation, *Soil Biol. Biochem.* **9**:247–251.

Osa-Afiana, L. O., and Alexander, M., 1981, Factors affecting predation by a microcrustacean (*Cypris* sp.) on nitrogen-fixing blue-green algae, *Soil Biol. Biochem.* **13**:27–32.

Osa-Afiana, L. O., and Alexander, M., 1982a, Differences among cowpea rhizobia in tolerance to high temperature and desiccation in soil, *Appl. Environ. Microbiol.* **43**:435–439.

Osa-Afiana, L. O., and Alexander, M., 1982b, Clays and the survival of *Rhizobium* in soil during desiccation, *Soil Sci. Soc. Am. J.* **46**:285–288.

Pant, S. D., and Iswaran, V., 1970, Survival of groundnut *Rhizobium* in Indian soils, *Mysore J. Agric. Sci.* **4**:19–25.

Pate, J. S., 1961, Temperature characteristics of bacterial variation in legume symbiosis, *Nature* **192**:637–639.

Pena-Cabriales, J. J., and Alexander, M., 1979, Survival of *Rhizobium* in soils undergoing drying, *Soil Sci. Soc. Am. J.* **43**:962–966.

Pena-Cabriales, J. J., and Alexander, M., 1983a, Growth of *Rhizobium* in unamended soils, *Soil Sci. Soc. Am. J.* **47**:81–84.

Pena-Cabriales, J. J., and Alexander, M., 1983b, Growth of *Rhizobium* in soil amended with organic matter, *Soil Sci. Soc. Am. J.* **47**:241–245.

Philpotts, H., 1977, Survival of *Rhizobium trifolii* strains on inoculated seed held at 35°C, *Aust. J. Exp. Agric. Anim. Husb.* **17**:995–997.

Rai, R., 1983, The salt tolerance of *Rhizobium* strains and lentil genotypes and the effects of salinity on aspects of symbiotic N-fixation, *J. Agric. Sci.* **100**:81–86.

Rains, D. W., and Talley, S. N., 1979, Use of azolla in North America, in: *Nitrogen and Rice,* pp. 419–430, International Rice Research Institute, Los Banos, Philippines.

Ramirez, C., and Alexander, M., 1980, Evidence suggesting protozoan predation on *Rhizobium* associated with germinating seeds and in the rhizosphere of beans (*Phaseolus vulgaris* L.), *Appl. Environ. Microbiol.* **40**:492–499.

Robson, A. D., and Loneragan, J. F., 1970, Nodulation and growth of *Medicago truncatula* on acid soil. I. Effect of calcium carbonate and inoculation level on the nodulation of *Medicago truncatula* on a moderately acid soil, *Aust. J. Agric. Res.* **21**:427–434.

Roger, P. A., and Reynaud, P. A., 1979, Ecology of blue-green algae in paddy fields, in: *Nitrogen and Rice,* pp. 287–309, International Rice Research Institute, Los Banos, Philippines.

Roughley, R. J., 1968, Some factors influencing the growth and survival of root nodule bacteria in peat cultures, *J. Appl. Bacteriol.* **31**:259–265.

Roughley, R. J., Blowes, W. M., and Herridge, D. F., 1976, Nodulation of *Trifolium subterraneum* by introduced rhizobia in competition with naturalized strains, *Soil Biol. Biochem.* **8**:403–407.

Roughley, R. J., Bromfield, E. S. P., Pulver, E. L., and Day, J. M., 1980, Competition between species of *Rhizobium* for nodulation of *Glycine max, Soil Biol. Biochem.* **12**:467–470.

Ryther, J. H., and Kramer, D. D., 1961, Relative iron requirement of some coastal and offshore plankton algae, *Ecology* **42**:444–446.

Salema, M. P., Parker, C. A., Kidby, D. K., and Chatel, D. L., 1982, Death of rhizobia on inoculated seed, *Soil Biol. Biochem.* **14**:13–14.

Singh, P. K., 1977, Multiplication and utilization of the fern "Azolla" containing nitrogen-fixing algal symbiont as green manure in rice cultivation, *Riso Rep* **26**:125–137.

Singh, P. K., 1979, Use of azolla in rice production in India, in: *Nitrogen and Rice,* pp. 407–417, International Rice Research Institute, Los Banos, Philippines.

Singleton, P. W., and Bohlool, B. B., 1984, Effect of salinity on nodule formation by soybean, *Plant Physiol.* **74**:72–76.

Singleton, P. W., El Swaify, S. A., and Bohlool, B. B., 1982, Effect of salinity on *Rhizobium* growth and survival, *Appl. Environ. Microbiol.* **44**:884–890.

Steinborn, J., Roughley, R. J., 1975, Toxicity of sodium and chloride ions to *Rhizobium* spp. in broth and peat culture, *J. Appl. Bacteriol.* **39**:133–138.

Stovold, G. E., and Evans, J., 1980, Fungicide seed dressings: Their effects on emergence of soybean and nodulation of pea and soybean, *Aust. J. Exp. Agric. Anim. Husb.* **20**:497–503.

Strivastava, J. S., Singh, B. D., Dhar, B., Singh, V. P., Singh, R. B., and Singh, R. M., 1980, Use of streptomycin resistance and phage sensitivity as markers in competition studies with *Rhizobium leguminosarum, Ind. J. Exp. Biol.* **18**:1171–1173.

Talley, S. N., Lim, E., and Rains, D. W., 1981, Application of *Azolla* in crop production, in: *Genetic Engineering of Symbiotic Nitrogen Fixation and Conservation of Fixed Nitrogen* (J. M. Lyons, R. C. Valentine, D. A. Phillips, D. W. Rains, and R. C. Huffaker, eds.), pp. 363–384, Plenum Press, New York.

Thompson, J. A., 1960, Inhibition of nodule bacteria by an antibiotic from legume seed coats, *Nature* **187**:619–620.

Thornton, F. C., and Davey, C. B., 1983, Acid tolerance of *Rhizobium trifolii* in culture media, *Soil Sci. Soc. Am. J.* **47**:496–501.

Tung, H. F., and Watanabe, I., 1983, Differential response of *Azolla-Anabaena* associations to high temperature and minus phosphorus treatments, *New Phytol.* **93**:423–431.

Vidor, C., and Miller, R. H., 1980, Relative saprophytic competence of *Rhizobium japonicum* strains in soils as determined by the quantitative fluorescent antibody technique (FA), *Soil Biol. Biochem.* **12**:483–487.

Watanabe, I., Berja, N. S., and Del Rosario, D. C., 1980, Growth of *Azolla* in paddy field as affected by phosphorus fertilizer, *Soil Sci. Plant Nutr. (Tokyo)* **26:**301–307.

Weber, D. F., and Miller, V. L., 1972, Effect of soil temperature on *Rhizobium japonicum* serogroup distribution in soybean nodules, *Agron. J.* **64:**796–798.

Wilkins, J., 1967, The effects of high temperatures on certain root-nodule bacteria, *Aust. J. Agric. Res.* **18:**299–304.

Wilson, D. O., and Trang, K. M., 1980, Effects of storage temperature and enumeration methods on *Rhizobium* spp. numbers in peat inoculants, *Trop. Agric. (Trin.)* **57:**233–238.

Wilson, J. T., and Alexander, M., 1979, Effect of soil nutrient status and pH on nitrogen-fixing algae in flooded soils, *Soil Sci. Soc. Am. J.* **43:**936–939.

Wilson, J. T., Greene, S., and Alexander, M., 1979, Effect of interactions among algae on nitrogen fixation by blue-green algae (cyanobacteria) in flooded soils, *Appl. Environ. Microbiol.* **38:**916–921.

Wilson, J. T., Greene, S., and Alexander, M., 1980, Effect of microcrustaceans on blue-green algae in flooded soils, *Soil Biol. Biochem.* **12:**237–240.

5

Ecological Aspects of Heavy Metal Responses in Microorganisms

TREVOR DUXBURY

1. Introduction

Much of the information concerning the influences of heavy metals on microorganisms, and the processes they mediate, is fragmentary and scattered over a wide range of scientific literature. This undoubtedly reflects the ubiquitous nature of this very important group of elements, which are characterized by metallic properties and a specific gravity greater than five (Gadd and Griffiths, 1978). Such elements include several essential for the growth, reproduction, and/or survival of all living things, some with no known biological function, and many with economic, industrial, and/or military uses. In all, about 65 elements comply with the definition of a heavy metal (Hammond, 1976). Wood (1974) considered that certain elements could be placed into three categories, each defined in terms of its environmental impact (Table I). For the purposes of this review similar guidelines have been adopted and most attention will be focused upon those metals in category 2. Little emphasis is placed here on interactions with Hg, since this metal was the subject of a recent review by Jeffries (1982). Lastly, although As has often been included in discussions of heavy metals, I have chosen to exclude it, as well as Se and Te, because it is a nonmetallic element (Hawley, 1977).

Some of the data pertaining to heavy metal interactions with microorganisms have been obtained from primarily physiological studies

TREVOR DUXBURY • Department of Microbiology, University of Sydney, Sydney, New South Wales 2006, Australia.

Table I. Classification of Heavy Metals According to Their Potential Toxicity[a]

Type 1: noncritical	Type 2: potentially toxic and relatively accessible	Type 3: potentially toxic but insoluble or rare
Fe	Co, Ag, Sb, Ni, Cd, Bi, Cu, Pt, Cr, Zn, Au, Mn, Sn, Hg, Mo, Pd, Pb, Tl	Hf, Ta, Os, Zr, Re, Rh, W, Ga, Ir, Nb, La, Ru

[a]Modified from Wood (1974).

aimed at elucidating their roles as trace nutrients. Manganese, Co, Cu, and Zn are elements frequently essential for growth, whereas Cr, Ni, Mo, and Sn may be required for specific metabolic processes by particular organisms (Pirt, 1975). However, when present in excess of normal requirements, even these elements can be toxic to living cells. Since the Minamata incident in the early 1950s (Ottaway, 1980), there has been a steady accumulation of information regarding the deleterious effects of heavy metals on microorganisms. Toxic effects of metals on pure cultures of microbes can be shown by altered cell morphology or metabolism, by bacteriostasis, or by cell death (Ehrlich, 1978). However, some organisms possess very effective tolerance mechanisms and some have the capacity to metabolize metals in various ways (Saxena and Howard, 1977; Iverson and Brinckman, 1978; Summers and Silver, 1978). There is a substantial amount of detailed information about the physiology, biochemistry, and, in some cases, the genetics of metal/microbe interactions, derived from investigations carried out under defined laboratory conditions. Regrettably, the same cannot be said when microorganisms have been considered in their natural environments.

Most ecological data have been characterized by considerable variability, which undoubtedly has arisen from the complex nature of many habitats. That the physicochemical characteristics of any environment influence the chemical form and mobility of heavy metals and hence their ability to interact with microorganisms is clear (Babich and Stotzky, 1978). Soil is not only the most complex of all environments, it is also one of the most important of the earth's resources (Gray and Williams, 1971a). Without the activity of microorganisms in soil it is doubtful whether any but the most meager plant populations could be sustained. Yet, only relatively recently have the effects of heavy metals on soil organisms and soil processes been reviewed (Tyler, 1981). Aquatic environments are less complex, but are equally prone to heavy metal contamination. The fertility of any aquatic ecosystem, whether it be marine or freshwater, ultimately depends upon the activities of the algae living within it (Russell-Hunter, 1970). Consequently, the review by Rai et al. (1981) on the effects of heavy metal pollution on algae was long overdue.

The reviews by Tyler and Rai *et al.* provide the backdrop against which some of the more recent literature will be considered in this chapter. However, the majority of the studies discussed by these authors were concerned with the deleterious effects of high concentrations of heavy metals arising from various forms of pollution. It is also reasonable to assume that, intermittently, locally high concentrations of such elements may arise naturally and, although considered insignificant alongside anthropogenic sources, may be of tremendous significance to an individual microorganism or small microbial community. Consequently, I will attempt to discuss not only the effects of heavy metals on microorganisms at the community or ecosystem level, but also their effects on some ecological interactions that may have particular relevance to microbes as individuals. Further, I will examine the means by which some of these data have been obtained, so that several of the pitfalls that have marred many previous studies may be avoided in the future.

2. Effects of Heavy Metals on Microbial Communities

2.1. Abundance

The effect of heavy metals on the size of microbial communities varies, depending upon which group of microorganisms is being considered, on the metal involved, and on the particular environment. For instance, on the phylloplane, bacteria appear to be more sensitive to metal pollution than fungi. Bewley (1980) showed that polluted oak leaves had fewer bacteria on them than uncontaminated controls, and a highly negative correlation was found between numbers of bacteria and the concentration of Pb on hawthorn leaves (Bewley and Campbell, 1980). On the other hand, the abundance of fungi appeared to be unaffected irrespective of whether perennial rye grass (Bewley, 1979), oak, or hawthorn leaves were considered. However, this particular environment is highly susceptible to other forms of pollution, such as sulfur dioxide, and these may contribute to the observed effects, making it difficult to draw any firm conclusions about the true significance of heavy metals in this type of environment. In contrast to the situation found on living leaves, contaminated leaf litter derived from *Agrostis tenuis* growing on Pb–Zn mine waste contained significantly lower numbers of fungi than litter from a control pasture site, but there were no differences in the abundance of bacteria and actinomycetes (Williams *et al.*, 1977). There was also a marked reduction in the numbers of mites, although springtails were more numerous. Strojan (1978a,b) recorded lower numbers of arthropods, particularly mites, in leaf litter at polluted sites adjacent to a Zn smelter. However, no such differences were found when polluted and clean litter were compared during incubation studies at uncontaminated sites, an observation later confirmed by Freedman and Hutchinson (1980a). They also recorded consistently lower numbers of fungi in soils very close to a Cu–Ni smelter, but these values were not significantly different from those at more distant sites. Neither could Nordgren *et al.* (1983) find a decrease in fungal viable counts along a heavy metal gradient, although, using a direct microscopic method, they did detect a reduction in total fungal biomass with increasing metal concentrations, particularly that of Cu. Bisessar (1982) also found that soil Cu concentrations correlated with changes only in bacterial numbers, whereas the concentrations of Pb and Cd showed a negative correlation with the abundance not only of bacteria, but also actinomycetes, fungi, nematodes, and earthworms. The overall impression gained from these reports is that heavy metals reduce the abundance of microorganisms, but this is not invariably so. In a Zn-polluted region of Dublin Bay, 15 times more bacteria per gram of dry sediment were found than at control sites (Pickaver and Lyes, 1981), but more will be said about this later (Section 4.1).

Such static figures unfortunately provide no information as to how such population changes might have occurred or whether changes were still occurring at the time of sampling. Of interest in this regard is the report by Vaccaro *et al.* (1977) that the addition of Cu (0.01 and 0.05 ppm) to two enclosed marine ecosystems caused an increase in the relative numbers of heterotrophic bacteria. This increase was thought to have occurred as a result of organic carbon having been released from Cu-sensitive organisms. It was also suggested that the bacteria that survived the prevailing Cu concentration developed tolerance to the metal over a period of time and in turn provided a source of inorganic nutrients that could then be used by later phytoplankton communities. Cycling of phytoplankton communities that showed some similarities to this was observed by Effler *et al.* (1980). When Cu was applied to a lake ecosystem on three occasions at about monthly intervals, there were substantial reductions in productivity initially, but after 5 or 6 days there appeared to be a recovery of activity. One definite conclusion that can be drawn from such data is that heavy metals display a differential toxic action, one of the consequences of which will be the alteration of the qualitative composition of microbial communities.

2.2. Diversity

One generalization that applies not only to heavy metals and microorganisms but also to other forms of environmental stress and higher organisms is that pollution causes a reduction in species diversity (Whittaker, 1975). This has been indicated by some of the data concerning the effects of heavy metals on the abundance of broad groups of microorganisms. However, this differential toxicity can also be seen in the presence or absence of particular genera of organisms, or even individual species, in their particular habitats. For example, *Sporobolomyces roseus* was absent from hawthorn leaves heavily contaminated with Cd, Pb, and Zn, although there were indications that its overall distribution was under the influence of several environmental factors, not just heavy metals (Bewley and Campbell, 1980). Conversely, *Aureobasidium pullulans* and *Cladosporium* spp. were found in greater numbers on polluted oak leaves (Bewley, 1980). Further, on hawthorn leaves both *A. pullulans* and nonpigmented yeasts showed a positive correlation with Pb concentration (Bewley and Campbell, 1980).

Williams *et al.* (1977) studied the fungal populations of mine waste and litter and found that *Mortierella* was present in both contaminated and uncontaminated litter, could be isolated with high frequency from mine waste, but was not present in a control pasture soil. Similarly, *Verticillium* was not isolated from pasture soil or from the above-lying litter, but was isolated very frequently from mine waste and contaminated litter. The genus *Penicillium* was found in the pasture soil, both uncontaminated and metal-containing litter, but not in mine waste. In contrast, in a different study, *P. waksmanii* was confined to polluted soils and represented 17–20% of total isolates in those habitats (Carter, 1978; quoted in Freedman and Hutchinson, 1980a). The data of Nordgren *et al.* (1983), however, appear to confirm the findings of Williams *et al.* (1977), in that the higher the soil metal concentration, the less frequently was *Penicillium* isolated (Fig. 1a). Some of their other data (Figs. 1b–1d) also showed how influential metal concentration could be on the distribution of particular organisms. *Oidiodendron* (Fig. 1b) showed a trend similar to that of *Penicillium,* whereas sterile forms (Fig. 1c) appeared to show the reverse. The most striking distribution was that of *Paecilomyces farinosus* (Fig. 1d), which was only isolated from soils contaminated with >1000 ppm Cu.

Another interesting distribution pattern was recorded by Williams *et al.* (1977) for *Arthrobacter,* which was isolated only from leaf litter at the mine waste site. The genus *Bacillus,* on the other hand, occurred in mine waste and pasture soil, as well as in litter at both sites. The genus *Streptomyces* dominated the isolates from the actinomycete group.

In their review, Rai *et al.* (1981) concluded that the presence of heavy metals caused changes in the qualitative composition of algal communities, and data reported since then tend to confirm this view. For instance, Effler *et al.* (1980) observed that application of Cu (0.008–0.014 ppm) favored the development of coccoid, colonial cyanobacteria but

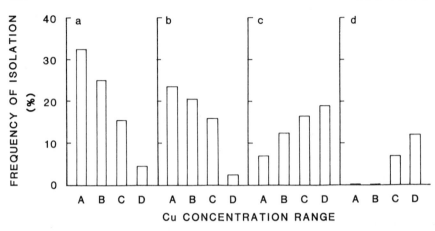

Figure 1. Effect of soil Cu concentration on the frequency of isolation of fungi. (a) *Penicillum*, (b) *Oidiodendron*, (c) sterile forms, (d) *Paecilomyces farinosus.* Concentration ranges: (A) <100 ppm, (B) 100–1000 ppm, (C) 1000–10,000 ppm, (D) >10,000 ppm. Data from Nordgren *et al.* (1983).

inhibited *Anabaena flos-aquae.* Sanders *et al.* (1981a) also observed that the addition of Cu to phytoplankton communities resulted in an overall reduction in centric diatoms, except *Skeletonema costatum,* and a complete absence of dinoflagellates. However, some groups, particularly pennate diatoms and small, unidentified flagellates, became much more abundant, and when very high Cu concentrations (0.015–0.02 ppm) were used it resulted in total domination of the community by the pennate diatom *Amphiprora paludosa* var. *hyalina* (Sanders *et al.,* 1981b). Mercury, even at very low concentrations (0.0005–0.005 ppm), also affects the species composition of periphyton communities, and at concentrations of around 0.05 ppm may inhibit their growth completely (Grollé and Kuiper, 1980). More recently Foster (1982a) studied algal communities in two rivers that drained from Cu and Pb mining regions of Cornwall, England. Even though the metal pollution had extended over a long period of time, it was clear that the diversity and abundance not only of algae, but also of invertebrates (Brown, 1977), were still affected. In addition, it was concluded that the degree of metal pollution, rather than the particular metal, determined the species composition of the algal communities. A similar reduction of species diversity in acid tundra ponds containing elevated concentrations of Zn and Ni was also recorded by Sheath *et al.* (1982). Lampkin and Sommerfeld (1982) recorded reduced species diversity at the point of entry of acid mine drainage containing high levels of Cu, Zn, Mn, and Fe. Unfortunately, in studies of natural ecosystems polluted by mine drainage or by burning bituminous shales, it is often difficult to separate the influences of heavy metals from those of pH changes. Nevertheless, it is clear that elevated levels of heavy metals can alter the qualitative as well as the quantitative structure of microbial communities. Unfortunately, we know very little about the environmental concentrations of metals that are required to bring about these changes. Nor do we know why certain organisms become dominant, although metal tolerance of one form or another undoubtedly plays a part.

2.3. Metal Tolerance

One of the major difficulties common to many studies of microbial metal tolerance lies in the criteria used to define tolerance. At the outset it could be argued that only stressful situations need be tolerated. Consequently, indicators of stress should provide a guide as to what concentration of heavy metal could distinguish tolerant from nontolerant organisms, whether it be in laboratory media or in natural environments. One parameter often used to indicate stress is the reduction in numbers or biomass of particular members of a community. With this in mind, the data of Hines and Jones (1982) (Table II) indicate that the medium they used, designed for the optimal recovery of aerobic heterotrophic bacteria from marine sediments (Litchfield *et al.,* 1975), was itself stressful, since, when amended with Cu or Ni (0.06–0.6 ppm), recoveries of up to 531% of the colony counts on unamended medium were recorded. The authors suggested that the reason for these enhanced recoveries was because the metals were displacing stimulatory cations from binding sites in the medium. However, Hines and Jones were really concerned with the metal tolerance of microbial communities at different depths in various carbonate sediments. The data obtained showed a considerable amount of variation depending on the depth, the metal, its concentration, and the particular sediment sample. In general, only at concentrations of 6 ppm or more of Cu or Ni did it become apparent that the percent recovery of bacteria on media supplemented to this extent decreased with increasing depth. No attempt was made to correlate metal tolerance of organisms with sediment metal concentrations.

Hallas and Cooney (1981) set out to test the hypothesis that Sn selected for Sn-resistant microbial communities. They concluded that there was no significant correlation between Sn concentrations in sediments and the proportion of organisms in a community resistant to Sn in laboratory media (Fig. 2). They also concluded that Sn alone did not select for Sn resistance and suggested that other factors were involved. A similar approach was taken by Olson and Thornton (1982), who attempted to correlate soil metal concentrations with metal-resistance

Table II. Recovery[a] of Aerobic Heterotrophic Bacteria from Carbonate Sediments on Metal-Supplemented Media[b]

Depth (cm)	Percent recovered on medium supplemented with (ppm)							
	Cu^{2+}				Ni^{2+}			
	0.064	0.64	6.4	64	0.059	0.59	5.9	59
	Coot Pond sediment							
0–3	531	114	57.0	<0.1	114	91	45.0	0.17
3–6	525	183	10.0	<0.1	258	225	7.0	0.09
6–8	135	85	25.0	<0.1	164	107	48.0	0.07
8–10	89	103	18.0	<0.1	128	103	39.0	0.01
12–15	12	16	12.0	<0.1	12	11	4.0	0.001
	Mills Creek sediment							
0–3	276	26	0.5	<0.1	300	30	0.8	<0.1
3–6	141	15	2.1	<0.1	171	10	1.3	<0.1
6–8	446	73	5.8	<0.1	367	80	18.0	<0.1
8–10	632	224	177.0	<0.1	ND[c]	110	161.0	<0.1
12–15	192	84	3.1	<0.1	192	46	0.8	0.08
	Devil's Hole sediment							
0–3	129	190	34.0					
3–6	80	32	7.0					
9–12	58	51	4.2					
15–18	59	29	0.41					
24–27	35	54	0.76					

[a]Calculated as a percentage of counts on unamended medium.
[b]Modified from Hines and Jones (1982).
[c]ND, Not determined.

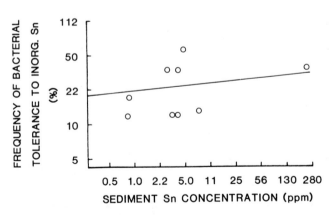

Figure 2. Percentage of bacteria from carbonate sediments able to grow on media supplemented with 75 ppm inorganic Sn. Data from Hallas and Cooney (1981).

patterns of bacteria isolated from them. Significant positive trends were found, but only when media were supplemented with particular metal concentrations (Fig. 3). This illustrates the importance of choosing a metal concentration that is discriminatory. With 10 ppm Cd in the medium there was virtually no increase in community tolerance with soil metal concentration (Fig. 3a) and this kind of response reflected that shown by the data of Hallas and Cooney (1981). Only at 100 ppm Cd did a trend begin to show (Fig. 3b), which became very marked when 200 ppm Cd (Fig. 3c) was incorporated in the plating medium. These results also raise another point of interest. Only when the metal concentration in the medium was relatively high and the proportion of resistant organisms very low (2.5%) did the correlation become apparent.

Duxbury and Bicknell (1983) also used a range of metal concentrations in their plating media and found that the Zn tolerance patterns of soil bacterial communities could be described by two superimposed responses (Figs. 4f and 4g). Each individual response is described by a single exponential equation,

$$y = a \exp(-bx) \tag{1}$$

where y is the number of bacteria able to grow in the presence of a metal at a concentration x (mM), and a is the number of bacteria able to grow on medium without added metals. The factor b is an exponent that is a measure of the metal toxicity, a combined expression of the inherent toxic qualities of the metal itself and the metal-binding properties of the medium. The whole bacterial community response can be described by combining the two individual equations,

$$y = a_1 \exp(-b_1 x) + a_2 \exp(-b_2 x) \tag{2}$$

where a_1 and b_1 refer to the curve covering the lower metal concentrations and a_2 and b_2 relate to the extended concentration range. The distribution of data points for the numbers of Co-, Cd-, Cu-, and Ni-tolerant bacteria indicated similar relationships for these metals (Figs. 4a–4e). Here, too, it was only when the metal concentrations in the media were in excess of 25 ppm and the metal-tolerant proportion of the total bacterial community was very low ($\sim$1% or less) did the two responses become apparent. Not only that, but when organisms belonging to the more tolerant community were characterized, 87% were found to be Gram-negative. Earlier, Doelman and Haanstra (1979c) had also recorded proportionally more Gram-negative rods in Pb-containing soils, although coryneforms were reduced in number. Olson and Thornton (1982) also found that, although on unsupplemented medium the bacteria belonged to a variety of genera

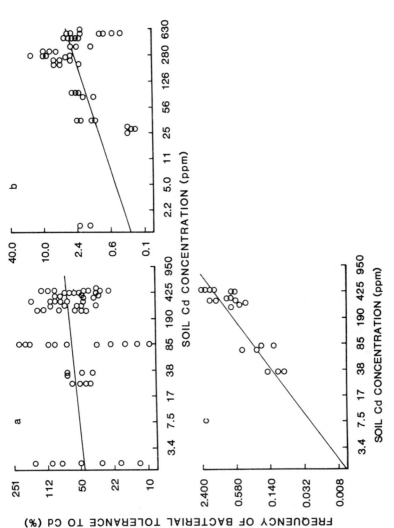

Figure 3. Percentage of bacteria from soils able to grow on media supplemented with (a) 10 ppm, (b) 100 ppm, or (c) 200 ppm Cd. Data from Olson and Thornton (1982).

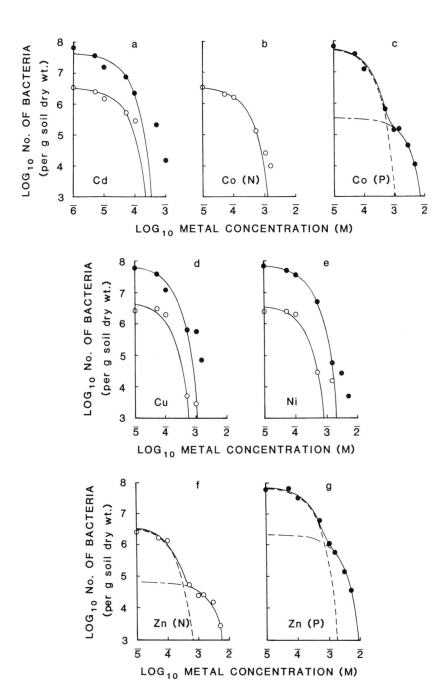

Figure 4. Numbers of soil bacteria able to grow on media supplemented with either (a) Cd, (b, c) Co, (d) Cu, (e) Ni, or (f, g) Zn. (—) Equation (1) for a, b, d, and e; Eq. (2) for c, f, and g. (- - -) Lower concentration range [Eq. (2), a_1, b_1 curve]; (- — -) extended concentration range [Eq. (2), a_2, b_2 curve]. (O, N) Natural soil, (●, P) polluted soil. Data from Duxbury and Bicknell (1983).

(Bacillus, Arthrobacter, Pseudomonas, Chromobacterium, Klebsiella, Alcaligenes, Flavobacterium), when 200 ppm Cd was present in the medium only four Gram-negative genera *(Flavobacterium, Pseudomonas, Acinetobacter, Alcaligenes)* were represented. The occurrence of metal-tolerant Gram-negative bacteria, particularly strains of *Pseudomonas,* in metal-polluted environments had earlier been reported by Austin *et al.* (1977) and Houba and Remacle (1980), although some Gram-positive organisms *(Bacillus,* coryneforms) may also be significant (Austin *et al.,* 1977; Timoney *et al.,* 1978). There may be particular significance in the data of Duxbury and Bicknell (1983) in that, although one of the soils used was metal-contaminated (in which it was expected to find metal-tolerant organisms), the other was a natural, unpolluted soil, which contained very low levels of the various metals tested, and here, too, the two community responses were detected (Fig. 4f). Foster (1982b) also recorded the isolation of metal-resistant algae from unpolluted habitats and often these displayed multiple tolerances.

So what is the significance of metal-tolerant microorganisms in natural environments? Alexander (1976) and Spain *et al.* (1980) have suggested that the ability of communities to adapt depends on the presence of specific microorganisms. Duxbury and Bicknell (1983) suggested that Gram-negative bacteria were tolerant because of their cell structure and that this group were better fitted to endure metal stress than Gram-positive organisms. In addition, they also suggested that plasmids might also be significant in the development of specifically metal-tolerant soil bacterial communities, but this could equally apply to other genetic elements.

2.4. Evolution and Gene Transfer

Transformation, conjugation, and transduction are generally accepted mechanisms of gene transfer among bacteria in laboratory cultures, but how widespread they are in natural environments is still a matter of conjecture. The transfer of drug-resistance mechanisms and the adaptation of microbial communities to the metabolism of pesticides and other unusual compounds have provided some evidence that DNA exchange processes do operate in nature (Reanney *et al.,* 1983). Although many have speculated that the existence of metal-tolerant organisms has been the result of adaptation in response to high environmental metal concentrations, no evidence exists to substantiate this. Despite the fact that many metal-tolerance mechanisms are coded for by plasmid-borne genes, antibiotic resistance markers have been the preferred investigative tool in the study of plasmid transfer. In the gastrointestinal tract the spread of such traits among the microbial community is of tremendous clinical significance and has undoubtedly helped to rationalize the use of broad-spectrum antibiotics. But what of the significance of Cu-tolerance plasmids in the porcine gut (Tetaz and Luke, 1983)? Pig feeds have been supplemented with $CuSO_4$ for many years because of its ability to promote animal growth, although the mechanism of its action is far from clear. However, as pointed out by Tetaz and Luke (1983), $CuSO_4$ in pig feed may select for Cu-tolerant strains of *Escherichia coli* and other bacteria, which may lead to a proliferation of resistance plasmids in general. Indeed, there is at least one report that the feeding of Cu to pigs does select for drug-resistant strains of *E. coli* (Gedek, 1981; quoted in Tetaz and Luke, 1983). The gastrointestinal tract of higher animals is a highly populated ecosystem, which may allow fairly efficient DNA transfer to take place. In soils microorganisms are often separated by relatively large distances, and, more often than not, probably exist under starvation conditions (Gray and Williams, 1971b). However, here, too, the possible role of animals in facilitating gene transfer should not be overlooked. Animals that graze on litter also ingest the associated microorganisms and this may play a significant role in the development of metal-tolerant microbial communities. Coughtrey *et al.* (1980) observed that, with the exception of Cd tolerance in fungi, the tolerance to Pb, Zn, and Cd of litter communities of bacteria and fungi decreased during grazing experiments involving woodlice *(Oniscus asellus)* (Fig. 5). On the other hand, when gut contents and fecal material were analyzed, the bacterial and fungal communities showed a dramatic increase in the proportions of organisms tolerant to the three metals.

Stanisich *et al.* (1977) showed that tolerance to Hg(II) ion, as well as being plasmid-borne, could also be located on transposable elements (Tn501). Radford *et al.* (1981) also found resistance not only to Hg(II) ion, but also to phenyl mercuric acetate on a transposon (Tn3401) in a soil isolate of *P. fluorescens*. Translocatable Hg(II) resistances were also found in a *Klebsiella* sp. (Tn3403) and a *Citrobacter* sp. (Tn3402). Restriction maps of these two transposons were identical but were different from those of Tn3401 and Tn501.

Unfortunately, metal-tolerance plasmids or transposons, once found and characterized, often become merely a tool of molecular biologists for genetic analysis or gene cloning (Glover and Hopwood, 1981). Rarely, if ever, have they been reintroduced into their natural environment as part of an ecological investigation. Once again their significance has been left to speculation, such as that of Radford *et al.* (1981), who suggested that soil could act as a reservoir for such resistance genes. Because the soil they used had not been subjected to direct Hg contamination, they further speculated that such transposable elements may have aided the diffusion of Hg resistance from bacteria in highly contaminated soils to

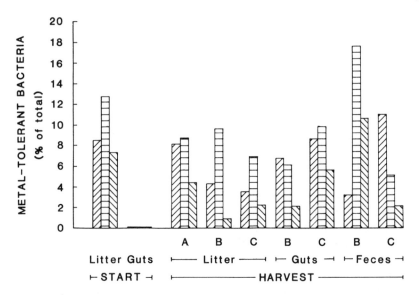

Figure 5. Tolerance of bacteria isolated from leaf litter and the guts and feces of *Oniscus asellus* to 800 ppm Pb ▨, 400 ppm Zn ☰, or 100 ppm Cd ▧ in growth media. (A) control litter bags, no animals present, shaken daily; (B) *O. asellus* added, feces and litter debris removed by daily shaking; (C) *O. asellus* added, feces not removed until harvest. Data from Coughtrey *et al.* (1980).

organisms in uncontaminated areas. The seemingly widespread occurrence of Hg resistance in bacteria from quite different habitats (Jeffries, 1982), many of which have apparently never been subjected to contamination by the metal, I find disquieting. Another explanation quite often put forward is that, because of its volatility, elemental Hg is ubiquitous and that, because of its extreme toxicity, only minute quantities might be required to select for tolerant organisms. I have yet to be convinced that these are indeed the reasons.

A rather different aspect in which heavy metals may be involved with DNA transfer in natural environments is by interacting with the processes that mediate such transfer. For instance, laboratory studies have shown that Cu, Zn, or Ni interfere with the process of transformation in *Bacillus subtilis* (Young and Spizizen, 1963; Dubes *et al.*, 1982). There is also evidence that Pt(II) complexes may block the entry of bacteriophage DNA into host cells (Kerszman *et al.*, 1979), which would consequently interfere with transduction. Until we know more about gene transfer processes in nature, however, the question as to the significance of heavy metal interactions with them must remain unresolved.

The mere presence of microorganisms or microbial communities in any environment is of little consequence when compared to the activities in which they may be involved. Many of these activities are of significance only to the microorganisms themselves in their fight for survival or success. Other activities, however, have a wider significance, particularly those involved in nutrient cycling, and it is by interfering with these that heavy metals may have profound effects on ecosystem functioning.

3. Effects of Heavy Metals on Microbially Mediated Processes

3.1. Carbon Fixation

Although Kepkay *et al.* (1979) suggested that bacterial autotrophic carbon fixation could make significant contributions to sediment activity, all attention so far has been focused on the role of eukaryotic algae in primary productivity in aquatic ecosystems and how heavy metals influence this process. Axler *et al.* (1980) showed that the addition of Mo could cause stimulation of both photosynthesis and nitrate uptake. The enhancement was observed only in spring, when nitrate was relatively abundant, and as the growing season progressed the response to added Mo disappeared. They proposed that Mo could limit phytoplankton productivity through its role in the assimilation of nitrate, since it was an essential component of the enzyme nitrate reductase. Huntsman and Sunda (1980) cited other work indicating that the addition of heavy metals, such as Fe, Mn, Zn, Cu, and Mo, to unenriched lake waters could enhance phytoplankton growth, although of these only Fe, Mn, and Cu were regarded as metals that occurred naturally at concentrations that could adversely affect the growth of algae. Iron and Mn were considered to be normally at growth-limiting concentrations, whereas Cu toxicity could occur even in unpolluted waters. Agreement with the latter observation came from Davies and Sleep (1980), who found that the lowest concentrations of Cu that could cause detectable inhibition of photosynthesis were in the range of 0.001–0.0025 ppm, which were well below levels reported for many natural marine localities. Inhibition of carbon fixation was also recorded at Zn concentrations of 0.01–0.015 ppm, levels observed in coastal waters of several European countries (Davies and Sleep, 1979a,b).

That the addition of synthetic chelators or of Fe to natural phytoplankton communities often stimulates their activity is well known (Huntsman and Sunda, 1980), yet the reasons for this are unclear. Chelators may enhance the availability of limiting trace nutrients or they may reduce the active concentrations of toxic metals. The mechanism by which added Fe enhances phytoplankton activity is less easy to explain. Although it is probably the most quantitatively important trace element for algae, Sunda *et al.* (1981) have proposed that one possible way in which Fe could stimulate algal activity is not, as might at first be thought, by satisfying an Fe requirement but by reducing the free Cu concentration by its adsorption onto hydrous iron oxides (Swallow *et al.,* 1980).

Data recently reported by Ortner *et al.* (1983) underlined the complex nature of heavy metal effects on carbon uptake by algae. It showed that carbon uptake by phytoplankton in both oligotrophic and eutrophic conditions could be inhibited by Cu and Zn, but not by Mn, over a range of concentrations up to ~0.4 ppm, and that the phytoplankton community from the oligotrophic site was affected at much lower concentrations of Cu and Zn than that from the eutrophic site. Also, the addition of organic matter, in the form of natural or synthetic fulvic acids, could completely alleviate Cu-induced inhibition of the oligotrophic community at low Cu concentrations (0.001 ppm) but not at a higher one (0.01 ppm). Further, the inhibition caused by Zn (0.04 and 0.4 ppm) in the oligotrophic samples and by both Zn (0.4 ppm) and Cu (0.01 and 0.1 ppm) in the eutrophic samples was not reversed by the addition of fulvic acid. Ortner *et al.* (1983) concluded that such differences in response may have been due to different water chemistries at the two sites or to a difference in metal sensitivities in geographically separate phytoplankton communities.

3.2. Methanogenesis

In marine sediments sulfate reduction is regarded as being much more important than methanogenesis, whereas in freshwater sediments growing evidence suggests the opposite to be the case (J. G. Jones, 1982). Nickel, Co, and Mo are essential elements for certain methanogens, and consequently some recent studies have been made on the effects of these metals on methane production in natural environments. J. G. Jones *et al.* (1982) could find no stimulation of methanogenesis in freshwater sediment slurries amended with 0.06 ppm Ni or Co or 0.096 ppm Mo, although slight stimulation was observed with some surface sediment samples. In sulfate-limited sediment samples, however, the addition of about 1900 ppm Mo resulted in a 60% or 80% decrease in methane production when the incubation atmosphere was hydrogen and carbon dioxide or nitrogen and carbon dioxide, respectively. Contrary to this, when adequate sulfate was present with a nitrogen and carbon dioxide atmosphere, stimulation of methanogenesis was observed. Enhanced methane production from salt marsh sediments after the addition of 1000 ppm sodium molybdate was also recorded by Capone *et al.* (1983). They suggested that the stimulation by sodium molybdate could be attributed to the inhibition of sulfate-respirers. J. G. Jones *et al.* (1982) had earlier come to a similar conclusion, but in addition proposed that bacteria other than sulfate-reducers might also be involved.

In an investigation of the effects of various pollutants on methanogenesis in freshwater sediments, Pedersen and Sayler (1981) found that 1 and 10 ppm of $HgCl_2$ had no significant effects on the process. Capone *et al.* (1983) similarly observed no effect with 10 and 100 ppm $HgCl_2$. They also included several other metals in their study and found that the effects were variable and depended not only on the metal itself, but also on its form. At concentrations of 1000 ppm methanogenesis was inhibited by CH_3HgCl, whereas $HgCl_2$, $PbCl_2$, and $KCrO_7$ caused an initial inhibition, followed by a period of stimulation. The chlorides of Ni, Cd, and Cu, as well as $ZnSO_4$, PbS, and HgS, caused short-term inhibition, but displayed no significant long-term effects.

3.3. Respiration

Working with respiration rates of soils polluted to varying degrees around a Ni–Cu smelter, Freedman and Hutchinson (1980a) recorded lower rates of carbon dioxide efflux at more contaminated sites, and statistical analysis revealed that Cu had a greater influence than equal amounts of Ni. Copper was also thought to be the most important heavy metal around the town of Gusum, South Sweden (Nordgren *et al.,* 1983), where again decreased soil respiration was characteristic of the more polluted soils, particularly when Cu was present in excess of 1000 ppm.

The addition of Pb (Doelman and Haanstra, 1979a) and of Cr, Cd, Cu, Zn, and Mn (Chang and Broadbent, 1981) to soil samples also caused a decrease in respiration rates. Doelman and Haanstra (1979a) showed that sandy soils exhibited about a 15% decrease in respiration when amended with 375 ppm Pb, the lowest concentration used, whereas a clay soil required 1500 ppm Pb to achieve the same inhibition. A peat soil showed no effects even at the highest concentration of Pb (7500 ppm). Thus, although Tyler (1981) concluded that impeded respiration was common in metal-polluted soils, it is clear that it is highly dependent upon the prevailing soil conditions.

3.4. Litter Decomposition

Soil respiration has been interpreted as providing an indication of overall activity (Macfadyen, 1971), but it is also highly correlated with organic matter decomposition (Witkamp, 1973). This is probably the reason why trends shown by data based on soil respiration have been reflected in studies of the effects of heavy metals on litter decomposition.

Williams *et al.* (1977) concluded that the inhibitory effects of Pb and Zn contained in leaf litter derived from metal-tolerant grasses growing on mine waste were indicated by greater accumulation of litter and less soil humus compared with a control pasture site. Later studies on forests and woodlands polluted by heavy metals from smelting operations (Strojan, 1978a; Coughtrey *et al.*, 1979; Freedman and Hutchinson, 1980a) also showed that increased litter standing crops, resulting from decreased decomposition and not an increase in litter production, were characteristic of areas nearer to the pollution sources compared to those more distant. Coughtrey *et al.* (1979) used statistical analysis to show that the strongest influence on litter decomposition was exerted by Cd and Zn and that there appeared to be very little effect of Pb, Cu, or pH. Another feature revealed by the analysis was that the influence of metal contamination was not at the initial stage of litter decomposition but at a later stage involving more fragmented material. Indeed, some of the data of Strojan (1978a) (Fig. 6) also indicated that the difference in rates of litter decomposition between contaminated and control sites only deviated markedly after 6 months. Further, in both the above studies it was also observed that there was no difference between the decomposition of contaminated versus uncontaminated litter. The lowered decomposition rates nearer the pollution sources were ascribed to site effects.

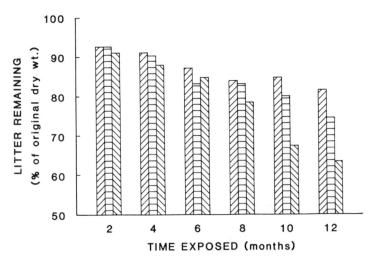

Figure 6. First-year weight loss of oak leaves at sites of low (◨ Fox Gap), intermediate (▤ Little Gap), and high (▨ Lehigh Gap) levels of heavy metal pollution. Data from Strojan (1978a).

3.5. Dinitrogen Fixation

Although in water and sediments almost all heavy metal studies have been directed toward various aspects of carbon cycling, in the case of soil the effects of heavy metals on nitrogen transformations have also received a considerable amount of attention. Tyler (1981) remarked that there had been few studies on the inhibition of dinitrogen fixation by heavy metals. Since then the situation has changed little, and this may also be extended to include other aspects of nitrogen cycling. Rother *et al.* (1982) could find little or no effect of various metals on nitrogenase activity (acetylene reduction) in a number of polluted and uncontaminated soils. In fact they concluded that in their investigation soil moisture was probably more limiting than heavy metal toxicity. The rates of dinitrogen fixation they recorded were low (0.6–8.0 g N/ha per day), but were within ranges recorded for similar habitats (Hardy *et al.,* 1973). The metals studied were Cd, Pb, and Zn, of which the latter two were present in particularly high concentrations in some soil samples (8000 and 26,000 ppm, respectively). Rother *et al.* (1982) suggested that the reason for the lack of effect was because the pollution was long-standing and only relatively small amounts of free metal might be present. In addition, the pH range (6.5–6.8) of the soils was such that there would have been very little Pb or Zn available, since these metals were present mostly in the form of carbonates and sulfides, which would be insoluble. Contrary to this, J. J. Slater and D. G. Capone (unpublished data) have observed that, in general, 1000 ppm of Hg, Pb, Zn, Cr, Mo, and Cd all caused inhibition of acetylene reduction activity in saltmarsh sediments, whereas Ni caused a stimulation at concentrations of 100 and 1000 ppm. They suggested that this enhancement may have been the result of a Ni requirement that has been found to be necessary for certain hydrogenases (Takakuwa and Wall, 1981).

Most studies of the inhibitory effects of heavy metals on symbiotic dinitrogen-fixing associations have placed greater emphasis on the whole plant response rather than on the process of dinitrogen fixation or the interactions with the microorganisms involved. In the alfalfa symbiotic dinitrogen-fixing system, Porter and Sheridan (1981) showed that Cd, Cu, and Zn were highly toxic. Lead, on the other hand, appeared to have little or no effect when included in nutrient solutions at concentrations up to 100 ppm. Earlier, Wickliff *et al.* (1980) had observed that $CdCl_2$ in nutrient solutions in which *Alnus rubra* (red alder) was growing could affect dinitrogen fixation. The presence of up to 15 ppm Cd decreased nitrogenase activity by up to 89%. Inhibition of the enzyme was found to occur when root nodules contained in excess of 3.4 ppm Cd. At the lower end of the concentration range (0.01–0.1 ppm), although nitrogenase activity was decreased, plant growth was not altered. In addition, when nodulated plants were exposed to 0.01–0.1 ppm Cd there was increased nodulation as the Cd concentration increased. They suggested that at these low concentrations more nodules were formed to compensate for the reduced enzyme activity. Because of the high negative correlation between acetylene-reduction activity and Cd concentration in the nodules, Wickliff and Evans (1980) suggested that one way the metal exerted its effect on the enzyme system was by causing increased resorption, or lysis, of the endophyte. However, Wickliff et al. (1980) also suggested that the inhibition may have arisen as a result of a reduction in available photosynthate since, at high Cu concentrations, chlorosis of the plant leaves indicated that chlorophyll biosynthesis and therefore photosynthesis were being impaired. Recently, Porter (1983) came to a similar conclusion with regard to the alfalfa system.

Earlier mention was made of the essentiality of several of the heavy metals for many biological processes. Consequently, the beneficial effects of many metals on dinitrogen fixation have been the subject of a few recent studies. Yatazawa et al. (1980) recorded unfavorable effects on the Azolla–Anabaena symbiosis when certain heavy metals were deficient. They found that the threshold levels of Mn and Mo for growth were 0.02 and 0.0003 ppm and for dinitrogen fixation were 0.01 and 0.001 ppm, respectively. Molybdenum was also studied by Gault and Brockwell (1980), but on legume dinitrogen-fixing systems. The survival of lucerne and clover rhizobia was adversely affected by sodium molybdate, but not by molybdic acid, ammonium molybdate, or molybdenum disulfide, and this was reflected in poorer nodulation of host plants. Even so, dinitrogen fixation, as measured by foliage nitrogen content, was always higher in Mo treatments when compared to treatments that did not incoporate Mo.

Skukla and Yadav (1982), working with chick peas *(Cicer arietinum),* observed that as Zn concentrations were increased up to 19 ppm the number of nodules, their dry weight and leghemoglobin contents, and the amount of nitrogen fixed increased, but beyond this concentration there was a decline in these parameters. However, the presence of 25–50 ppm phosphorus could counteract the effects of 40–100 ppm Zn and maximum nodulation and dinitrogen fixation was observed when 25–50 ppm phosphorus was combined with 5–10 ppm Zn. The observation of these effects with Zn over a relatively narrow range of low concentrations could well be a reflection of their choice of a Zn-deficient loamy sand as experimental material. Rhoden and Allen (1982) also studied the effects of Zn along with Mn on dinitrogen fixation, but quantitative comparison with other studies was made difficult by their choice of units of addition. They amended a Norfolk sandy loam with Mn and Zn at levels of 0, 5, 10, and 20 kg/ha, with pH regimes of 5.5, 6.0, and 6.5, and measured the responses of various cultivars of *Vigna unguiculata*. They found that the effects of Mn and Zn on nodulation and dinitrogen fixation depended on the cultivar and soil pH, but, in general, maximum nodulation was observed with 5 kg Mn/ha or 20 kg Zn/ha. Dinitrogen fixation rates responded similarly, except that only 10 kg Zn/ha was required for optimum activity.

3.6. Other Nitrogen Transformations

There have been virtually no new data on the effects of heavy metals on other aspects of nitrogen cycling in natural environments since Tyler's (1981) review. Flowers and O'Callaghan (1983) made a passing comment to the effect that Cu and Zn contained in pig slurry appeared to have no inhibitory effects on nitrification in metal-amended soil. J. J. Slater and D. G. Capone (unpublished data) found that the effects of several heavy metals on denitrification in saltmarsh sediments were difficult to assess because of variability between samples. In general, Cr, Pb, Zn, and Cu caused an inhibition of the initial rate of nitrous oxide evolution but an increase in the maximum amount of gas produced. Nickel, on the other hand, tended to cause a reduction in both parameters. Results for Cd, Hg, and Mo were inconclusive.

The foregoing, albeit brief, review emphasizes that much of the information about the effects of heavy metals on microbially mediated processes is quite conflicting—so much so that Doelman and Haanstra (1979a) expressed doubts about general values for heavy metal concentrations above which deleterious effects may be expected. Similarly, Coughtrey *et al.* (1979) thought it unwise to speculate about the long-term effects of heavy metals on ecosystem functioning until more detailed studies had been undertaken. Considering the data that have become available since 1979, I can only agree and further recommend caution in any attempts at generalizing about the effects of heavy metals on microbially mediated processes in natural environments.

4. Effects of Heavy Metals on Various Interactions

In order that students might begin to understand what is meant by microbial ecology, I often advise them to envisage themselves in the place of a microorganism, to try to imagine what the physical, chemical, and biological aspects of their environment might be like, and to consider what aspects of themselves (as microorganisms) and of their environment might be significant in contributing to their success in a particular ecosystem. Using such an approach, this section is devoted to those aspects of the lifestyle of a microorganism that might be affected by increased concentrations of heavy metals, and to some of the strategies that a microorganism might avail itself of in dealing with them.

4.1. Substrate Interactions

The effects of heavy metals on substrate decomposition have received little recent attention. Doelman and Haanstra (1979b) found that the addition of Pb to sandy soil retarded the decomposition of various compounds, particularly cellulose, and millipede excrement. However, the inhibitory effects became less apparent over a long period of time (2 years) and the authors suggested that adaptation of individual organisms to Pb or the selection of Pb-resistant members of the microbial community could have been the reason. Contrary to this, Pickaver and Lyes (1981) recorded higher rates of cellulose degradation in sediment containing 300 ppm Zn compared with one containing less (30 ppm), whether measured *in vivo* or *in vitro*. This stimulation of microbial activity was also reflected in respiration rates. However, Pickaver and Lyes (1981) maintained that heavy metals were having detrimental effects in the sediment. They argued that, because bacterial numbers were 15 times higher in the more polluted site, which showed 2–4 times the metabolic rate using sediment dry weight as a basis, the true metabolic rate was about five times lower when based on bacterial numbers. Although, as microbiologists, we should perhaps consider the plight of the individual microbe, as ecologists, should not whole ecosystem functioning also be of prime concern?

A rather different aspect of substrate interactions concerns the effects of metal-contaminated substrates on microorganisms. Swift *et al.* (1979) discussed the importance of resource quality on decomposition processes and the microbial communities involved. The small amount of information available from natural ecosystems suggested that micronutrient heavy metals would not limit decomposition. Not only were they usually present in above-limiting concentrations in the orginal substrate, but they also tended to be more concentrated in the tissues of the decomposer organisms, although this need not be invariably so (Table III). Indeed, some of the decomposers have quite high requirements for certain metals. Isopods, for example, store large quantities of Cu in the hepatopancreas and their feeding rates are linked to the amount of available Cu (Wieser and Wiest, 1968). Consequently, lowered feeding rates and therefore reduced decomposition rates could well result from the presence of abnormally high Cu concentrations in polluted litter. There is no direct evidence to support this suggestion, but litter accumulation was found to be a characteristic of polluted sites (see Section 3.4). In an earlier study, Coughtrey *et al.* (1979) did raise the point that, if the soil fauna or micro-

Table III. Heavy Metal Composition of Leaf Litter and Decomposer Organisms[a]

	Metal content (ppm)				
	Cu	Zn	Mn	Pb	Cd
Leaf litter[b]	17	67	858	—	—
Basidiocarps[b]	44	108	157	—	—
Insecta[c]	50	150	30	—	—
Isopoda[c]	50	130	40	—	—
Leaf litter[d,e]	157	2169	—	2418	33
Isopoda *(Oniscus asellus)*[d,e]	459	462	—	568	232
Isopoda *(Oniscus asellus)*[d,f]	289	329	—	231	19

[a]Modified from Swift *et al.* (1979).
[b]Cromack *et al.* (1975).
[c]McBrayer *et al.* (1974).
[d]Coughtrey *et al.* (1980).
[e]Contaminated woodland site.
[f]Garden site.

biota were to develop metal tolerance, this might lead to increased litter decomposition, thereby releasing previously unavailable metals. If this were to happen, what would be the consequences for the microorganisms involved?

Lighthart (1980) studied the effects of Cd on *E. coli* and *P. aeruginosa* when citrate was present in the growth medium. Citrate was not metabolizable by *E. coli* and, by virtue of its chelating capacity, alleviated the metal toxicity throughout the experiment (Fig. 7). On the other hand, *P. aeruginosa* could use citrate as a nutrient. Consequently, the organism exhibited an initial period of growth followed by a decline, which, according to Lighthart, corresponded to the time when the organism began to use citrate. Was Cd entering the cells combined with the citrate, thereby resulting in intracellular accumulation, or, where *P. aeruginosa* utilized citrate, was it releasing ionic Cd into the external medium, resulting in a gradual increase in the extracellular concentration that ultimately proved toxic? These kinds of questions are important because in soils and many other environments heavy metals may be associated in a similar way with organic matter. Strojan (1978a) and Freedman and Hutchinson (1980a) showed that some components, e.g., K, were lost very rapidly from decomposing leaf litter, whereas the loss of others, such as Ca, paralleled litter weight loss. A third class of elements appeared to increase in concentration as decomposition ensued, and Strojan found that the smelter pollutants in particular belonged to this group. Coughtrey *et al.* (1979) also observed higher metal concentrations in the smaller particle-size classes of litter at contaminated sites.

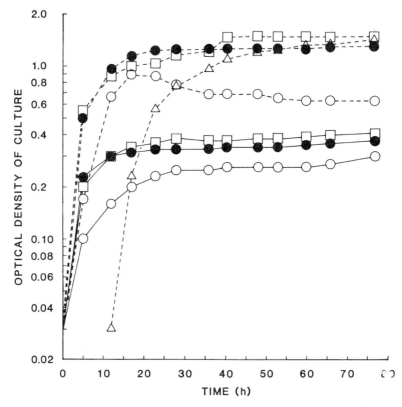

Figure 7. Effect of sodium citrate on the growth of (—) *E. coli* and (- - -) *Pseudomonas* sp. (●) Control (no citrate, no Cd), (△) with 11 ppm Cd, (□) with sodium citrate, (○) with 11 ppm Cd and sodium citrate. No growth of *E. coli* in the presence of Cd without sodium citrate. Data from Lighthart (1980).

From this fragmentary information several questions may be posed. How was this metal concentration occurring? Were the metal-containing litter components being avoided by the "early" decomposers or were the metals accumulating in the decomposers themselves? If so, what effects would this have on the decomposer community as a whole?

In the above studies fungi appeared to be less abundant in contaminated litter. Hanlon (1978; quoted in Swift *et al.,* 1979) showed that fungal catabolism of litter was related to particle size and that the most significant effects occurred with sizes less than 300 μm. Perhaps the high heavy metal concentrations associated with the small particle sizes recorded by Coughtrey *et al.* (1979) were predominantly affecting decomposer fungi. If so, this might be expected to have some effect on organisms that feed on fungi. Mites are usually regarded as not being involved with litter comminution, but rather with the control of microbial communities (Swift *et al.*, 1979). Williams *et al.* (1977), Strojan (1978a,b), and Freedman and Hutchinson (1980a) all recorded reduced populations of mites at metal-contaminated sites. In a quite different study, Drifmeyer and Rublee (1981) also observed that, during decomposition of *Spartina alterniflora* detritus, the concentrations of Mn, Fe, Cu, and Zn increased substantially. In an attempt to localize the site of accumulation, the metal contents and standing crops of bacteria and fungi were determined but were found to account for only part of the accumulated metal. Clearly, there is a need for more detailed studies on this intriguing aspect of substrate decomposition.

4.2. Predator–Prey Interactions

While discussing substrate relationships it should also be recognized that microorganisms themselves may be substrates for other organisms. Microbes are well known for their capacity to not only bind heavy metals, but also to accumulate them in their cells (Hassett *et al.*, 1981; Blair *et al.*, 1982; Kurek *et al.*, 1982; Macaskie and Dean, 1982). One consequence of this is the very real danger of biomagnification of metals, particularly through food chains (Wright, 1978; Leland *et al.*, 1979). Duddridge and Wainwright (1980) used the hemp seed baiting technique to isolate aquatic fungi. Three of the isolates were a *Phythium* sp., *Dictyuchus sterile*, and *Scytalidum lignicola*, and these were able to accumulate Zn, Pb, and Cd, mostly by surface adsorption. When the Cd-loaded *Phythium* sp. was used as food for *Gammarus pulex* there was a marked reduction in shrimp viability and the dead animals were found to contain significant quantities of Cd.

Similar movement of metals through a food chain was demonstrated by Berk and Colwell (1981). They observed that Hg readily accumulated in a *Pseudomonas* sp. and a *Vibrio* sp. and that this could be subsequently transferred to ciliates *(Uronema nigricans)* and possibly to copepods *(Eurytemora affinis)* through predation. However, they only recorded magnification of the metal up to the ciliate trophic level and suggested that the failure to detect magnification from ciliate to copepod may have been due to the large differences in biomass between these organisms or that the experimental period may have been too short to allow detectable Hg accumulation to occur. Other effects of feeding *U. nigricans* with Hg-laden bacteria had previously been investigated by Berk *et al.* (1978). They found that the ciliate, normally sensitive to 0.01 ppm Hg in solution, could ingest bacteria containing 1 ppm of the metal without apparent ill effects. Indeed, it appeared to acquire Hg tolerance, but this was shown not to be due to the activities of the bacteria, since their retention time in the protozoan was considered too low. In addition, when Hg-free bacteria of the same strain were used as the food source the ciliate was still found to be Hg sensitive. It was suggested that the tolerance acquired during the ingestion of Hg-laden bacteria was a consequence of the metal already being safely complexed with organic cellular contents. However, *U. nigricans* could become tolerant if it was exposed to increasing levels of Hg in solution, but the basis of this form of tolerance could not be explained. Berk *et al.* (1978) also considered whether bacteria grown in the presence of Hg might have been altered in such a way that they may not have been as easily recognized as food by the predator. While they could find no evidence for this, considering the number of reports of heavy metal-induced morphological and/or biochemical changes in microorganisms, it is certainly a distinct possibility.

That heavy metals may interfere in other predator–prey situations was shown by Rosenzweig and Pramer (1980), who found that the effects of Cd, Pb, and Zn on trap formation in several nematode-trapping fungi varied according to the metal and the species of fungus involved. In general, Cd was the most toxic and Pb the least. There appeared to be a direct correlation between growth inhibition and decreased trap formation. However, there were exceptions, such as *Monascosporium eudermatum*, which grew almost as well in the presence of 50 ppm Zn as when unsupplemented medium was used, but trap formation was completely inhibited. On the other hand, 300 ppm Pb reduced the growth of *Arthrobotrys oligospora* by 36%, but did not affect trap formation. Whatever the degree of inhibition, traps that were formed were observed to be functional. The production of collagenase, an enzyme implicated in the penetration of nematode cuticle (Schenk *et al.*, 1980), was much more resistant to inhibition by the three metals. For example, 5 ppm Cd reduced the growth of *A. amerospora* by 67%, but the enzyme was unaffected. Up to 300 ppm Pb similarly had no effect, but 50 ppm Zn produced a significant increase in enzyme activity.

The size of nematode-trapping fungi and their prey makes them amenable to observation under the light microscope. Interactions involving bdellovibrios are less easy to observe directly, but Varon and Shilo (1981) adopted an intriguing approach by using the light emission from photobacteria as a measure of the penetration by the predator and showed that Cd could produce very striking effects (Fig. 8). Although no deleterious effects were observed on *Photobacterium leiognathi* until it was suspended in a Cd concentration of 100 ppm, 0.1 ppm caused a marked reduction in the predatory activity of the bdellovibrio. Copper sulfate showed similar effects. One observation that they omitted from their discussion was the lack of effect of 100 ppm Cd on the photobacterium when the bdellovibrio was present. Further, no explanations or suggestions

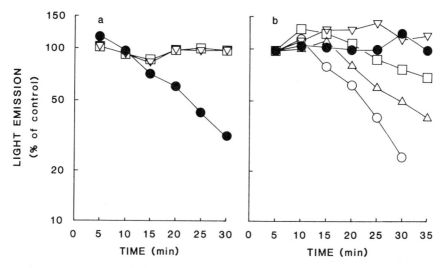

Figure 8. Effect of CdCl₂ on (a) the light emission from *Photobacterium leiognathi* and (b) the bioluminescence decay in the presence of bdellovibrios. Concentrations of CdCl₂: (○) no Cd, (△) 0.1 ppm, (□) 1 ppm, (▽) 10 ppm, (●) 100 ppm. Data from Varon and Shilo (1981).

were offered as to which particular aspects of the interaction were being affected by the heavy metals. Surely a fascinating interaction worthy of more detailed study!

Of even smaller dimensions are bacteriophages, but here, too, there is some evidence of heavy metal interference in the host–parasite relationship. Kerszman *et al.* (1979) suggested that Pt(II) complexes blocked the entry of T4 phage DNA into host cells. They postulated that conformational changes in the DNA, cross-linking between DNA molecules or DNA–protein molecules, or possibly damage to proteins participating in the injection process might have been responsible.

One way in which predator–prey relationships might also be affected by the presence of heavy metals is by interference with chemotactic responses. Walsh and Mitchell (1974) suggested that changes in the kill rates of *E. coli* when exposed to natural marine predators were the result of the inhibition or excitation of chemotactic responses in the predators. Whether heavy metals interfere with chemotaxis is largely unknown. Tso and Adler (1974) observed that concentrations in excess of 1.2 ppm NiSO₄ or 11.6 ppm CoSO₄ could elicit a negative chemotactic response in *E. coli,* but that up to 5.2 ppm CrCl₂, 0.6 ppm CuCl₂, or 0.6 ppm NiCl₂ was nonrepellent. Apart from this report and that of Daniels *et al.* (1980), who found that concentrations of CoCl₂ above 0.6 ppm repelled a spiroplasma, there is almost no information regarding heavy metal influences on chemotaxis, particularly in an ecological context. However, one of the effects of some heavy metals is the inhibition of motility (Adler and Templeton, 1967), which will obviously translate into altered chemotactic responses.

4.3. Interfacial Interactions

Among other things, predator–prey relationships must involve a consideration of the interfacial interactions between the two partners involved. However, interfaces play significant roles in many other aspects of microbial ecology (Marshall, 1976). Because of the propensity of ions, macromolecules, and microorganisms to associate with various kinds of interface, it would be expected that heavy metals could significantly influence the ecology of microorganisms in many ways. Marshall (1978) reported that the proportion of bacteria associated with particulates tolerant to 10 ppm Zn or Cd was higher near the source of a sewage outfall than at sites further downstream. Because particulates provided a large surface area for metal concentration, organisms attached to such particulate material could be expected not only to exhibit greater metal tolerance, but also to accumulate the metals in their cells, which may lead to bioconcentration in higher trophic levels. In addition, Remacle et al. (1982) suggested that bacteria attached to surfaces may act as secondary sources of metal contamination. How significant this would be in aquatic ecosystems is difficult to judge, but microbial communities in soils, where movement of the metal away from such a source would be restricted, might possibly be influenced by such localized release. That surface-associated microorganisms do concentrate heavy metals was shown by Johnson et al. (1981), who observed that periphytic bacteria attached to the crab Helice crassa contained high levels of Cr. Such attached communities have also been found to influence the accumulation of Hg in gastropods (Titus et al., 1980). In this case, feeding habits played a rather important role, in that Helisoma trivolivis accumulated Hg only after Hg had appeared in the periplankton film adhering to the aquarium glass used in the investigation. On the other hand, Campeloma decisa only began to accumulate Hg after the appearance of Hg-laden organic material that had been deposited by fish on the sediment.

There are other, possibly more important ecological consequences of such surface-associated metal–microbe interactions. For example, Harvey et al. (1982) found significant quantities of bacteria and Pb associated with surface layer particles in a saltmarsh ecosystem. Kirchman and Mitchell (1982) observed that in certain aquatic ecosystems surface-associated bacteria were metabolically more active per cell than unattached bacteria. Further, Dawson *et al.* (1981) and Kjelleberg *et al.* (1982) suggested that marine bacteria associated with surfaces as part of their survival strategy. The presence of heavy metals could detrimentally affect this enhanced activity and survival of such organisms, and consequently alter the flow of nutrients in such ecosystems.

One aspect of microbial interactions with surfaces that is of interest to all but the very young and possibly the very old is in the formation of dental plaque. Dental plaque is a complex microbial community found adhering to the tooth surface, embedded in a matrix of polymeric material (Marsh, 1980). Mouth rinses have been introduced to combat plaque formation and some of these incorporate SnF_2. Although probably not the only mechanism involved, it is thought that Sn ions may interfere with plaque formation by reducing bacterial affinity for tooth surfaces (Rolla, 1980).

The effects of several metal salts on microbial adhesion to other animal surfaces, this time human buccal cells, were studied by Sugarman (1980). At Zn concentrations of 0.4–65 ppm there was an increase in adhesion of *E. coli, Klebsiella pneumoniae,* and *Proteus mirabilis,* but at concentrations in excess of 65 ppm no further increase was observed (Fig. 9). Preincubation experiments with *K. pneumoniae* revealed that the increased adhesion was due to some interaction between the bacteria and the metal (Fig. 10) apparently within an initial 5-min period. Bhattercherjee and Saxena (1983) also recorded enhanced adhesion in the presence of Zn (3.4 ppm) of *E. coli* and *Streptococcus faecalis* to human buccal cells but not of *Candida albicans.* Cadmium (5.6 ppm), on the other hand, caused a decrease in adhesion in all three organisms. In addition, when *S. faecalis* was grown in broth containing Zn or Cd it also showed a decrease in adhesion. However, preincubation in the presence of Zn eliminated the inhibitory effects of Cd. Because many infections are caused by autochthonous bacteria normally associated with particular tissues or organs that colonize adjacent, normally sterile tissues (Cheng *et al.,* 1981), metal-induced changes to the patterns of adhesion exhibited by various microorganisms could influence the ability of such pathogens to establish infection. However, heavy metals may alter the balance between host and parasite in other ways.

4.4. Pathogenic Interactions

D. G. Jones and Suttle (1983) recently showed that the susceptibility of a host to pathogenic bacteria could be influenced by heavy metal deficiencies. Mice were fed on normal or Cu-deficient food and then subjected to experimental infection by intraperitoneal injection with *Pasteurella haemolytica.* The results of LD_{50} tests indicated that subclinical Cu

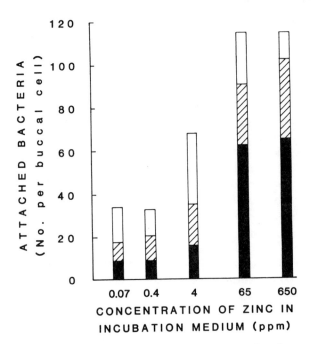

Figure 9. Effect of Zn concentration on the adhesion of bacteria to human buccal cells. □ *K. pneumoniae;* ▨ *P. mirabilis;* ■ *E. coli.* Data from Sugarman (1980).

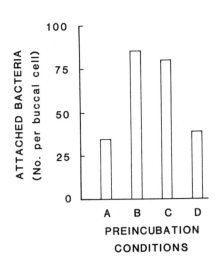

Figure 10. Effect of preincubation conditions on the adhesion of bacteria to human buccal cells in the presence of Zn. (A) Bacteria and buccal cells preincubated separately in buffer, (B) bacteria and buccal cells preincubated separately in 20 ppm Zn, (C) bacteria preincubated in 20 ppm Zn, buccal cells in buffer, (D) bacteria preincubated in buffer, buccal cells in 20 ppm Zn. Data from Sugarman (1980).

deficiency could be associated with increased susceptibility to the pathogen. The cause of the altered response was not investigated, but indirect evidence indicated that impaired reticuloendothelial system activity might have been involved. Earlier, however, Hart *et al.* (1982) had shown that the extent of metal deficiency was also important. They found that moderately Fe-deficient rats were less susceptible to experimentally induced *Proteus mirabilis* pyelonephritis than were Fe-sufficient or severely Fe-deficient littermates. They suggested that this was due to a balance between the requirement for sufficient Fe to maintain the host's defense mechanisms, with the restricting effect that limitation of Fe had on bacterial growth. When Fe was at normal levels in the host there would be sufficient Fe for bacterial proliferation, whereas in severely deficient rats, although there would be less Fe available for microbial growth, defects in the immunocompetence of the host would also occur.

Bouley *et al.* (1982) suggested that Cd altered host susceptibility to infection by interfering with the immune system. In their investigation they exposed mice to airborne CdO microparticles for 15 min and challenged the mice 48 hr later with either *Pasteurella multocida* or *Orthomyxovirus influenzae* via the respiratory route. The death rate resulting from bacterial challenge was significantly increased by exposure to Cd, but the opposite effect was observed with the viral challenge. This difference in response was attributed to the detrimental effects of Cd on the immune system. Cadmium impairs macrophage activity and antibody production, which would result in lower rates of bacterial removal. No explanation could be made for the decreased death rate following the viral treatment, although a similar observation had been reported earlier (Exon *et al.*, 1979). Hatch *et al.* (1981) also showed that the particular heavy metal and the dose rate were important factors when considering host–pathogen interactions. Of 22 chemicals studied, they observed that, when administered by intratracheal injection at lung concentrations of 1 μg/animal, $CdSO_4$, $CuSO_4$, and $ZnSO_4$ caused significant increases in mortality due to *Streptococcus pyogenes* infection. Less, but more variable, elevations in mortalities were caused by 2.16 μg/animal of $Zn(NH_4)_2(SO_4)_2$, whereas salts of Fe, Pb, Mn, or Ni in general caused no significant changes in the death rate at lung concentrations in excess of 2 μg/animal.

A less obvious way in which heavy metals might affect the balance between a parasite and its host was shown by Newsome and Wilhelm (1983). They found that Fe chelators of microbial origin severely inhibited *Naegleria fowleri,* the causative organism of primary amebic meningoencephalitis. Because pathogenic strains of *Naegleria* are most often found in warm water, particularly in excess of 30°C, it was suggested that in these environments the production of microbial Fe-chelators may be suppressed, thereby allowing *Naegleria,* regarded as one of the most virulent protozoa, to proliferate.

Interactions between viruses and plants are also affected by heavy metals. For instance, Harkov and Brennan (1981) observed that in the presence of Cd in bean and tobacco leaf tissue at concentrations that normally did not produce visible symptoms of metal toxicity there was an increase in the number of lesions compared with controls following inoculation with tobacco mosaic virus. Although this could not be explained, it was noticed that the amount of Cd present in the tissues when the effect was observed depended on the plant species, in that tobacco tissue contained 47.9 ppm, whereas beans contained only 5 ppm.

Looking back at this and previous sections (Sections 4.1–4.3), I find it most dissatisfying to see so many ecologically important interactions between microbes and heavy metals for which the reasons remain either completely unknown or, at best, obscure. Of course, it is quite possible that these and others of my criticisms may be premature, and perhaps in the not too distant future some explanations will be forthcoming based upon reliable experimental data. With this in mind, Section 5 is devoted to a discussion of some of the problems commonly associated with heavy metal experimentation.

5. Common Problems

5.1. Field-Oriented Studies

Field observations are invaluable and should be the first step in any study of the ecology of a microorganism in its natural habitat (Lynch and Poole, 1979). Studies involving natural environments frequently require multiple sampling of one particular site. In field-oriented pollution studies, however, it has often been customary to compare contaminated sites with control sites. But what is a control site? In terms of metal concentrations, the data from some of the soil studies already discussed (Table IV) indicate that what was a control site in one investigation could be considered quite contaminated in another. For example, Williams *et al.* (1977) measured 406 ppm Zn in their contaminated litter, whereas Strojan (1978a) considered 676 ppm Zn to be normal. The range of concentrations used for comparison within each investigation also varied quite markedly. Freedman and Hutchinson (1980a) compared respiration in soil samples containing 1400–2600 ppm Cu against control soils containing 230–470 ppm Cu. On the other hand, Nordgren *et al.* (1983) used a metal gradient approach, the extremes of which were 20,000 ppm Cu compared with 15 ppm Cu, which was regarded as a background level.

Table IV. Metal Contents of Leaf Litter and Soils Used in Various Investigations

Site and study	Distance from source (km)	Metal concentration (ppm)				
		Cd	Cu	Pb	Zn	Ni
Leaf litter						
Williams *et al.* (1977)						
Control	0.5	—	—	172	84	—
Contaminated	0	—	—	14,207	406	—
Strojan (1978a)						
Control	40	9	47	258	676	20
Highly contaminated	6	256	172	971	14,600	20
Very highly contaminated	1	885	340	2,333	25,750	21
Coughtrey *et al.* (1979)						
Control	23	2	15	44	80	—
Contaminated	3	98	135	2,179	2,814	—
Freedman and Hutchinson (1980a)						
Control	50	—	8–16	—	—	4–37
Contaminated	3	—	31–94	—	—	60–207
Soils						
Williams *et al.* (1977)						
Control	0.5	—	—	274	79	—
Mine Waste	0	—	—	21,320	1,273	—
Freedman and Hutchinson (1980a)						
Control	50	—	230–470	—	—	270–590
Contaminated	3	—	1400–2600	—	—	1200–1900
Bisessar (1982)						
Control	1	1	73	703	—	—
Contaminated	0.02	151	599	28,000	—	—
Nordgren *et al.* (1983)						
Control	10	—	15	50	100	—
Contaminated	0.1	—	20,000	1,000	20,000	—

Considering these very large differences, it is little wonder that many of the conclusions from field investigations have been at variance with each other.

Little is known about the effects of different metal concentrations in natural environments on microbial communities. It has been suggested that Gram-negative bacteria may be able to withstand moderate levels of pollution better than Gram-positive organisms (Duxbury and Bicknell,

1983), and some authors have reported observing threshold effects with regard to certain microbial responses to heavy metals (Williams and Wollum, 1981). It would not be surprising, therefore, to find that the microbial community present in a soil containing 20,000 ppm Cu was very different from one found in soil containing 2600 ppm Cu (see Section 2.2). This also introduces another potential pitfall in field studies, particularly those involving soil, of whether the contaminated sites were of the same type as the controls. In the case of the study by Williams *et al.* (1977) this was not so much of a problem, since they were knowingly comparing mine waste with pasture soil. However, other studies have focused on comparisons of supposedly the same soil types, while often, to obtain high versus low metal-containing sites, using large distances beween sampling sites (Table IV). For example, Freedman and Hutchinson (1980a) compared soils up to 47 km apart. Fortunately, in this particular study extreme care was exercised in selecting their transects to ensure that the physiographic characteristics of the soils remained the same (Freedman and Hutchinson, 1980b). But what of other field investigations? The drastic effects different soil types can have on heavy metal responses was well shown by Liang and Tabatabai (1977) (Table V). Their data showed that Cu, in particular, could be very inhibitory toward nitrogen mineralization in Okoboji soil and yet show apparently no inhibition in Judson soil.

Table V. Inhibition of Nitrogen Mineralization in Four Soils during Incubation with Added Metals[a]

| Heavy metal | Oxidation state | Percent inhibition of nitrogen mineralization in soil | | | | |
		Webster	Judson	Harps	Okoboji	Average
Ag	I	73	41	59	52	56.3
Hg	II	73	39	35	32	44.8
Cu	II	20	0	7	82	27.3
Cd	II	17	27	39	18	25.3
Pb	II	17	19	28	10	18.5
Mn	II	12	15	26	18	17.8
Zn	II	14	12	15	14	13.8
Ni	II	17	7	17	14	13.8
Sn	II	8	10	20	11	12.3
Co	II	12	2	7	6	6.8
Cr	III	20	15	13	24	18.0
Mo	VI	10	22	22	54	27.0
Soil pH		5.8	6.6	7.8	7.4	
Percent organic C		2.58	2.95	3.74	5.45	
Percent clay		23	45	30	34	

[a]Modified from Liang and Tabatabai (1977).

Further, Doelman and Haanstra (1979a) concluded that the parameters they used to measure the toxic effects of Pb could only be considered reliable when an unpolluted sample of the same soil was used as a control. With environments already polluted this may be impossible to achieve.

5.2. Laboratory-Oriented Studies

To determine the effects of heavy metals on many microbially mediated processes, samples usually have to be removed from field sites and transported to the laboratory. Sample manipulation introduces many problems, particularly with soils. Because of its heterogeneous nature, soil has often been prepared by sieving, removing roots or other debris, and occasionally by air drying. This preparation reduces some of the variability within and between samples to allow significant differences to become more apparent. Metals have then been added, either as powders or in solution, and the soil moisture adjusted to some ecologically acceptable level. Parkinson *et al.* (1971) pointed out that such preparation would result in a biased sample. I often wonder how realistic such samples are. Such drastic manipulations of a soil must destroy its micropedological characteristics. The significance of sorptive interactions between microorganisms and soil components was emphasized by Marshall (1971, 1975). Marshall and Marshman (1978) further stressed the importance of the relationships between a microorganism and a variety of environmental factors, physical, chemical, and biological. Unfortunately, the plea by Marshall (1971) for more realistic studies of microbial activity in natural soils has for the most part gone unheeded, particularly in heavy metal studies. In some investigations even the nutritional status of the soil has been changed. For example, Chang and Broadbent (1981) amended their soil with 1% alfalfa to provide a high level of microbial activity as a background for comparison with heavy metal effects. The growing realization that soils and other natural environments are relatively poor in nutrients and that under normal conditions microbial growth rates are probably low or at best only transiently high (Gray, 1976) casts some doubt on the validity of the data derived from such soil amendment studies.

In a few studies the metal has been added to the sample in a form similar to that associated with a particular source of pollution to introduce realism into the investigation (Castignetti and Klein, 1979; Doelman and Haanstra, 1979a,b,c). One unrealistic aspect of some investigations has been the range of metal concentrations used, since the higher values often have exceeded levels expected to occur in nature. Castignetti and Klein (1979) realized this during their investigation into the effects of high concentrations of Ag that might arise from the use of cloud nucleating agents. However, they concluded that even if such high levels were achieved in aquatic or sewage treatment systems, Ag would not have any significant effects on methanogenesis. Capone *et al.* (1983) similarly acknowledged that the 1000 ppm metal level they used for the majority of their experiments was exceptionally high. However, they argued that such concentrations had been previously recorded in polluted sediments and went on to suggest that, because of the inhibitory effects on methanogenesis they observed, chronic pollution might alter the flow of carbon in sediments.

Coughtrey *et al.* (1979) expressed doubts about making predictions and also highlighted what I believe to be one of the greatest oversights perpetrated in pollution-oriented heavy metal investigations, that of the inescapable involvement of time. "Chronic" describes conditions that have developed over a long period of time, and this would apply to the majority of polluted environments. Yet, almost every pollution-oriented study in microbial ecology in which soils, sediments, or water have been amended with heavy metals has been set up to simulate acute situations. The reason for such an approach has often been to determine how much pollution might cause a detrimental change in ecosystem functioning. But how useful is such data? Doelman and Haanstra (1979a) did include long-term effects in their experimental design. Unfortunately, they began a 3-year trial period by dosing a soil with relatively large amounts of Pb (up to 1500 ppm). At the end of the experimental period they observed that in Pb-amended soil, respiration was still reduced compared to a control soil. Such findings are not surprising. The initial impact of the Pb may have caused irreparable damage to some important groups of microorganisms, such that reestablishment of the original community structure could not take place. Time allows for two important processes to take place. Pollutants normally enter ecosystems as small, probably continuous, additions. In natural environments microorganisms may never experience sudden increases in accessible metal concentrations, because the incoming metal may be immediately sequestered by various environmental components. Of greater significance is the opportunity that time allows for microorganisms to adapt, either physiologically or, perhaps more importantly, genetically, to a gradually changing environment (see Section 2.4).

Perhaps the most common problem associated with laboratory-oriented heavy metal investigations is that of media design. This may be looked at from two viewpoints. The first is in the selection of the metal concentrations used. Because of various problems it is almost impossible to relate metal concentrations in soils or most other environments to metal concentrations in laboratory media. Nonetheless, this has been tried. Pickaver and Lyes (1981) remarked that only 1% of bacteria isolated from sediment containing 250 ppm Zn were resistant to the same concentration in agar, whereas 14% of bacteria from a less polluted sediment (50 ppm Zn) were resistant to 50 ppm Zn in agar medium. They apparently expected the majority of the bacterial communities at each site to be resistant to environmental Zn concentrations. Beck (1981) concluded that heavy metals had a more toxic effect on soil bacteria when they were incorporated into laboratory media than when they were added to soils in the same concentrations. How it is possible to equate the concentration of a metal in a particulate environment such as soil to a concentration in liquid or solidified laboratory media is beyond my comprehension.

The second aspect of media design that is often overlooked is its composition. Metal–media interactions were highlighted by Ramamoorthy and Kushner (1975). Even so, investigations are still being described that incorporate the use of high concentrations of peptones and other organic nutrients in media in which it is impossible to estimate the actual amount of heavy metal causing the observed effects. Even the particular gelling agent used may influence the response of microorganisms subsequently inoculated onto the medium (Hallas *et al.*, 1982). Ion-selective electrodes may be of some value in defining metal concentrations (Zevenhuizen *et al.*, 1979), but media design should be extended even further. Most organic media presently used contain nutrients far in excess of environmental concentrations. Tempest *et al.* (1983) discussed the relevance of laboratory cultures to the growth of microorganisms in natural ecosystems, and even though there is a wide acceptance that microorganisms in many environments, especially soils (Gray, 1976) and water (Stevenson, 1978), live under starvation conditions, very few studies have used either starved cells or even slow-growing populations. Chai (1983) reported that *E. coli* grown in bay water was more sensitive to heavy metals than when it was grown in rich medium. The altered susceptibility was ascribed to a change in the cell envelope composition. Cell envelope damage following short-term stress in water had earlier been observed by Zaske *et al.* (1980) also for *E. coli.* That alterations in the cell envelope of Gram-negative organisms may modify their response to heavy metals was reported by Lutkenhaus (1977), who observed that when an outer membrane protein was missing from *E. coli* the organism exhibited an increase in Cu tolerance. On the other hand, Pan-Hou *et al.* (1981) showed that an increase in sensitivity toward Hg by *Enterobacter aerogenes* was correlated with the loss of a plasmid concomitantly with the loss of two outer membrane proteins. In *P. aeruginosa,* Calcott (1981) recorded a transient loss of plasmid-mediated Hg(II) ion resistance associated with freeze–thaw damage to the outer membrane. In the light of these findings, caution clearly should be exercised in extrapolating many of the observations of microbial responses to heavy metals obtained under laboratory conditions often quite different from those prevailing in natural environments.

5.3. Problems in General

5.3.1. Relative Metal Toxicities

One of the problems that has characterized both field- and laboratory-oriented studies is the attempt in many to express relative metal toxicities. For example, Wood (1974) categorized elements based on their relative environmental impact (Table I), but he appreciated that not all elements could be so neatly compartmentalized, and pointed out that, for example, Mn could fit into more than one category. Unfortunately, the temptation to compare the relative toxicity of one heavy metal to another has often proved too strong. Even Tyler (1981), in summarizing some of the data of Liang and Tabatabai (1977) (Table V), maintained that of 19 elements tested, only Ag and Hg were more inhibitory toward nitrogen mineralization than Cu. This conclusion was based on an average 27% inhibition measured in four different soil types. However, the data showed that in only one of these, Okoboji, was Cu very inhibitory (82%). In Judson soil, in fact, Cu caused no inhibition at all, and in another (Harps) it was among the least inhibitory metals. Such a generalization could lead to the misconception that Cu was the metal most toxic toward nitrogen mineralization regardless of soil type.

Average percent inhibition is only one of several parameters used to indicate the relative toxicities of heavy metals. In general, pollution-oriented field studies have used loading rates, usually as ppm, whereas in laboratory investigations concentrations have also been expressed in terms of molarity. Duxbury (1981) derived an order of toxicity of various heavy metals toward soil bacteria when they were plated onto laboratory media (Fig. 11). The decrease in the number of bacteria able to grow in the presence of increasing concentrations of heavy metals can be described by an exponential equation [Eq. (1), Section 2.3]. Based on the values of b, it was suggested that there were three broad categories of toxicity: one of extreme toxicity, containing metals such as Hg, another of intermediate toxicity, containing metals such as Cd, and a third containing elements such as Cu, Ni, and Zn, regarded as being of only relatively low toxicity.

Unfortunately, the complexity of natural environments introduces a great deal of uncertainty, and relative toxicities vary quite markedly, depending upon the parameter chosen as the basis for comparison. The data of Chang and Broadbent (1981) illustrate this extremely well (Table

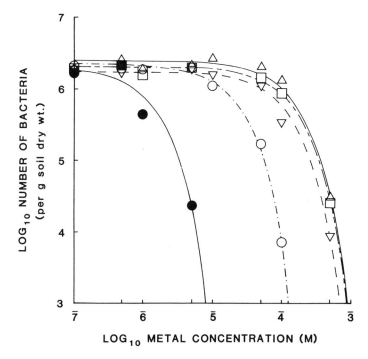

Figure 11. Number of soil bacteria able to grow on media containing either (O) Cd, (▽) Cu, (●) Hg, (□) Ni, or (△) Zn. Lines are the best fit according to Eq. (1) (Section 2.3). Data from Duxbury (1981).

VI). In investigating the effects of various metals on soil respiration, they expressed their data in terms of a "threshold concentration" C_{10}, which was defined as the concentration of metal required for a 10% inhibition of carbon dioxide production, and a "toxicity index" T, calculated from a curve of percent inhibition of respiration versus the concentration of metal extractable with diethylenetriaminepentaacetic acid (DTPA) or nitric acid. The toxicity index T_{10} or T_{50} was the slope of the curve at 10% or 50% inhibition, respectively. Considering their data, a variety of relative toxicities can be derived (Table VII), depending upon which extractant was used, whether T_{10} or T_{50} values were used, or whether ppm or molar concentrations were used. It may prove difficult to standardize on extractants, or to have universal agreement as to what percent inhibition should be regarded as significant, but there should be few problems in using molar values for future heavy metal studies. From a biochemical viewpoint, molar values must be the parameter of choice. The range of atomic weights throughout the heavy metals is quite large. For example, in the Zn triad, 65 ppm is equivalent to 1.0 mM Zn, 0.57 mM Cd, and

Table VI. Threshold Concentrations (C_{10}), Toxicity Indexes (T_{10}, T_{50}), and Loading Rates of Soils That Would Produce Threshold Concentrations[a]

Metal	Loading rate to produce C_{10}		C_{10} (nM/g)		Toxicity index[b]	
	ppm	nM/g	DTPA	HNO$_3$	T_{10}	T_{50}
Cd	8.7	77.4	22.1	48.0	0.456	0.235
Cr	8.6	165	14.5	73.4	6.46	1.35
Cu	11.8	186	65.6	339	0.607	0.0491
Pb	26.8	129	13.6	98.6	0.916	ND
Zn	11.7	179	96.2	266	1.04	0.0141

[a]Modified from Chang and Broadbent (1981). See text for definitions.
[b]ND, Not determined.

0.32 mM Hg, a threefold difference in molar concentration. Having advocated the use of molarity, I must point out that in this chapter, because of the necessity for standardization, ppm has been used. This has been necessary because of the ambiguity of some of the quantitative values in the literature, which makes it inappropriate for them to be converted into molar values.

5.3.2. Effective Metal Concentrations

Many environmental factors may affect the availability of heavy metals to microorganisms (Babich and Stotzky, 1980). The chemical form and mobility of heavy metals can be affected both by pH and *Eh*.

Table VII. Relative Toxicities of Heavy Metals

Basis for comparison	Relative toxicity[a]				
	Cd	Cr	Cu	Pb	Zn
Concentration					
ppm	2	1	4	5	3
molar	1	3	5	2	4
Extractant					
DPTA	3	2	4	1	5
HNO$_3$	1	2	5	3	4
Toxicity index					
T_{10}	5	1	4	3	2
T_{50}	2	1	3	ND	4

[a]1, most toxic; 5, least toxic; ND, not determined.

They may be adsorbed onto hydrous metal oxides; or anions, such as carbonates or phosphates, may interact with them to form insoluble compounds. Negatively charged clay minerals have charge-compensating cations adsorbed to them and these may be able to exchange with heavy metal ions. Cation exchange, as well as chelation and covalent bonding, can also be responsible for the sorption of some heavy metal ions to organic matter. Even microbial cells may effectively bind heavy metals. Kurek *et al.* (1982) found that bacterial cells, particularly dead ones, sorbed more Cd than either montmorillonite clay or sand. Natural waters are considered to be dilute environments (Huntsman and Sunda, 1980), yet, even so, there may be sufficient organic matter present to interact significantly with heavy metals. Although not invariably the case, there appears to be a general consensus that it is the free ion concentration of heavy metals that controls their influence on primary productivity and phytoplankton populations (Huntsman and Sunda, 1980). However, numerous reports indicate that metal toxicity may be a function of the metal content of the organism (Foster, 1977; Fisher and Frood, 1980), and Davies and Sleep (1979a,b, 1980) proposed that it was the intracellular levels of rate-limiting constituents, rather than their concentration in the surrounding water, that determined the growth kinetics of phytoplankton populations. They found that the Cu or Zn contents of natural phytoplankton communities correlated with the concentrations of these metals in the water, and that the rates of carbon fixation decreased as the intracellular Cu or Zn content, measured as metal:chlorophyll *a* ratio, increased. In the case of Zn, it appeared that there was a saturation level at about 0.03 ppm. Above this level there was very little alteration in the Zn:chlorophyll *a* ratio, and no further increase in the inhibition of carbon fixation.

In quite a different study, Laegreid *et al.* (1983) compared the effects of Cd on *Selenastrum capricornutum* suspended in water samples from two lakes. Lake Lille Bakketjern was a small, dystrophic bog lake [low inorganic, high organic content (Whittaker, 1975)], whereas Lake Gjersjøen was eutrophic (high inorganic content) and contained only moderate amounts of humic material. Using an ion-selective electrode with Lille Bakketjern water, Laegreid *et al.* (1983) observed that the toxic response of the algae was a function of the free metal concentration. With Gjersjøen water the ion-selective electrode gave unreliable results. However, they recorded increasing metal toxicity toward the algae during the growing season, which they suggested was due to chemical changes in the lake water rather than to a reduction in metal binding. They indicated that low-molecular-weight compounds arising from algal blooms by extracellular production, autolysis, and decomposition may have increased metal uptake and hence the toxicity.

It would appear that even in relatively simple aquatic environments the concentrations of heavy metals that actually interact with microorganisms are not easy to ascertain, but in more complex environments, such as soils, it is even more difficult to determine how much metal might be accessible to microorganisms. Different extraction techniques have produced very different values for metal contents of soils. Some of the data of Chang and Broadbent (1981) serve as an example (Table VIII). Sequential extraction with water, potassium nitrate, DPTA, and nitric acid removed increasing amounts of heavy metals from a Yolo soil, although the efficiency of each extractant varied, depending on the metal involved. DTPA- and nitric acid-extractable forms of metal were more closely correlated with the quantitites of metals added to the soil than were water- or potassium nitrate-extractable forms. In addition, they could not find a linear correlation between the inhibition of soil respiration and the concentration of metals soluble in any of the extractants.

Of all the problems discussed in this section, this last one is probably the most signficant. Because in the majority of cases it has not been possible to determine the actual effective concentration of a heavy metal causing an observed effect, we are repeatedly confronted with the most undesirable situation of not knowing the quantitative parameter fundamental to all heavy metal–microbe interactions. But what of the future?

Table VIII. Concentrations of Metals Extractable with Various Reagents in Relation to the Quantity of Metal Added to Soil[a]

Metal added (ppm)		Extractant			Percent recovery from four sequential extractions
	H_2O	KNO_3	DTPA	HNO_3	
Cd 0	0.0028	0.0093	0.0035	0.0583	
50	0.0078	0.0561	11.0	23.9	47.7
100	0.0141	0.294	17.9	38.3	38.3
200	0.521	1.55	29.2	56.5	28.2
300	0.539	1.92	40.5	73.2	24.4
400	0.570	2.22	46.8	85.7	21.4
Cu 0	0.0288	0.537	5.52	14.9	
50	0.0129	0.0897	6.15	32.0	34.3
100	0.114	0.267	14.0	58.8	43.9
200	0.192	0.466	43.0	123	54.3
300	0.324	0.687	66.6	180	55.0
400	0.425	0.812	85.9	226	52.8

[a]Modified from Chang and Broadbent (1981).

6. Future Prospects

In this chapter I have attempted to show the many ways in which heavy metals interact with microorganisms and the ecological consequences of those interactions. Unfortunately, for many of the interactions we know very little about the underlying mechanisms responsible for the organismal or ecosystem response. Admittedly, some of the problems and pitfalls to which I have alluded may have precluded more detailed investigations, but it also could mean that different approaches are necessary. Perhaps the use of mathematical models, such as that used by Khummongkol *et al.* (1982) to describe the uptake of heavy metals by *Chlorella vulgaris,* may provide a way of predicting the behavior of that most elusive of parameters, the "effective metal concentration." Other models, such as chemostats, could also be employed to reproduce particular aspects of natural ecosystems (Mayfield *et al.,* 1980; Tan, 1980) or to produce more physiologically realistic cultures for detailed laboratory studies.

With regard to environmentally oriented studies, there has been an overwhelming emphasis on the negative aspects of heavy metal pollution. Apart from the study by Van Hook *et al.* (1977), there is almost no information about how heavy metals are partitioned and cycled in natural, unpolluted environments, nor do we know what potential damaged ecosystems possess to enable them to recover when a pollution source is removed. What we do know is that, although some of the data are conflicting, high concentrations of certain heavy metals do interfere with ecosystem functioning. It would be reassuring, although naive, to think that heavy metal pollution was a thing of the past—a leftover from a less-enlightened era. But even now the contents of Wood's table (see Table I) may require additions as the environmental concentrations of transuranic elements begin to increase due to fallout from nuclear testing and the increased reliance on the nuclear generation of electricity. Perhaps the future will see more reports similar to those of Wildung and Garland (1982) and Fisher *et al.* (1983a,b) involving plutonium, neptunium, americium, and californium.

Finally, a plea. In my introduction I described heavy metals as ubiquitous because they find mention in a diverse array of literature. In preparing this chapter I was dismayed to find heavy metal interactions with microorganisms mentioned in over 250 journals. For heavy metal research to be better coordinated and to facilitate more rapid progress, this is one instance in which a reduction in diversity would be beneficial. A *Journal of Heavy Metal Research* would be useful.

Throughout this chapter I have interpreted microbial ecology in its broadest sense, which has necessitated my encroaching into some unfamiliar territory. Consequently, I apologize for any misinterpretations of the data. The views I have expressed and the criticisims I have raised are personal ones, but if they have provided some food for thought or stimulated a degree of discussion, I will have achieved my aim.

References

Adler, J., and Templeton, B., 1967, The effect of environmental conditions on the motility of *Escherichia coli, J. Gen. Microbiol.* **46**:175–184.

Alexander, M., 1976, Natural selection and the ecology of microbial adaptation in a biosphere, in: *Extreme Environments: Mechanisms of Microbial Adaptation* (M. R. Heinrich, ed.), pp. 3–25, Academic Press, New York.

Austin, B., Allen, D. A., Mills, A. L., and Colwell, R. R., 1977, Numerical taxonomy of heavy metal-tolerant bacteria isolated from an estuary, *Can. J. Microbiol.* **23**:1433–1447.

Axler, R. P., Gersberg, R. M., and Goldman, C. R., 1980, Stimulation of nitrate uptake and photosynthesis by molybdenum in Castle Lake, California, *Can. J. Fish. Aquat. Sci.* **37**:707–712.

Babich, H., and Stotzky, G., 1978, Effects of cadmium on the biota: Influence of environmental factors, *Adv. Appl. Microbiol.* **23**:55–117.

Babich, H., and Stotzky, G., 1980, Environmental factors that influence the toxicity of heavy metal and gaseous pollutants to microorganisms, *Crit. Rev. Microbiol.* **8**:99–145.

Beck, T., 1981, Untersuchungen über die toxische Wirkung der in Siedlungsabfällen häufigen Schwermetalle auf die Bodenmikroflora, *Z. Pflanzenernaehr. Bodenkd.* **144**:613–627.

Berk, S. G., and Colwell, R. R., 1981, Transfer of mercury through a marine microbial food web, *J. Exp. Mar. Biol. Ecol.* **52**:157–172.

Berk, S. G., Mills, A. L., Hendricks, D. L., and Colwell, R. R., 1978, Effects of ingesting mercury-containing bacteria on mercury tolerance and growth rates of ciliates, *Microb. Ecol.* **4**:319–330.

Bewley, R. J. F., 1979, The effects of zinc, lead, and cadmium pollution on the leaf surface microflora of *Lolium perenne* L., *J. Gen. Microbiol.* **110**:247–254.

Bewley, R. J. F., 1980, Effects of heavy metal pollution on oak leaf microorganisms, *Appl. Environ. Microbiol.* **40**:1053–1059.

Bewley, R. J. F., and Campbell, R., 1980, Influence of zinc, lead, and cadmium pollutants on the microflora of hawthorn leaves, *Microb. Ecol.* **6**:227–240.

Bhattacherjee, J. W., and Saxena, R. P., 1983, Effect of cadmium and zinc on microbial adherence, *Toxicol. Lett.* **15**:139–145.

Bisessar, S., 1982, Effect of heavy metals on microorganisms in soils near a secondary lead smelter, *Water Air Soil Pollut.* **17**:305–308.

Blair, W. R., Olson, G. J., Brinckman, F. E., and Iverson, W. P., 1982, Accumulation and fate of tri-*n*-butyltin cation in estuarine bacteria, *Microb. Ecol.* **8**:241–251.

Bouley, F., Chaumard, C., Quero, A.-M., Girard, F., and Boudene, C., 1982, Opposite effects of inhaled cadmium microparticles on mouse susceptibility to an airborne bacterial and an airborne viral infection, *Sci. Total Environ.* **23**:185–188.

Brown, B. E., 1977, Effects of mine drainage on the River Hayle, Cornwall. A: Factors affecting concentrations of copper, zinc and iron in water, sediments and dominant invertebrate fauna, *Hydrobiologia* **52**:221–233.

Calcott, P. H., 1981, Transient loss of plasmid-mediated mercuric ion resistance after stress in *Pseudomonas aeruginosa, Appl. Environ. Microbiol.* **41**:1348–1354.

Capone, D. G., Reese, D. D., and Kiene, R. P., 1983, Effects of metals on methanogenesis, sulfate reduction, carbon dioxide evolution, and microbial biomass in anoxic saltmarsh sediments, *Appl. Environ. Microbiol.* **45**:1586–1591.

Carter, A., 1978, Some aspects of the fungal flora in nickel-contaminated and non-contaminated soils near Sudbury, Ontario, Canada, M.Sc. thesis, University of Ontario, Toronto.

Castignetti, D., and Klein, D. A., 1979, Silver iodide burn complex and silver phosphate effects on methanogenesis, *J. Environ. Sci. Health Part A Environ. Sci. Eng.* **14**:529–546.

Chai, T.-J., 1983, Characteristics of *Escherichia coli* grown in bay water as compared with rich medium, *Appl. Environ. Microbiol.* **45**:1316–1323.

Chang, F.-H., and Broadbent, F. E., 1981, Influence of trace metals on carbon dioxide evolution from a Yolo soil, *Soil Sci.* **132**:416–421.

Cheng, K.-J., Irvin, R. T., and Costerton, J. W., 1981, Autochthonous and pathogenic colonization of animal tissues by bacteria, *Can. J. Microbiol.* **27**:461–490.

Coughtrey, P. J., Jones, C. H., Martin, M. H., and Shales, S. W., 1979, Litter accumulation in woodlands contaminated by Pb, Zn, Cd and Cu, *Oecologia (Berl.)* **39**:51–60.

Coughtrey, P. J., Martin, M. H., Chard, J., and Shales, S. W., 1980, Microorganisms and metal retention in the woodlouse, *Oniscus asellus, Soil Biol. Biochem.* **12**:23–27.

Cromack, K., Jr., Todd, R. L., and Monk, C. D., 1975, Patterns of basidiomycete nutrient accumulation in conifer and deciduous forest litter, *Soil Biol. Biochem.* **7**:265–268.

Daniels, M. J., Longland, J. M., and Gilbart, J., 1980, Aspects of motility and chemotaxis in spiroplasmas, *J. Gen. Microbiol.* **118**:429–436.

Davies, A. G., and Sleep, J. A., 1979a, Photosynthesis in some British coastal waters may be inhibited by zinc pollution, *Nature* **277**:292–293.

Davies, A. G., and Sleep, J. A., 1979b, Inhibition of carbon fixation as a function of zinc uptake in natural phytoplankton assemblages, *J. Mar. Biol. Assoc. U. K.* **59**:937–949.

Davies, A. G., and Sleep, J. A., 1980, Copper inhibition of carbon fixation in coastal phytoplankton assemblages, *J. Mar. Biol. Assoc. U. K.* **60**:841–850.

Dawson, M. P., Humphrey, B. A., and Marshall, K. C., 1981, Adhesion: A tactic in the survival strategy of a marine vibrio during starvation, *Curr. Microbiol.* **6**:195–199.

Doelman, P., and Haanstra, L., 1979a, Effect of lead on soil respiration and dehydrogenase activity, *Soil Biol. Biochem.* **11**:475–479.

Doelman, P., and Haanstra, L., 1979b, Effects of lead on the decomposition of organic matter, *Soil Biol. Biochem.* **11**:481–485.

Doelman, P., and Haanstra, L., 1979c, Effect of lead on the soil bacterial microflora, *Soil Biol. Biochem.* **11**:487–491.

Drifmeyer, J. E., and Rublee, P. A., 1981, Mn, Fe, Cu and Zn in *Spartina alterniflora* detritus and microorganisms, *Bot. Mar.* **24**:251–256.

Dubes, G. R., Sambol, A. R., Al-Moslih, M. I., and Bradshaw, G. L., 1982, Inactivation of transforming DNA by a transient product of the interaction of Cu^{2+} and hydroquinone, *Biochem. Biophys. Res. Commun.* **109**:888–894.

Duddridge, J. E., and Wainwright, M., 1980, Heavy metal accumulation by aquatic fungi and reduction in viability of *Gammarus pulex* fed Cd^{2+} contaminated mycelium, *Water Res.* **14**:1605–1611.

Duxbury, T., 1981, Toxicity of heavy metals to soil bacteria, *FEMS Microbiol. Lett.* **11**:217–220.

Duxbury, T., and Bicknell, B., 1983, Metal-tolerant bacterial populations from natural and metal-polluted soils, *Soil Biol. Biochem.* **15**:243–250.

Effler, S. W., Litten, S., Field, S. D., Tong-Ngork, T., Hale, F., Meyer, M., and Quirk, M., 1980, Whole lake responses to low level copper sulfate treatment, *Water Res.* **14**:1489–1499.

Ehrlich, H. L., 1978, How microbes cope with heavy metals, arsenic and antimony in their environment, in: *Microbial Life in Extreme Environments* (D. J. Kushner, ed.), pp. 381–408, Academic Press, London.

Exon, J. H., Koller, L. D., and Isaacson-Kerkvliet, N., 1979, Lead–cadmium interaction: Effects on viral-induced mortality and tissue residues in mice, *Arch. Environ. Health* **34**:469–475.

Fisher, N. S., and Frood, D., 1980, Heavy metals and marine diatoms: Influence of dissolved organic compounds on toxicity and selection for metal tolerance among four species, *Mar. Biol.* **59**:85–93.

Fisher, N. S., Bjerregaard, P., and Fowler, S. W., 1983a, Interactions of marine plankton with transuranic elements I. Biokinetics of neptunium, plutonium, americium, and californium in phytoplankton, *Limnol. Oceanogr.* **28**:432–447.

Fisher, N. S., Bjerregaard, P., and Huynh-Ngoc, L., 1983b, Interactions of marine plankton with transuranic elements II. Influence of dissolved organic compounds on americium and plutonium accumulation in a diatom, *Mar. Chem.* **13**:45–56.

Flowers, T. H., and O'Callaghan, J. R., 1983, Nitrification in soils incubated with pig slurry or ammonium sulphate, *Soil Biol. Biochem.* **15**:337–342.

Foster, P. L., 1977, Copper exclusion as a mechanism of heavy metal tolerance in a green alga, *Nature* **269**:322–323.

Foster, P. L., 1982a, Species associations and metal contents of algae from rivers polluted by heavy metals, *Freshwater Biol.* **12**:17–39.

Foster, P. L., 1982b, Metal resistances of Chlorophyta from rivers polluted by heavy metals, *Freshwater Biol.* **12**:41–61.

Freedman, B., and Hutchinson, T. C., 1980a, Effects of smelter pollutants on forest leaf litter decomposition near a nickel–copper smelter at Sudbury, Ontario, *Can. J. Bot.* **58**:1722–1736.

Freedman, B., and Hutchinson, T. C., 1980b, Long-term effects of smelter pollution at Sudbury, Ontario, on forest community composition, *Can. J. Bot.* **58**:2123–2140.

Gadd, G. M., and Griffiths, A. J., 1978, Microorganisms and heavy metal toxicity, *Microb. Ecol.* **4**:303–317.

Gault, R. R., and Brockwell, J., 1980, Studies on seed pelleting as an aid to legume inoculation 5. Effects of incorporation of molybdenum compounds in the seed pellet on inoculant survival, seedling nodulation and plant growth of lucerne and subterranean clover, *Aust. J. Exp. Agric. Anim. Husb.* **20**:63–71.

Gedek, B., 1981, Zur Wirkung von Kupfer im Tierfutter als Selektor antibiotikaresistenter *E. coli*-Keime beim Schwein, *Tieraerztl. Umsch.* **36**:6–21.

Glover, S. W., and Hopwood, D. A. (eds.), 1981, *Genetics as a Tool in Microbiology*, Cambridge University Press.

Gray, T. R. G., 1976, Survival of vegetative microbes in soil, in: *The Survival of Vegetative Microbes* (T. R. G. Gray and J. R. Postgate eds.), pp. 327–364, Cambridge University Press.

Gray, T. R. G., and Williams, S. T., 1971a, *Soil Microorganisms*, Oliver and Boyd, Edinburgh.

Gray, T. R. G., and Williams, S. T., 1971b, Microbial productivity in soil, in: *Microbes and Biological Productivity* (D. F. Hughes and A. H. Rose, eds.), pp. 255–286, Cambridge University Press.

Grollé, T., and Kuiper, J., 1980, Development of marine periphyton under mercury stress in a controlled ecosystem experiment, *Bull. Environ. Contam. Toxicol.* **24**:858–865.

Hallas, L. E., and Cooney, J. J., 1981, Tin and tin-resistant microorganisms in Chesapeake Bay, *Appl. Environ. Microbiol.* **41:**446–471.

Hallas, L. E., Thayer, J. S., and Cooney, J. J., 1982, Factors affecting the toxic effect of tin on estuarine microorganisms, *Appl. Environ. Microbiol.* **44:**193–197.

Hammond, C. R., 1976, The elements, in: *Handbook of Chemistry and Physics,* 57th ed. (R. C. Weast, ed.), pp. B5–60, Chemical Rubber Co., Cleveland, Ohio.

Hanlon, D., 1978, Soil animal/microbial interactions during litter decomposition, Ph.D. thesis, University of Exeter.

Hardy, R. W. F., Burns, R. C., and Holsten, R. D., 1973. Applications of the acetylene–ethylene assay for measurement of nitrogen fixation, *Soil Biol. Biochem.* **5:**47–81.

Harkov, R., and Brennan, E., 1981, Cadmium in foliage alters plant response to tobacco mosaic virus, *J. Air. Pollut. Control Assoc.* **31:**166–167.

Hart, R. C., Kadis, S., and Chapman, W. L., Jr., 1982, Nutritional iron status and susceptibility to *Proteus mirabilis* pyelonephritis in the rat, *Can. J. Microbiol.* **28:**713–717.

Harvey, R. W., Lion, L. W., Yong, L. Y., and Leckie, J. O., 1982, Enrichment and association of lead and bacteria at particulate surfaces in a salt-marsh surface layer, *J. Mar. Res.* **40:**1201–1212.

Hassett, J. M., Jennett, J. C., and Smith, J. E., 1981, Microplate technique for determining accumulation of metals by algae, *Appl. Environ. Microbiol.* **41:**1097–1106.

Hatch, G. E., Slade, R., Boykin, E., Hu, P. C., Miller, F. J., and Gardner, D. E., 1981, Correlation of effects of inhaled versus intra-tracheally injected metals on susceptibility to respiratory infection in mice, *Am. Rev. Respir. Dis.* **124:**167–173.

Hawley, G. G., 1977, *The Condensed Chemical Dictionary,* 9th ed., Van Nostrand Reinhold, New York.

Hines, M. E., and Jones, G. E., 1982, Microbial metal tolerance in Bermuda carbonate sediments, *Appl. Environ. Microbiol.* **44:**502–505.

Houba, C., and Remacle, J., 1980, Composition of the saprophytic bacterial communities in freshwater systems contaminated by heavy metals, *Microb. Ecol.* **6:**55–69.

Huntsman, S. A., and Sunda, W. G., 1980, The role of trace metals in regulating phytoplankton growth with emphasis on Fe, Mn and Cu, in: *Studies in Ecology,* Vol. 7, *The Physiological Ecology of Phytoplankton* (I. Morris, ed.), pp. 285–328, Blackwell Scientific Press, Oxford.

Iverson, W. P., and Brinckman, F. E., 1978, Microbial metabolism of heavy metals, in: *Water Pollution Microbiology,* Vol. 2 (R. Mitchell, ed.), pp. 201–232, Wiley, New York.

Jeffries, T. W., 1982, The microbiology of mercury, *Prog. Indust. Microbiol.* **16:**21–75.

Johnson, I., Flower, N., and Loutit, M. W., 1981, Contribution of periphytic bacteria to the concentration of chromium in the crab, *Helice crassa, Microb. Ecol.* **7:**245–252.

Jones, D. G., and Suttle, N. F., 1983, The effect of copper deficiency on the resistance of mice to infection with *Pasteurella haemolytica, J. Comp. Pathol.* **93:**143–149.

Jones, J. G., 1982, Activities of aerobic and anaerobic bacteria in lake sediments and their effect on the water column, in: *Sediment Microbiology* (D. B. Nedwell and C. M. Brown, eds.), pp. 107–145, Academic Press, London.

Jones, J. G., Simon, B. M., and Gardener, S., 1982, Factors affecting methanogenesis and associated anaerobic processes in the sediments of a stratified eutrophic lake, *J. Gen. Microbiol.* **128:**1–11.

Kepkay, P. E., Cooke, R. C., and Novitsky, J. A., 1979, Microbial autotrophy: A primary source of organic carbon in marine sediments, *Science* **204:**68–69.

Kerszman, G., Josephsen, J., and Fernholm, B., 1979, Platinum(II) complexes block the entry of T4 phage DNA into the host cell, *Chem. Biol. Interact.* **28:**259–268.

Khummongkol, D., Canterford, G. S., and Fryer, C., 1982, Accumulation of heavy metals in unicellular algae, *Biotechnol. Bioeng.* **24:**2643–2660.

Kirchman, D., and Mitchell, R., 1982, Contribution of particle-bound bacteria to total microheterotrophic activity in five ponds and two marshes, *Appl. Environ. Microbiol.* **43**:200–209.

Kjelleberg, S., Humphrey, B.A., and Marshall, K. C., 1982, Effect of interfaces on small, starved marine bacteria, *Appl. Environ. Microbiol.* **43**:1166–1172.

Kurek, E., Czaban, J., and Bollag, J.-M., 1982, Sorption of cadmium by microorganisms in competition with other soil constituents, *Appl. Environ. Microbiol.* **43**:1011–1015.

Laegreid, M., Alstad, J., Klaveness, D., and Seip, H. M., 1983, Seasonal variation of cadmium toxicity toward the alga *Selenastrum capricornutum* Printz in two lakes with different humus content, *Environ. Sci. Technol.* **17**:357–361.

Lampkin, A. J., III, and Sommerfeld, M. R., 1982, Algal distribution in a small, intermittent stream receiving acid mine-drainage, *J. Phycol.* **18**:196–199.

Leland, H. V., Luoma, S. N., and Fielden, J. M., 1979, Bioaccumulation and toxicity of heavy metals and related trace elements, *J. Water Pollut. Control Fed.* **51**:1592–1616.

Liang, C. N., and Tabatabai, M. A., 1977, Effects of trace elements on nitrogen mineralization in soils, *Environ. Pollut.* **12**:141–147.

Lighthart, B., 1980, Effects of certain cadmium species on pure and litter populations of microorganisms, *Antonie van Leeuwenhoek J. Microbiol. Serol.* **46**:161–167.

Litchfield, C. D., Rake, J. B., Zindulis, J., Watanabe, R. T., and Stein, D. J., 1975, Optimization of procedures for recovery of heterotrophic bacteria from marine sediments, *Microb. Ecol.* **1**:219–233.

Lutkenhaus, J. F., 1977, Role of a major outer membrane protein in *Escherichia coli, J. Bacteriol.* **131**:631–637.

Lynch, J. M., and Poole, N. J. (eds.), 1979, *Microbial Ecology: A Conceptual Approach,* Blackwell Scientific Publications, Oxford.

Macaskie, L. E., and Dean, A. C. R., 1982, Cadmium accumulation by microorganisms, *Environ. Technol. Lett.* **3**:49–56.

MacFadyen, A., 1971, The soil and its total metabolism, in: *Methods of Study in Quantitative Soil Ecology: Population, Production and Energy Flow* (J. Phillipson, ed.), pp. 1–13, Blackwell Scientific Publications, Oxford.

Marsh, P., 1980, *Oral Microbiology,* Thomas Nelson and Sons, Walton-on-Thames.

Marshall, K. C., 1971, Sorptive interactions between soil particles and microorganisms, in: *Soil Biochemistry,* Vol. 2 (A. D. McLaren and J. Skujins, eds.), pp. 409–445, Marcel Dekker, New York.

Marshall, K. C., 1975, Clay mineralogy in relation to survival of soil bacteria, *Annu. Rev. Phytopathol.* **13**:357–373.

Marshall, K. C., 1976, *Interfaces in Microbial Ecology,* Harvard University Press, Cambridge, Massachusetts.

Marshall, K. C., 1978, The effects of surfaces on microbial activity, in: *Water Pollution Microbiology,* Vol. 2 (R. Mitchell, ed.), pp. 51–70, Wiley, New York.

Marshall, K. C., and Marshman, N. A., 1978, The role of interfaces in soil microenvironments, in: *Environmental Biogeochemistry and Geomicrobiology,* Vol. 2: *The Terrestrial Environment* (W. E. Krumbein, ed.), pp. 611–618, Ann Arbor Science Publishers, Ann Arbor, Michigan.

Mayfield, C. I., Inniss, W. E., and Sain, P., 1980, Continuous culture of mixed sediment bacteria in the presence of mercury, *Water Air Soil Pollut.* **13**:335–349.

McBrayer, J. F., Reichle, D. E., and Witkamp, M., 1974, Energy flow and nutrient cycling in a cryptozoan food-web, Oak Ridge National Laboratory.

Newsome, A. L., and Wilhelm, W. E., 1983, Inhibiton of *Naegleria fowleri* by microbial iron-chelating agents: Ecological implications, *Appl. Environ. Microbiol.* **45**:665–668.

Nordgren, A., Bååth, E., and Söderström, B., 1983, Microfungi and microbial activity along a heavy metal gradient, *Appl. Environ. Microbiol.* **45**:1829–1837.

Olson, B. H., and Thornton, I., 1982, The resistance patterns to metals of bacterial populations in contaminated land, *J. Soil Sci.* **33**:271–277.

Ortner, P. B., Kreader, C., and Harvey, G. R., 1983, Interactive effects of metals and humus on marine phytoplankton carbon uptake, *Nature* **301**:57–59.

Ottaway, J. H., 1980, *The Biochemistry of Pollution,* Edward Arnold, London.

Pan-Hou, H. S., Nishimoto, M., and Imura, N., 1981, Possible role of membrane proteins in mercury resistance of *Enterobacter aerogenes, Arch. Microbiol.* **130**:93–95.

Parkinson, D., Gray, T. R. G., and Williams, S. T., 1971, *Methods for Studying the Ecology of Soil Microorganisms,* Blackwell Scientific Publications, Oxford.

Pedersen, D., and Sayler, G. S., 1981, Methanogenesis in freshwater sediments: Inherent variability and effects of environmental contaminants, *Can. J. Microbiol.* **27**:198–205.

Pickaver, A. H., and Lyes, M. C., 1981, Aerobic microbial activity in surface sediments containing high or low concentrations of zinc taken from Dublin Bay, Ireland, *Estuarine Coastal Shelf Sci.* **12**:13–22.

Pirt, S. J., 1975, *Principles of Microbe and Cell Cultivation,* Blackwell Scientific Publications, Oxford.

Porter, J. R., 1983, Variation in the relationship between nitrogen fixation, leghaemoglobin, nodule numbers and plant biomass in alfalfa *(Medicago sativa)* caused by treatment with arsenate, heavy metals and fluoride, *Physiol. Plant.* **57**:579–583.

Porter, J. R., and Sheridan, R. P., 1981, Inhibition of nitrogen fixation in alfalfa by arsenate, heavy metals, fluoride, and simulated acid rain, *Plant Physiol.* **68**:143–148.

Radford, A. J., Oliver, J., Kelly, W. J., and Reanney, D. C., 1981, Translocatable resistance to mercuric and phenyl mercuric ions in soil bacteria, *J. Bacteriol.* **147**:1110–1112.

Rai, L. C., Gaur, J. P., and Kumar, H. D., 1981, Phycology and heavy metal pollution, *Biol. Rev. Camb. Philos. Soc.* **56**:99–151.

Ramamoorthy, S., and Kushner, D. J., 1975, Binding of mercuric and other heavy metal ions by microbial growth media, *Microb. Ecol.* **2**:162–176.

Reanney, D. C., Gowland, P. C., and Slater, J. H., 1983, Genetic interactions among microbial communities, in: *Microbes in Their Natural Environments* (J. H. Slater, R. Whittenbury, and J. W. T. Wimpenny, eds.), pp. 379–421, Cambridge University Press.

Remacle, J., Houba, C., and Ninane, J., 1982, Cadmium fate in bacterial microcosms, *Water Air Soil Pollut.* **18**:455–466.

Rhoden, E. G., and Allen, J. R., 1982, Effect of B, Mn and Zn on nodulation and N_2-fixation in southern peas, *Commun. Soil Sci. Plant Anal.* **13**:243–258.

Rolla, G., 1980, On the chemistry of the matrix of dental plaque, in: *Microbial Adhesion to Surfaces* (R. C. W. Berkeley, J. M. Lynch, J. Melling, P. R. Rutter, and B. Vincent, eds.), pp. 425–439, Ellis Horwood, Chichester, England.

Rosenzweig, W. D., and Pramer, D., 1980, Influence of cadmium, zinc, and lead on growth, trap formation, and collagenase activity of nematode-trapping fungi, *Appl. Environ. Microbiol.* **40**:694–696.

Rother, J. A., Millbank, J. W., and Thornton, I., 1982, Seasonal fluctuations in nitrogen fixation (acetylene reduction) by free-living bacteria in soils contaminated with cadmium, lead and zinc, *J. Soil Sci.* **33**:101–113.

Russell-Hunter, W. D., 1970, *Aquatic Productivity: An Introduction to Some Basic Aspects of Biological Oceanography and Limnology,* Collier-Macmillan, London.

Sanders, J. G., Ryther, J. H., and Batchelder, J. H., 1981a, Effects of copper, chlorine, and thermal addition on the species composition of marine phytoplankton, *J. Exp. Mar. Biol. Ecol.* **49**:81–102.

Sanders, J. G., Batchelder, J. H., and Ryther, J. H., 1981b, Dominance of a stressed marine phytoplankton assemblage by a copper-tolerant pennate diatom, *Bot. Mar.* **24**:39–41.

Saxena, J., and Howard, P. H., 1977, Environmental transformation of alkylated and inorganic forms of certain metals, *Adv. Appl. Microbiol.* **21**:185–226.

Schenck, S., Chase, T., Jr., Rosenzweig, W. D., and Pramer, D., 1980, Collagenase production by nematode-trapping fungi, *Appl. Environ. Microbiol.* **40**:567–570.

Sheath, R. G., Havas, M., Hellebust, J. A., and Hutchinson, T. C., 1982, Effects of long-term natural acidification on the algal communities of tundra ponds and Smoking Hills, N.W.T., Canada, *Can. J. Bot.* **60**:58–72.

Shukla, U. C., and Yadav, O. P., 1982, Effect of phosphorus and zinc on nodulation and nitrogen fixation in chickpea (*Cicer arietinum* L.), *Plant Soil* **65**:239–248.

Spain, J. C., Pritchard, P. H., and Bourquin, A. W., 1980, Effects of adaptation on biodegradation rates in sediment/water cores from estuarine and freshwater environments, *Appl. Environ. Microbiol.* **40**:726–734.

Stanisich, V. A., Bennett, P. M., and Richmond, M. H., 1977, Characterization of a translocation unit encoding resistance to mercuric ions that occurs on a nonconjugative plasmid in *Pseudomonas aeruginosa, J. Bacteriol.* **129**:1227–1233.

Stevenson, L. H., 1978, A case for bacterial dormancy in aquatic systems, *Microb. Ecol.* **4**:127–133.

Strojan, C. L., 1978a, Forest leaf litter decomposition in the vicinity of a zinc smelter, *Oecologia (Berl.)* **32**:203–212.

Strojan, C. L., 1978b, The impact of zinc smelter emissions on forest litter arthropods, *Oikos* **31**:41–46.

Sugarman, B., 1980, Effect of heavy metals on bacterial adherence, *J. Med. Microbiol.* **13**:351–354.

Summers, A. O., and Silver, S., 1978, Microbial transformations of metals, *Annu. Rev. Microbiol.* **32**:637–672.

Sunda, W. G., Barber, R. T., and Huntsman, S. A., 1981, Phytoplankton growth in nutrient rich seawater: Importance of copper–manganese cellular interactions, *J. Mar. Res.* **39**:567–586.

Swallow, K. C., Hume, D. N., and Morel, F. M. M., 1980, Sorption of copper and lead by hydrous ferric oxide, *Environ. Sci. Technol.* **14**:1326–1331.

Swift, M. J., Heal, O. W., and Anderson, J. M., 1979, *Decomposition in Terrestrial Ecosystems,* Blackwell Scientific Publications, Oxford.

Takakuwa, S., and Wall, J. D., 1981, Enhancement of hydrogenase activity in *Rhodopseudomonas capsulata* by nickel, *FEMS Microbiol. Lett.* **12**:359–363.

Tan, T. L., 1980, Effect of long-term lead exposure on the seawater and sediment bacteria from heterogeneous continuous flow cultures, *Microb. Ecol.* **5**:295–311.

Tempest, D. W., Neijssel, O. M., and Zevenboom, W., 1983, Properties and performance of microorganisms in laboratory culture: Their relevance to growth in natural ecosystems, in: *Microbes in Their Natural Environments* (J. H. Slater, R. Whittenbury, and J. W. T. Wimpenny, eds.), pp. 119–152, Cambridge University Press.

Tetaz, T. J., and Luke, R. K. J., 1983, Plasmid-controlled resistance to copper in *Escherichia coli, J. Bacteriol.* **154**:1263–1268.

Timoney, J. F., Port, J., Giles, J., and Spanier, J., 1978, Heavy-metal and antibiotic resistance in the bacterial flora of sediments of New York Bight, *Appl. Environ. Microbiol.* **36**:465–472.

Titus, J. A., Parsons, J. E., and Pfister, R. M., 1980, Translocation of mercury and microbial adaptation in a model aquatic system, *Bull. Environ. Contam. Toxicol.* **25**:456–464.

Tso, W.-W., and Adler, J., 1974, Negative chemotaxis in *Escherichia coli, J. Bacteriol.* **118**:560–576.

Tyler, G., 1981, Heavy metals in soil biology and biochemistry, in: *Soil Biochemistry,* Vol. 5 (A. E. Paul and J. N. Ladd, eds.), pp. 371–414, Marcel Dekker, New York.

Vaccaro, R. F., Azam, F., and Hodson, R. E., 1977, Response of natural marine bacterial populations to copper: Controlled ecosystem pollution experiment, *Bull. Mar. Sci.* **27**:17–22.

Van Hook, R. I., Harris, W. F., and Henderson, G. S., 1977, Cadmium, lead and zinc distributions and cycling in a mixed deciduous forest, *Ambio* **6:**281–286.

Varon, M., and Shilo, M., 1981, Inhibition of the predatory activity of *Bdellovibrio* by various environmental pollutants, *Microb. Ecol.* **7:**107–111.

Walsh, F., and Mitchell, R., 1974, Inhibition of intermicrobial predation by chlorinated hydrocarbons, *Nature* **249:**673–674.

Whittaker, R. H., 1975, *Communities and Ecosystems,* 2nd ed., Macmillan, New York.

Wickliff, C., and Evans, H. J., 1980, Effect of cadmium on the root and nodule ultrastructure of *Alnus rubra, Environ. Pollut. (Series A)* **21:**287–306.

Wickliff, C., Evans, H. J., Carter, K. R., and Russell, S. A., 1980, Cadmium effects on the nitrogen fixation system of red alder, *J. Environ. Qual.* **9:**180–184.

Wieser, W., and Wiest, C., 1968, Ökologische Aspekte des Kupferstoffwechsels terrestrischer Isopoden, *Oecologia (Berl.)* **1:**38–48.

Wildung, R. E., and Garland, T. R., 1982, Effects of plutonium on soil microorganisms, *Appl. Environ. Microbiol.* **43:**418–423.

Williams, S. E., and Wollum, A. G., II, 1981, Effect of cadmium on soil bacteria and actinomycetes, *J. Environ. Qual.* **10:**142–144.

Williams, S. T., McNeilly, T., and Wellington, E. M. H., 1977, The decomposition of vegetation growing on metal mine waste, *Soil Biol. Biochem.* **9:**271–275.

Witkamp, M., 1973, Compatibility of microbial measurements, *Bull. Ecol. Res. Commun. (Stockholm)* **17:**179–188.

Wood, J. M., 1974, Biological cycles for toxic elements in the environment, *Science* **183:**1049–1052.

Wright, D. A., 1978, Heavy metal accumulation by aquatic invertebrates, *Appl. Biol.* **3:**331–394.

Yatazawa, M., Tomomatsu, N., Hosoda, N., and Nunome, K., 1980, Nitrogen fixation in *Azolla–Anabaena* symbiosis as affected by mineral nutrient status, *Soil Sci. Plant Nutr. (Tokyo)* **26:**415–426.

Young, F. E., and Spizizen, J., 1963, Incorporation of deoxyribonucleic acid in the *Bacillus subtilis* transformation system, *J. Bacteriol.* **86:**392–400.

Zaske, S. K., Dockins, W. S., and McFeters, G. A., 1980, Cell envelope damage in *Escherichia coli* caused by short-term stress in water, *Appl. Environ. Microbiol.* **40:**386–390.

Zevenhuizen, L. P. T. M., Dolfing, J., Eshuis, E. J., Scholten-Koerselman, I. J., 1979, Inhibitory effects of copper on bacteria related to the free ion concentration, *Microb. Ecol.* **5:**139–146.

6

Ecology of Microbial Cellulose Degradation

LARS G. LJUNGDAHL and KARL-ERIK ERIKSSON

1. Introduction

Worldwide photosynthetic fixation of carbon dioxide is estimated to yield annually up to 150×10^9 tons of dry plant material (biomass) (Lieth, 1973; Whittaker and Likens, 1973; Bassham, 1975; Stephens and Heichel, 1975). Almost half of this material consists of cellulose (28–50%); other major components are hemicelluloses (20–30%) and lignin (18–30%) (Thompson, 1983). Additional important but minor constituents of biomass are proteins, lipids, and carbohydrates such as chitin, starch, and pectin. A list of the amounts of cellulose in various plants has been compiled by Stephens and Heichel (1975). The biomass is eventually degraded and oxidized to CO_2 and returned to the atmosphere. Thus, we have a carbon cycle. The gain of the cycling process is the capture of solar energy, which is then used by living cells for growth and maintenance.

The degradation of cellulose is almost exclusively by microbiological processes (Swift, 1977; Fenchel and Jørgensen, 1977). However, it must be understood that cellulose is closely associated with hemicellulose and lignin as well as other substances, and that interaction occurs among the degradative processes of different constituents of biomass (Kirk, 1973; Rosenberg, 1978; Zeikus, 1981; Eriksson, 1981a). These processes are

LARS G. LJUNGDAHL • Center for Biological Resource Recovery, Department of Biochemistry, University of Georgia, Athens, Georgia 30602. KARL-ERIK ERIKS-SON • Swedish Forest Products Research Laboratory, S-11486, Stockholm, Sweden.

also dependent on minerals and compounds such as sulfate, nitrate, and ammonia present in the environment of the biomass degradation (Jeffries *et al.*, 1981; Reid, 1983a,b; Aumen *et al.*, 1983). The microbial degradation of cellulose is a complicated process. It is now being studied extensively due to the possibility of converting cellulose with the aid of microorganisms to useful industrial products such as ethanol, acetic acid, acetone, butanol, and protein.

Most of the cellulose in the natural environment is oxidized by aerobic microorganisms to CO_2 according to the reaction

$$C_6H_{12}O_6 + 6O_2 \rightarrow 6CO_2 + 6H_2O \tag{1}$$

However, a substantial amount of cellulose (5–10%) is converted in the global methane cycle (Ehhalt, 1976; Vogels, 1979) to methane and CO_2, as summarized in the reaction

$$C_6H_{12}O_6 \rightarrow 3CH_4 + 3CO_2 \tag{2}$$

According to Vogels (1979), $(0.55–1.2) \times 10^9$ tons/year of methane is formed from cellulose in the anaerobic ecosystem and enters the atmosphere, where it is oxidized in reactions with ozone to CO_2 and water.

In this chapter we discuss microorganisms involved in the aerobic and anaerobic degradations of cellulose, interactions among these microorganisms, and cellulolytic enzyme systems. We begin with a brief introductory discussion of cellulose structure.

2. Structure of Cellulose

The structure and biosynthesis of cellulose are not completely known. Both subjects are being intensely investigated in several laboratories [see the books edited by Brown (1982) and Bikales and Segal (1971)]. Although these subjects are somewhat outside the realm of this chapter, it will be useful to review some salient points.

Cellulose is a homopolymer consisting of glucose moieties joined together in β-1,4 linkages, as illustrated for the disaccharide cellobiose in Fig. 1. However, in cellobiose the two glucose residues are joined not only by the β-1,4 linkage, but also by a hydrogen bond between the C-3 hydroxyl group of one glucose residue and the ring oxygen of the other glucose moiety (Chu and Jeffrey, 1968). In methyl β-cellobioside, the C-6 hydroxyl as well as the ring oxygen are involved in a bifurcated hydrogen bond to the hydroxyl group of C-3 of the other glucose residue (Ham and Williams, 1970). Thus, cellobiose and methyl β-cellobioside (Fig. 2)

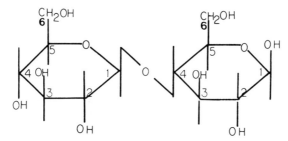

Figure 1. Structure of cellobiose.

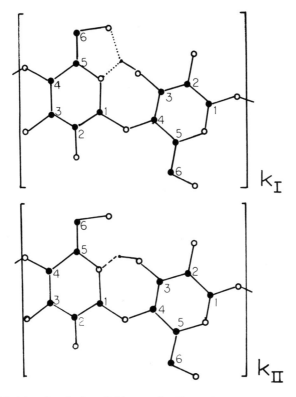

Figure 2. Models of anhydrocellobiose units K_I and K_{II}, according to Atalla and VanderHart (1984), based on stable conformations of cellobiose with one isolated intramolecular hydrogen bond K_{II} (Chu and Jeffrey, 1968) and of methyl β-cellobioside with a bifurcated intramolecular hydrogen bond K_I. [Ham and Williams (1970).]

represent two types of stable linkage between glucose moieties. It should be pointed out that the C-6 involved in the bifurcated hydrogen bond in methyl β-cellobioside is not equivalent to the same C-6 of cellobiose. The movement of the former is restricted in comparison to the latter (Atalla, 1979).

The size of the cellulose molecule is often given in terms of the degree of polymerization (DP), which is the number of glucose moieties in the molecule. Native cellulose from higher plants seems to have a DP of 14,000 (Marx-Figini and Schulz, 1966). This value appears to be remarkably constant for plant cellulose, which indicates that the DP of the cellulose (glucan) chain is controlled by the plant when it is synthesized (Marx-Figini, 1982). The DP in cellulose from bacteria is lower; a value of 3470 has been obtained for cellulose synthesized by *Acetobacter xylinum* (Marx-Figini, 1982). Again, a control mechanism is evident. Commercially available celluloses have DP varying from 50 to 5000.

Based on x-ray and infrared data, the anhydrocellobiose ($C_{12}H_{20}O_{10}$) rather than the anhydroglucose unit ($C_6H_{10}O_5$) seems to be the basic structure of cellulose (Tønesen and Ellefsen, 1971). The shortest cellulose-like molecule would then be represented by cellotetrose, DP = 4,

$$(C_6H_{11}O_5)\cdot(C_{12}H_{20}O_{10})\cdot(C_6H_{11}O_6)$$

in which ($C_6H_{11}O_5$) represents the glucose moiety with a free secondary hydroxyl group on C-4, ($C_{12}H_{20}O_{10}$) is an anhydrocellobiose unit, and ($C_6H_{11}O_6$) is the glucose molecule with the reducing group. Blackwell (1982) and Atalla (1983) also consider anhydrocellobiose to be the basic building unit of cellulose. Interestingly, cellobiose is the principal product or an intermediate in the formation of glucose from hydrolysis of cellulose by both fungal and bacterial cellulase systems. Thus, the cellulolytic enzymes also appear to recognize the cellobiose unit of cellulose.

Cellobiose, cellotriose, and cellotetrose are soluble in water (13.3, 11.7, and 7.5 g/100 ml of water at 25°C, respectively). However, the solubility of cellopentose is only 0.59 g and that of celloheptose is 0.18 g/100 ml of water (Tønnesen and Ellefsen, 1971). Cellulose molecules with a DP higher than 6 can be considered insoluble. This has been interpreted as a result of hydrogen bond formation between cellulose molecules with exclusion of water. Thus, hydrogen bonding in cellulose occurs between glucose moieties in the same glucan chain as well as between chains. Hydrophobic (van der Waals) interactions between chains also occur.

Cellulose is polymorphic and at least four crystalline forms (cellulose I–IV) have been described. They differ in x-ray diffraction patterns and in spectra (Blackwell, 1982; Atalla and VanderHart, 1984). Cellulose I is native cellulose from plants. Cellulose II is formed from cellulose I by mercerizing, which involves treatment of cellulose I with NaOH for 1–2 h at 18°C followed by removal of the alkali. Treatment of celluloses I and II with liquid ammonia or ammonia vapor at 50°C yields cellulose III, and exposure of celluloses I and II to glycerol at 270–280°C for 1–2 hr results in the formation of cellulose IV (Wellard, 1954). Celluloses III and IV derived from cellulose I are often designated celluloses III_I and IV_I. Similarly, when derived from cellulose II they are designated celluloses III_{II} and IV_{II} (Blackwell, 1982). Transitions are possible between the different types of cellulose (Ellefsen and Tønnesen, 1971). However, the formation of cellulose I from cellulose II has not been observed, which is an indication that the structure of cellulose II is more stable than that of cellulose I (Blackwell, 1982). This, in turn, implies that during the synthesis of cellulose, cells direct the crystallization in such a way that cellulose I is formed. This process and the formation of cellulose fibers are discussed by Mueller (1982), Vian (1982), Marx-Figini (1982), Haigler and Benziman (1982), and Carpita (1982) and in a summary by Brown (1981).

In addition to crystalline forms, there is amorphic cellulose, and it has been suggested that in indigenous cellulose fibers, amorphic regions exist, although the fiber cores are highly crystalline, as indicated by early observations of Norkrans (1950) and Cowling (1963) and several more recent observations (Caulfied and Moore, 1974; Hurst *et al.,* 1978; Fan *et al.,* 1980; Ohmine *et al.,* 1983) that hydrolysis of amorphous cellulose occurs faster than that of crystalline portions with cellulase preparations.

Cellulose molecules form elementary fibrils through intramolecular hydrogen bonding, and these fibrils in turn form microfibrils and then fibers. The structures of and interactions among cellulose molecules in celluloses I and II have been intensely investigated [for reviews see, e.g., Tønnesen and Ellefsen (1971), Blackwell (1982), and Atalla (1983)]. There is a great deal of controversy about the subject. Blackwell (1982), referring to work in his laboratory (Gardner and Blackwell, 1974; Kolpak and Blackwell, 1976) and by Sarko and Muggli (1974) and Stipanovic and Sarko (1976), proposed that cellulose I consists of staggered parallel (polarity in the same direction) chains, whereas in cellulose II the chains are antiparallel (chains have polarity "up" or "down"). However, Atalla (1983) and Atalla and VanderHart (1984) proposed, based on the cross-polarization, magic angle spinning (CP-MAS) technique, that native celluloses are composites of two distinct crystalline forms named I_α and I_β. In these forms the basic anhydrocellobiose unit may have the same bindings as that of methyl β-cellobioside (Fig. 2). The I_α crystalline form is dominant in celluloses produced by bacteria, such as *Acetobacter xylinum,* and *Valonia venticosa,* whereas the I_β form is dominant in plant cellulose. Cellulose II, on the other hand, may consist of anhydrocellobiose units of the structure corresponding to cellobiose (Fig. 2). It is well known that celluloses of different origins are attacked by cellulolytic enzymes at different rates. This suggest that celluloses have different crystalline forms or structures, which are recognized by the enzymes. The understanding of cellulose structure is therefore of paramount importance for the understanding of functions of cellulolytic enzymes.

3. Cellulolytic Microorganisms

Microorganisms that degrade cellulose are abundant in nature. They include both aerobic and anaerobic fungi and bacteria, many of which grow under extreme conditions of temperature and pH. Thus, most habitats include cellulolytic microorganisms. In this section we present representatives of cellulose-degraders of different types and environments. The discussion begins with the fungi, which are being focused upon by many investigators because of the potential industrial use of the fungal cellulolytic enzymes for the degradation of cellulose to produce fermentable carbohydrates. This enormous interest is evidenced in Table I, which is a partial list of fungal species being considered in studies of cellulases.

Table I. Examples of Fungi Considered in Cellulase Studies

Agaricus bisporus	Manning and Wood (1983)
Ascobolus furfuraceus	Wicklow *et al.* (1980)
Aspergillus fumigatus	Rogers *et al.* (1972), Jain *et al.* (1979)
Aspergillus niger	Ikeda *et al.* (1973), Sternberg *et al.* (1977), Hurst *et al.* (1978)
Aspergillus phoenicis	Sternberg *et al.* (1977)
Aspergillus terreus	Sinha *et al.* (1981), Garg and Neelakantan (1982)
Botryodiplodia theosbromae	Umezurike (1979, 1981)
Chaetomium cellulolyticum	Chahal and Hawksworth (1976), Moo-Young *et al.* (1978), Fähnrich and Irrgang (1982)
Chaetomium thermophile	Fergus (1969), Romanelli *et al.* (1975)
Cladosporium cladosporides	Lynch *et al.* (1981)
Coriolus versicolor	Highley (1975a)
Eupenicillium javanicum	Tanaka *et al.* (1981)
Fomes fomentarius	Kozlik and Schanel (1974)
Fusarium sp.	Trivedi and Rao (1980, 1981)
Humicola grisea	Fergus (1969), Araujo *et al.* (1983)
Humicola insolens	Fergus (1969), Jain *et al.* (1979)
Hypocopra merdaria	Wicklow *et al.* (1980)
Irpex lacteus	Kanda *et al.* (1983)
Myceliophthora thermophila	Sen *et al.* (1981)

(*continued*)

Table I. (*Continued*)

Myriococcum albomyces	Fergus (1969)
Myrothecium verrucaria	Whitaker and Thomas (1963)
Neocallimastix frontalis	Mountfort and Asher (1983)
Neurospora crassa	Eberhart *et al.* (1977), Rao *et al.* (1983)
Paecilomyces fusisporus	Mishra *et al.* (1981)
Paecilomyces varioti	Mishra *et al.* (1981)
Papulaspora thermophila	Chapman *et al.* (1975)
Penicillium chrysogenum	Mishra *et al.* (1981)
Penicillium funiculosum	T. M. Wood *et al.* (1980), Deshpande *et al.* (1983a), Joglekar *et al.* (1983)
Penicillium verruculosum	Szakács *et al.* (1981)
Pestalotiopsis versicolor	Thakur and Sastry (1981), Rao *et al.* (1983b)
Phanerochaete chrysosporium	Reid (1983a,b)
Phialophora malorum	Berg (1978)
Phoma hibernica	Urbanek *et al.* (1978)
Physarum polycephalum	Koevenig and Liu (1981)
Pleurotus ostreatus	Kozlik and Schanel (1974)
Pleurotus sajor-caju	Madan and Bisaria (1983)
Podospora decipiens	Wicklow *et al.* (1980)
Polyporus versicolor	Highley (1973)
Poria placenta	Highley (1980), Highley *et al.* (1981)
Poronia punctata	Wicklow *et al.* (1980)
Saccobolus truncatus	Wicklow *et al.* (1980)
Schizophyllum commune	Desrochers *et al.* (1981), Rho *et al.* (1982), Yaguchi *et al.* (1983)
Sclerotinia libertiana	Tanaka *et al.* (1982)
Sclerotium rolfsii	Mishra *et al.* (1981)
Sordaria fimicola	Wicklow *et al.* (1980)
Sporotrichum pulverulentum	Streamer *et al.* (1975), Deshpande *et al.* (1978), Eriksson and Hamp (1978), Ruel *et al.* (1981), Tanaka *et al.* (1982)
Sporotrichum thermophile	Fergus (1969), Romanelli *et al.* (1975), Margaritis and Merchant (1983)
Talaromyces emersonii	Folan and Coughlan (1981), McHale and Coughlan (1981b), Moloney *et al.* (1983)
Thermoascus aurantiacus	Romanelli *et al.* (1975), Tong *et al.* (1980), Shepherd *et al.* (1981)
Thraustotheca clavata	Berner and Chapman (1977)
Torula thermophile	Fergus (1969), Jain *et al.* (1979)
Trichoderma koningii	Iwasaki *et al.* (1964), T. M. Wood and McCrae (1978), Halliwell and Vincent (1981)
Trichoderma pseudokoningii	Zhu *et al.* (1982), Harrer *et al.* (1983)
Trichoderma reesei (viride)	Mandels and Reese (1960), Berghem *et al.* (1976), Sternberg and Mandels (1979), White and Brown (1981), Montenecourt *et al.* (1981)
Trichurus spiralis	Mishra *et al.* (1981)
Verticillium albo-atrum	D. P. Gupta and Heale (1971), Heale and Gupta (1971)

3.1. Properties of Wood-Rotting Fungi

In nature there is a continuous degradation of dead plant material by saprophytic microorganisms. The most important of these organisms for the degradation of wood are three different types of rot fungi, namely the white-rot, brown-rot, and soft-rot fungi. However, rarely, if ever, is wood degradation in nature caused by a monoculture of a fungus. Instead there is a succession of microorganisms colonizing dead wood (Käärik, 1974a). Garrett (1963) divided this succession into three stages: (1) The wood tissues are attacked by so-called primary saprophytic fungi, which metabolize the low-molecular-weight sugars and other compounds in wood that are easier to degrade than cellulose; (2) cellulose-decomposing fungi dominate, associated with so-called secondary saprophytic fungi, which live on cellulose degradation products; (3) lignin-degrading and associated fungi become active [for a discussion of lignin degradation see Zeikus (1981)].

As was pointed out by Käärik (1974a), however, the succession of microorganisms in decaying wood is often not this simple, and is strongly influenced by the way in which the wood is killed and exposed to microorganisms. What particularly influences both the degradation rate and the succession of microorganisms is whether the dead wood is in contact with the soil or not. There is also a marked difference between fungi attacking the above-soil and subsoil sections of decaying wood.

Wood, as such, is not a homogeneous material. It can, for instance, be divided into heartwood and sapwood. It is generally recognized that heartwood is more resistant to microbial attack than is sapwood. This resistance of the heartwood is due to the presence of chemical compounds that, at least in certain wood species, are inhibitory to microbial activity. Erdtman (1939), Rennerfelt (1944), and Rudman and DaCosta (1958) have isolated a large number of phenolic and quinonic compounds as well as tropolones from conifer woods. Käärik (1974b) gives a list of substances toxic to wood-rotting fungi.

The three main components of wood are cellulose, hemicellulose, and lignin. These components are attacked in different ways by different microorganisms. Fungi are the main degraders, and, as already noted, wood-degrading fungi are divided into different groups of rot fungi. A description of the way in which these groups of fungi attack wood and the various wood components is given below.

3.1.1. White-Rot Fungi

All white-rot fungi have the capacity to produce extracellular phenol oxidases. Bavendamm (1928) used gallic and tannic acids to differentiate between white-rot and brown-rot fungi. White-rot fungi are a heterogeneous group of fungi and are the only microorganisms that degrade all the wood components to a substantial degree, even the lignin. Most fungi giving a positive Bavendamm test response degrade lignin according to the test method of Sundman and Nase (1971).

White-rot fungi have often been separated into two groups: (1) Fungi degrading the different wood components at approximately the same rate are called simultaneous rot fungi (Liese, 1970), and (2) fungi preferentially degrading lignin, leaving considerable amounts of cellulose, are called white-pocket rot fungi (Blanchette, 1980a,b, 1982; Otjen and Blanchette, 1982). Wood surrounding the white pockets is sound (Blanchette, 1980a, 1982; Otjen and Blanchette, 1982). Formation of these pockets and patterns of selective delignification have been described for *Phellinus pini* in conifers (Blanchette, 1980a, 1982) and for *Ionotus dryophilus* in oaks (Otjen and Blanchette, 1982).

White-rot fungi, at least the simultaneous rot fungi, in contrast to brown-rot fungi, depolymerize wood polysaccharides only to the extent that the products are simultaneously utilized. This indicates that energy to degrade lignin must be derived from more easily accessible energy sources, such as wood polysaccharides and low-molecular-weight sugars (Ander and Eriksson, 1976a,b; Hiroi and Eriksson, 1976; Kirk *et al.*, 1976; Jeffries *et al.*, 1981). Alternatively, since hydrogen peroxide seems to be necessary for lignin degradation, it may be that the sugars are required as substrates only for the production of hydrogen peroxide, which is formed for instance when glucose is oxidized to gluconic acid by glucose oxidase.

The extracellular enzymes produced by white-rot fungi seem to carry out their attack in the immediate vicinity of the fungal hyphae and cause erosion troughs (Blanchette *et al.*, 1978; Eriksson *et al.*, 1980a,b; Ruel *et al.*, 1981).

Recent studies by Blanchette and Shaw (1978) indicate that yeasts as well as bacteria influence the rate of decay by wood-destroying basidiomycetes. When the yeast *Cryptococcus albidus* var *albidus* was combined with the white-rot fungus *Coriolus versicolor* wood decay was substantially increased. After 7 months of combined action the weight loss was 20.8% compared to 7.5% for the fungus alone (Blanchette *et al.*, 1978).

3.1.2. Brown-Rot Fungi

Brown-rot fungi generally belong to the *Basidiomycetes* and mainly decompose the wood polysaccharides. Kirk and Highley (1973) studied the degradation of five conifer woods by three different brown-rot fungi.

They found that the lignin content was generally only slightly reduced by these fungi.

The brown-rot fungal hyphae are normally localized in the wood cell lumen and penetrate adjacent cells either through existing openings or by producing bore holes in the wood cell walls. In early stages of decay, and contrary to white-rot fungi, brown-rot fungi depolymerize cellulose faster than the degradation products are utilized.

During the decay of wood the removal of cell wall substances by brown-rot fungi begins in the S_2 layer of the secondary wall. Due to their high lignin content (Sarkanen and Hegert, 1971), the primary wall and the middle lamella are very resistant to degradation by brown-rot fungi (Wilcox, 1970). In advanced stages of decay, when most of the polysaccharides are consumed, the cell wall collapses.

Brown-rot fungi growing in wood seem to produce a diffusable factor that is involved in cellulose degradation at considerable distances from the fungal hyphae (Cowling, 1961; Liese, 1970; Wilcox, 1970). Although brown-rot fungi degrade cellulose in wood, they do not seem to degrade pure, isolated cellulose (Nilsson, 1974a; Highley, 1977).

It was suggested by Eriksson (1981a,b) that brown-rot fungi must have access to a low-molecular-weight sugar before the attack on the cellulose can be started. This sugar is oxidized under formation of H_2O_2, which seems necessary for degradation of at least crystalline cellulose by these organisms (Koenigs, 1974a,b).

3.1.3. Soft-Rot Fungi

The term soft-rot derives from the softening of wood caused by this group of fungi, which are found among the *Ascomycetes* and *Fungi Imperfecti* species. Soft-rot fungi principally attack the carbohydrates in wood; the lignin is modified only to a limited extent (Seifert, 1968). The soft-rot fungi first grow in the wood cell lumen, but then invade the secondary wall, where they grow parallel to the fiber axes. During this growth, cylindrical cavities with conical ends are developed (Nilsson, 1974b). Some soft-rot fungi also cause an erosion of the cell walls starting from the cell lumen.

For detailed information concerning wood decay by soft-rot fungi, see Wilcox (1973) and Käärik (1974b).

3.2. Anaerobic Cellulolytic Fungi

It has long been recognized that the rumen contains protozoa [see reviews by Hungate (1966) and Wolin (1979)]; the rumen flagellates *Neocallimastix frontalis, Sphaeromonas communis,* and *Piromonas commu-*

nis have been isolated. Recently, these flagellates have been found to be species of anaerobic phycomycete fungi (Orpin, 1975, 1977a,b; Heath *et al.,* 1983). These or similar anaerobic fungi have been found to degrade carbohydrates, including cellulose (Orpin and Letcher, 1979; Bauchop and Mountfort, 1981; Orpin, 1981).

The fermentation of cellulose by a rumen anaerobic fungus resulted in the formation of acetate, carbon dioxide, formate, ethanol, lactate, and hydrogen. This fungus was also cocultured with a rumen methanogen and with the stable coculture the major products were acetate, carbon dioxide, and methane, with no or very low production of hydrogen, formate, or lactate (Bauchop and Mountfort, 1981). In coculture of the anaerobic fungus with *Methanosarcina barkeri* and *Methanobacter* sp., methane and carbon dioxide were the major products (Bauchop and Mountfort, 1981). These results show that the methanogens utilize electrons produced in the fermentation of cellulose by the fungus to form methane and also that the acetate is converted to carbon dioxide and methane by *Methanosarcina barkeri.* These results also indicate that cellulolytic anaerobic fungi may play an important role in cellulose degradation in the rumen. However, the cellulolytic enzyme systems in these fungi have not yet been studied.

That these fungi are involved in degradation of fiber material in the rumen is substantiated by electron microscopic investigations, which demonstrate that the fungi develop preferentially on lignocellulosic materials to which they adhere (Orpin, 1977c; Bauchop, 1981; Akin *et al.,* 1983; Akin and Barton, 1983). A review of these studies has been prepared by Akin (1985).

Clearly, the demonstration of the degradation of lignocellulosic material in the rumen by anaerobic fungi is a significant finding. It leads to the question of whether anaerobic fungi are also present in other anaerobic habitats, such as soil and mud.

3.3. Cellulolytic Enzyme Systems in Fungi

3.3.1. White-Rot Fungi

The enzyme mechanisms involved in cellulose degradation have been particularly well studied in two fungi, namely the white-rot fungus *Sporotrichum pulverulentum* (Eriksson, 1981a) and the mold *Trichoderma reesei* (Ryu and Mandels, 1980) (the fungus *T. viride* QM6a and strains derived from it are now referred to as *T. reesei*). The pattern of hydrolytic enzymes used by these two organisms for the hydrolysis of cellulose is similar. The cellulolytic enzyme system is composed of: (1) endo-1,4-β-glucanases, which randomly split 1,4-β-glucosidic linkages along the cellulose chain, (2) exo-1,4-β-glucanases, which split off either cellobiose or glucose from the nonreducing end of the cellulose, and (3) 1,4-β-glucosidases, which hydrolyze cellobiose and water-soluble cellodextrins to glucose. Amorphous cellulose can be degraded by both endo- and exoglucanases separately (Eriksson and Johnsrud, 1982). However, to degrade crystalline cellulose cooperative action between these two types of enzymes seems necessary (Streamer *et al.,* 1975). Crystalline cellulose is not attacked by either of these enzymes separately.

In *S. pulverulentum* an oxidative enzyme of importance in cellulose degradation has been discovered in addition to the hydrolytic enzymes described above (A. R. Ayers *et al.,* 1978). The enzyme has been purified and characterized and found to be a cellobiose oxidase, which oxidizes cellobiose and higher cellodextrins to their corresponding onic acids using molecular oxygen. It was recently reported by Vaheri (1982) that culture solutions of *T. reesei* grown on cellulose contain gluconic and cellobionic acids. These findings indicate that *T. reesei* also produces an oxidative enzyme involved in cellulose degradation.

In addition to the described cellobiose oxidase, *S. pulverulentum* has another pathway for the oxidation of cellobiose, namely through the enzyme cellobiose:quinone oxidoreductase (cellobiose dehydrogenase) (Westermark and Eriksson, 1974a,b). This enzyme reduces quinones and phenoxy radicals in the presence of cellobiose and is involved in the degradation of both cellulose and lignin. The quinones and phenoxy radicals are formed by oxidation of phenols from the lignin by phenol oxidases, and cellobiose is formed from cellulose by the hydrolytic enzymes described above. The cellobiose dehydrogenase enzyme is relatively specific for its disaccharide substrate, whereas the requirements on the quinone structure are less specific. A reaction scheme for the enzyme is presented in Figure 3.

The enzyme cellobiose dehydrogenase has been studied in many laboratories and exists in all white-rot fungi so far investigated (Ander and Eriksson, 1977; Kelleher, 1981; Hutterman and Noelle, 1982), the imperfect fungus *Monilia* (Dekker, 1980), and the fungus *Sporotrichum thermophile* (Coudray *et al.,* 1982). Phylogenetically, the latter belongs to the *Ascomycetes.* The properties of all of these enzymes seem to be very similar to those of the well-characterized cellobiose dehydrogenase from *S. pulverulentum* (Westermark and Eriksson, 1974a,b).

Two acidic proteases, proteases I and II, from culture solutions of *S. pulverulentum* have been purified and partly characterized (Eriksson and Pettersson, 1982). These proteases seem to activate the endoglucanases. If *S. pulverulentum* is grown on cellulose and the culture solution is treated with the individual proteases or a mixture of both, the endoglucanase activity in the solution increases approximately tenfold (Fig. 4). A

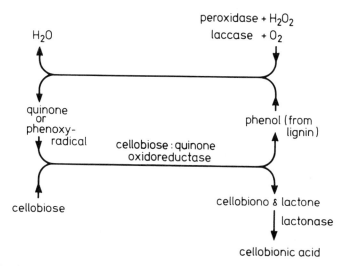

Figure 3. Oxidation of cellobiose through phenol from lignin with cellobiose:quinone oxidoreductase as described for *S. pulverulentum*. [Westermark and Eriksson (1974a,b).]

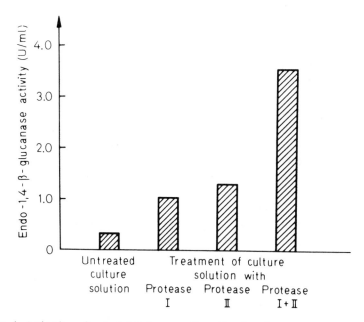

Figure 4. Activation of endo-1,4-β-glucanase from *S. pulverulentum* by acidic proteases I and II from *S. pulverulentum*. [Eriksson and Pettersson (1982).]

possible explanation is that the endoglucanases are activated by the proteases or that the proteases destroy a proteinaceous inhibitor of the endoglucanase. The proteases from *S. pulverulentum* also can increase the activity of endoglucanases from other fungi (Eriksson and Pettersson, 1982).

3.3.2. Brown-Rot Fungi

Brown-rot fungi, which can degrade cellulose extensively, seem to utilize different mechanisms for this degradation than those described for *S. pulverulentum* and *T. reesei*. Brown-rot fungi produce endo-1,4-β-glucanases, but seem to lack exo-1,4-β-glucanases (Highley, 1975a,b). Thus, in brown-rot fungi, crystalline cellulose is not degraded by a synergistic action between endo- and exoglucanases as is the case in *T. reesei* and in *S. pulverulentum*. Koenigs (1974a) found that brown-rot fungi are strong producers of H_2O_2, much more so than white-rot fungi (Koenigs, 1974b). He therefore suggested (Koenigs, 1974a) that the initial attack on crystalline cellulose by brown-rot fungi is via a H_2O_2/Fe^{2+} system. It has been reported by many authors that brown-rot fungi oxidize cellulose (Koenigs, 1974a) and also, when growing in wood, attack the cellulose at a considerable distance from the fungal cell wall (Ander and Eriksson, 1977). It is therefore an attractive hypothesis that the initial attack on cellulose takes place via a chemical agent of low molecular weight that can easily diffuse through the wood fiber walls rather than via an enzyme with limited diffusibility.

In recent experiments at the Swedish Forest Products Research Laboratory it was found that H_2O_2/Fe^{2+} pretreatment of cotton cellulose does not enhance the action of an endoglucanase attack on cellulose. If a mixture of H_2O_2/Fe^{2+} and oxalic acid was used as pretreating agent, however, an increased production of reducing sugars from a thus pretreated cotton cellulose was obtained when endoglucanases were added to a pretreated cotton fiber suspension.

Recently, Cobb (1982) demonstrated that a nonenzymatic cellulose decay mechanism may exist in brown-rot fungi. An ultrafiltration membrane that did not allow passage of enzymes was placed between a fungal culture and its cellulose substrate. A fungal culture grown on hemicellulose or monosaccharides decomposed the cellulose substrate to a greater extent than did a fungal culture grown on ball-milled pine, which is a poor substrate for H_2O_2 production. However, much more work is needed before the mechanism for cellulose degradation by brown-rot fungi is understood.

3.3.3. Soft-Rot Fungi

Although soft-rot fungi are capable of degrading crystalline, highly ordered cellulose, only a few of the studied species seem to release, at least under laboratory conditions, an extracellular enzyme system with this capability. Species in this group of fungi with a capability to produce a complete cellulase system, i.e., endo- and exoglucanases and β-glucosidases, are *Trichoderma viride, T. reesei* (Ryu and Mandels, 1980), *T. koningii* (Halliwell, 1965; T. M. Wood, 1968), *Penicillium funiculosum* (Selby, 1968; T. M. Wood *et al.,* 1980), and *Fusarium solani* (T. M. Wood and Phillips, 1969).

Cell-free culture filtrates of most other soft-rot fungi contain appreciable amounts of endoglucanases and variable amounts of β-glucosidases, but little, if any, exoglucanase. These fungi are therefore capable of degrading only the more disordered amorphous celluloses. It is not clear why some soft-rot fungi produce a more complete enzyme complex than others; this may be of crucial importance for understanding the mechanism of breakdown of cellulose by these fungi.

3.3.4. Synergistic Action among Different Cellulases

Degradation of highly ordered cellulose in native cotton fibers is almost completely dependent on the synergistic action of the three hydrolytic enzymes endoglucanase, exoglucanase, and β-glucosidase. None of the enzymes acting alone can solubilize cotton to any significant extent (Eriksson and Wood, 1984). Synergism is evident between exoglucanases and endoglucanases, but for maximum rate and extent of hydrolysis all three hydrolytic enzymes must be present at the same time. Synergism between endo- and exoglucanases has been reported in many studies (Eriksson and Wood, 1984).

Hydrolysis of crystalline cellulose is suggested to occur in a sequential attack on the substrate by the different hydrolytic enzymes. Endoglucanases, which attack randomly over the cellulose chain, function first, creating end groups on which the endwise-acting exoglucanases can attack from the nonreducing end of the chains. Since the mode of attack of an endoglucanase from one fungus is similar to the mode of attack of an endoglucanase from another fungus, it seems reasonable to expect that an endoglucanase from one fungus should act synergistically with an exoglucanase from another fungus. This phenomenon has been studied in particular by T. M. Wood and McCrae (1975) and by T. M. Wood (1980), who found that such cooperation is very variable. Synergism is highest between enzymes (endo- and exoglucanases) from different fungi when both fungi produce a so-called complete cellulase complex, i.e., an enzyme complex capable of extensive hydrolysis of crystalline cellulose. Thus, while synergism is high between any combination of the exo- and endoglucanases of *T. koningii, F. solani,* and *P. funiculosum,* it is low when the exoglucanase from any of these fungi is mixed with an endoglucanase from culture filtrates of fungi that do not apparently produce an exoglucanase, for instance, *Myrothecium verrucaria, Stachybotrus atra, Gleocladium roseum,* and *Memnoniella echinata* (T. M. Wood, 1980). In keeping with this is the recent observation (T. M. Wood, 1983) that synergism between the exoglucanases of *T. koningii, F. solani,* and *P. funiculosum* and an endoglucanase from the white-rot fungus *S. pulverulentum* is high. However, there is no such synergism when the exoglucanases act together with endoglucanases of rumen anaerobic bacteria such as *Ruminococcus albus, R. flavefaciens,* and *Bacteroides succinogenes.*

The cellulase systems of rumen bacteria appear to differ substantially from those of fungi, and synergistic interaction between fungal and bacterial cellulase systems may not be feasible. The bacterial cellulases will be described later in this chapter. Thus, it is apparent that not all endo- and exoglucanases are completely interchangeable in their synergistic actions. T. M. Wood (1980) discusses several possible explanations and suggests that the problem can be a stereochemical one. This suggestion has not been verified experimentally. It seems clear that the enzyme degradation of highly ordered crystalline cellulose is a quite complicated and not completely understood process, which not only requires several enzymes of different properties, but also that these enzymes act in synergy.

3.3.5. Regulation of Cellulolytic Enzymes of Fungi

The induction and repression phenomena of enzymes involved in cellulose hydrolysis have, like the properties of the enzymes, been best studied in the white-rot fungus *S. pulverulentum* and the soft-rot fungus *T. reesei.* In particular, the endoglucanases have been studied in this respect (Eriksson and Hamp, 1978). It has been found that in growth media containing carboxymethylcellulose (CMC), *S. pulverulentum* forms endoglucanases. This is not the case, however, with a *T. reesei* strain (Eriksson and Hamp, 1978). In the presence of cellobiose at a concentration as low as 1 mg/liter a much more rapid induction of endoglucanases takes place in *S. pulverulentum* compared to that of the induction in the CMC solution alone. Cellobiose at the same or even high concentrations does not induce any endoglucanase activity in *T. reesei* (Eriksson and Hamp, 1978). However, sophorose, a disaccharide with a structure different from cellobiose, induces endoglucanase activity in this fungus. Glucose is the end product of cellulose hydrolysis and causes catabolite repression of endoglucanase production in *S. pulverulentum* at concentrations as low as 50 mg/liter. Similarly, glucose represses endoglucanase production in *T. reesei.*

Production of cellulases in fungi is also regulated by phenomena other than induction and catabolite repression. Thus, Varadi (1972) demonstrated that various phenols repress the production of both cellulases and xylaneses in the fungi *Schizophyllum commune* and *Chaetomium globosum,* and that they do this at concentrations of less than 1 mM. Ander and Eriksson (1976b) demonstrated that endoglucanase production is drastically repressed in a phenol oxidase-less mutant (Phe 3) of *S. pulverulentum* in the presence of kraft lignin and phenols at concentrations of about 10^{-3} M. However, both the wild type (WT) and a phenol-oxidase-positive revertant (Rev 9) produced endoglucanases without significant repression by phenols. Furthermore, if a highly purified laccase preparation was added to the growth medium of Phe 3 in the presence of phenols, the endoglucanase production became normal. These results indicate that the influence of both kraft lignin and phenols on the synthesis of endoglucanase in Phe 3 is due to the absence of phenol-oxidizing and phenol-polymerizing enzymes. Phenol oxidases may thus function in regulating the production of cellulases, and most likely also xylanases, by oxidizing (polymerizing) lignin-related phenols, which may act as repressors of enzyme production when the fungus is growing in wood.

However, the activity of cellulose-hydrolyzing enzymes in culture filtrates of fungi is dependent not only on mechanisms regulating their biosynthesis, but also on the presence of specific inhibitors regulating the activity of the enzymes themselves. One such inhibitor is gluconolactone, which is produced by the oxidation of glucose by the enzyme glucose oxidase or by hydrolytic cleavage of cellobionolactone (cellobionic acid) formed by either cellobiose oxidase or cellobiose:quinone oxidoreductase (Eriksson, 1981a). Regulation of β-glucosidases from *S. pulverulentum* by gluconolactone has been studied by Deshpande *et al.* (1978). The *S. pulverulentum* seems to produce two β-glucosidases, one free and one bound to the fungal cell wall. That gluconolactone is a very powerful inhibitor is demonstrated by the fact that the affinity of the enzyme for the lactone is about 10,000 times higher than the affinity of the enzyme for its substrate, cellobiose.

3.4. Cellulolytic Bacteria and Their Cellulase Enzyme Systems

The degradation of cellulose by bacterial systems is a well-known phenomenon. It occurs both aerobically and anaerobically. It has already been mentioned that a substantial amount of biomass is converted to methane and carbon dioxide in the so-called methane cycle by anaerobic bacteria in the natural environment, such as in swamps, marshes, soil, river and lake sediments, and composts. Of special interest is the anaerobic digestion of cellulose in herbivorous animals and in insects. Like the degradation of cellulose by fungi, the degradation of cellulose by bacteria occurs by a consortium of several types of bacteria that interact. This will be discussed later. In this section we present the aerobic and anaerobic bacteria whose primary function is to hydrolyze the cellulose to simple sugars, and we discuss their cellulolytic enzyme systems. A list of such bacteria is given in Table II.

The bacterial cellulase systems have only recently been seriously investigated. There has been a tendency to compare them with the cellulase systems of fungi; thus, the bacterial cellulolytic enzymes are often characterized in terms of endo-β-1,4-glucanases, and exo-β-1,4-glucanases. As will be evident, the bacteria have cellulase systems that do not strictly allow them to be neatly assigned in this way. Furthermore, many bacteria do not have β-glucosidase (cellobiase), or, if they do have this enzyme, they may still metabolize cellodextrins and, more importantly, cellobiose by phosphorylases.

Studies of anaerobic bacteria have accelerated tremendously during the last decade, partly due to the interest in using them in industrial fermentations. In addition, convenient routine methods have been developed that make it possible to study these bacteria with almost the same ease as aerobic microorganisms (Bryant and Robinson, 1961; Hungate, 1950, 1969a; Miller and Wolin, 1974).

3.4.1. Properties of Some Anaerobic Cellulolytic Bacteria

This section will cover some mesophilic and thermophilic cellulolytic bacteria, some of which have been covered in earlier reviews by Hungate (1950) and McBec (1950), respectively.

3.4.1a. Acetivibrio cellulolyticus. This bacterium was isolated from sewage sludge by Saddler and Khan (1979) and later described by Patel *et al.* (1980) and Colvin *et al.* (1982). It is an obligate anaerobe, and grows at pH 6.5–7.7 at temperatures from 20 to 40°C. Of about 30 polysaccharides and sugars tested, it ferments only cellulose, cellobiose, and the glucoside salicin. Products of fermentation are hydrogen, carbon dioxide, acetate, and traces of ethanol, propanol, and butanol. Glucose and cellobiose may accumulate in the medium in fermentations of cellulose (Saddler and Khan, 1979). In the presence of excess cellobiose an iodophilic polysaccharide (glycogen-like) accumulates in the medium (Patel and Breuil, 1981). On starvation this saccharide is metabolized. When

Table II. Anaerobic and Aerobic Bacteria with Cellulolytic Activity[a]

Anaerobic
  Acetivibrio cellulolyticus
  Clostridium cellobioparum
  Clostridium papyrosolvens
  Clostridium stercorarium
  Clostridium acetobutylicum
  Bacteroides succinogenes
  Butyrivibrio fibrisolvens
  Ruminococcus albus
  Ruminococcus flavefaciens
  Eubacterium cellusolvens
  Micromonospora ruminantium
  Micromonospora propionici

Aerobic
  Cellulomonas flavigena
  Cellulomonas biazotea
  Cellulomonas cellasea
  Cellolomonas fimi
  Cellulomonas gelida
  Cellulomonas cartae
  Cellulomonas uda
  Cellulomonas turbata
  Bacillus brevis
  Bacillus firmus
  Bacillus licheniformis
  Bacillus pumilus
  Bacillus subtilis
  Bacillus polymyxa
  Bacillus cereus
  Serrata marcescens
  'Pseudomonas fluorescens var. *cellulosa'*
  'Cellvibrio viridus'
  'Cellvibrio flavescens'
  'Cellvibrio ochraceus'
  'Cellvibrio fulvus'
  'Cellvibrio vulgaris'
  'Cellvibrio gilvus'
  Cytophaga hutchinsonii
  Cytophaga aurantiaca
  Cytophaga rubra
  Cytophaga tenuissima
  Cytophaga winogradskii
  Cytophaga krzemieniewskae
  Herpetosiphon geysericolus
  Sporocytophaga myxococcoides
  Streptomyces flavorgriseus
  'Thermoactinomyces species'
  Thermomonospora curvata

[a]References to these bacteria are found in the text. The bacteria within prime signs are not validly classified.

growing on cellulose, *A. cellulolyticus* forms a yellow pigment, which stains the cellulose particles (Patel *et al.,* 1980).

The cellulase system of *A. cellulolyticus* has been studied by Khan (1980), Saddler and Khan (1981), and MacKenzie and Bilous (1982). It contains three extracellular cellulolytic enzymes, one of which has a molecular weight of 38,000. This enzyme is considered to be an exoglucanase, yielding cellobiose when Avicel is the substrate. The other two proteins, designated C2 and C3, have been characterized as endoglucanases (Saddler and Khan, 1981). The molecular weights of C2 and C3 are 33,000 and 10,400, respectively. Although some glucose and cellotriose are found, the major product of the cellulolytic system is cellobiose, which is further metabolized by a cell-associated β-glucosidase (Saddler and Khan, 1980). The cellulase system of *A. cellulolyticus* is induced by cellulose, cellobiose, and salicin. However, the induction by cellobiose is inhibited by glucose (Saddler *et al.,* 1980), although glucose is not fermented by *A. cellulolyticus.*

3.4.1b. Cellulolytic Clostridia. Clostridia are strictly anaerobic spore-forming rods, which are commonly found in soil, marine, and freshwater sediments, compost piles, and in the intestinal tracts of animals and man. Although they are anaerobic, their spores often survive in air for long periods. These spores are transported around the earth with the winds. An example of a ubiquitous *Clostridium* is the thermophile *C. thermohydrosulfuricum,* which has an optimum growth temperature of 70°C. Wiegel *et al.* (1979) found that this bacterium can be isolated from any types of soil and mud samples obtained from anaerobic and microanaerobic environments, and from places of low to high temperatures.

Both mesophilic and thermophilic cellulolytic clostridia are also commonly found in diverse environments, as is evidenced by many reported isolates. Among these are the validly described mesophiles *C. cellobioparum,* obtained from rumen (Hungate, 1944), and *C. papyrosolvens,* isolated from river mud (Madden *et al.,* 1982), as well as the thermophiles *C. thermocellum,* found in manure (Viljoen *et al.,* 1926), and *C. stercorarium,* purified from a compost (Madden, 1983). Allcock and Woods (1981) also indicate that *C. acetobutylicum* produces cellulase and cellobiase; this bacterium had not previously been considered to be cellulolytic. Other isolates of mesophilic cellulolytic clostridial strains are from decayed grass (Giallo *et al.,* 1983) and freshwater environments (Leschine and Canale-Parola, 1983). Several isolates of thermophilic cellulolytic clostridia, considered as strains of *C. thermocellum,* have been described (McBee, 1950, 1954), including strains LQR1 from sewage sludge (Lamed and Zeikus, 1980), JW20 from cotton bales (Ljungdahl *et al.,* 1981b) and M-7 from manure (Lee and Blackburn, 1975). In addition, Enebo (1951) obtained an isolate he called *Clostridium thermocellula-*

seum from decaying grass and leaves. Unfortunately, this isolate is lost and it is now considered a strain of *C. thermocellum.*

Attempts to elucidate the cellulolytic enzyme system in cellulolytic clostridia have been moderately successful. Most studies have been performed with *C. thermocellum* and they have recently been discussed in reviews by Carreira and Ljungdahl (1983) and Duong *et al.* (1983). *Clostridium thermocellum* excretes enzymes that degrade cellulose to cellodextrins and cellobiose (Ng and Zeikus, 1981a; Johnson *et al.,* 1982). Endoglucanases of different properties have been isolated from culture media of *C. thermocellum* (Ng and Zeikus, 1981b; Petre *et al.,* 1981) and *C. stercorarium* (Creuzet and Frixon, 1983), but an exoglucanase has not been obtained. Recently, it has been shown (Lamed *et al.,* 1983; Bayer *et al.,* 1983) that *C. thermocellum* produces a cellulase enzyme complex with a molecular weight of about 2.1 million that contains as many as 14 polypeptides and binds to cellulose. In agreement with the findings by Lamed and his group, Ljungdahl *et al.* (1983) observed that the cellulase system of *C. thermocellum* is tightly bound to cellulose and that this binding appears to be facilitated by a low-molecular-weight yellow substance. This substance is characteristically formed by *C. thermocellum* during cellulose fermentations and colors the cellulose brightly yellow. The cellulase complex is easily separated from the yellow cellulose by extraction with water, but not by salt or buffer solutions. During fermentations of cellulose, *C. thermocellum* cells also adhere to the cellulose, as shown in Fig. 5 (Wiegel and Dykstra, 1984). This also supports the findings by Lamed *et al.* (1983), Bayer *et al.* (1983), and Ljungdahl *et al.* (1983). The previously isolated endoglucanases (Ng and Zeikus, 1981b; Petre *et al.,* 1981) may well be components of the larger cellulase enzyme complex described by Lamed *et al.* (1983).

As mentioned, cellobiose and cellodextrins are the products of cellulose degradation by *C. thermocellum.* In most fungi the cellobiose is hydrolyzed by a cellobiase. A cellobiase or β-glucosidase has been isolated from *C. thermocellum* (Ait *et al.,* 1982). However, the affinity of cellobiose for the enzyme is very low and it is unlikely that it functions as a cellobiase; instead, it may aid in hydrolyzing saccharides having β-1,3 and β-1,2 linkages. Cellobiose and cellodextrins may instead enter the cell, where they are metabolized by specific cellodextrin and cellobiose phosphorylases to yield glucose-1-phosphate and free glucose (Alexander, 1968; Sheth and Alexander, 1969). The formation of glucose-1-phosphate is an advantage for an organism, since the energy of the glucosidic bond is then preserved. Glucose-1-phosphate and also glucose are fermented by clostridia via the Embden–Meyerhof pathway to pyruvate, which is further metabolized to hydrogen, CO_2, ethanol, acetate, lactate, and, in some cases, also to lesser amounts of formate, butyrate, and succinate.

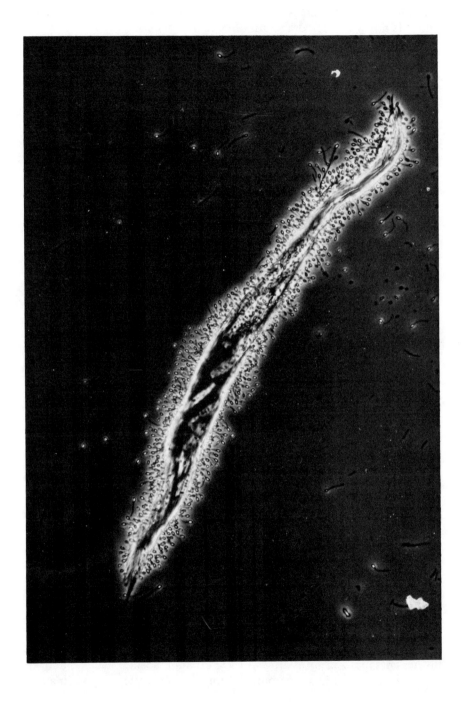

3.4.1c. Anaerobic Cellulolytic Bacteria of Rumen and Intestinal Tracts. One of the most studied anaerobic ecological systems is that of the rumen. This system has been the subject of many reviews (e.g., Hungate, 1966, 1969b, 1975; Hobson, 1971; Wolin, 1979; Wolin and Miller, 1983; Akin and Barton, 1983). Since cellulose is the major part of the ruminant diet, cellulolytic bacteria have been the targets of much interest. Several methods have been developed to determine the number of cellulolytic microorganisms in the rumen (Bryant and Robinson, 1961; van Gylswyk, 1970; Teather and Wood, 1982). Their nutritional requirements have been discussed in many papers [see the above-mentioned reviews and Bryant (1973)]. The predominant rumen cellulolytic bacteria seem to be *Bacteroides succinogenes, Butyrivibrio fibrisolvens, Ruminococcus albus,* and *Ruminococcus flavefaciens.* In addition, clostridial species, *Eubacterium cellulosolvens, Micromonospora ruminantium,* and *Micromonospora propionici,* have been isolated from rumen fluid. The last two bacteria are described by Maluszynska and Janota-Bassalik (1974) and Hungate (1946), respectively. *Micromonospora propionici* was first obtained from termites. These bacteria have not been the subject of intensive investigation. As already mentioned, anaerobic cellulolytic fungi are also present in the rumen (see Section 3.2).

Bacteroides succinogenes was first isolated from bovine rumen by Hungate (1950). He found that it ferments cellulose, cellobiose, glucose, starch, dextrin, maltose, and trehalose to succinate and acetate as the main products and that carbon dioxide is fixed. A valid description of *B. succinogenes* has been published by Cato *et al.* (1978). Other strains of *B. succinogenes* have been isolated from the bovine rumen (Bryant and Doetsch, 1954; Stewart *et al.,* 1981), rat cecum (Montgomergy and Macy, 1982), pig intestine (Varel *et al.,* 1984), and buffalo rumen (Sinha and Ranganathan, 1983). Cellulolytic *Bacteroides* strains (species type not indicated) have also been isolated from anaerobic marine environments (Miyoshi, 1978) and municipal sewage sludge (Khan *et al.,* 1980). These latter isolates differ substantially from *B. succinogenes* in that they do not produce succinate, whereas acetate is a major product.

Halliwell and Bryant (1963) reported that growing cultures of *B. succinogenes* degrade filter paper and cotton fibers. However, cell-free cul-

<hr>

Figure 5. Micrograph using phase contrast and 800-fold magnification of a filter paper fiber with attached cells of *C. thermocellum* strain JW20. The strong attachment was observed when the cellulose became colored yellow by a compound excreted by the bacterial cells. Note that almost all cells sporulated while adhering to the cellulose fiber. This occurred especially when the medium became slightly acid (pH 6.3 or below). Similar micrographs were obtained using Solka floc-cellulose and enrichment cultures from soil or mud samples. [Wiegel and Dykstra (1984).]

tural filtrates were almost void of cellulolytic activity. Recently, Forsberg and Groleau have more intensely studied the cellulolytic enzyme system of *B. succinogenes*. They noticed that filter paper cellulose fibers were completely covered by bacteria and were degraded within a few days (Groleau and Forsberg, 1981). However, cell extracts or culture fluid had very low activity toward filter paper or crystalline cellulose, although amorphorous cellulose and carboxymethylcellulose (CMC) were readily hydrolyzed, which indicated free endoglucanase activity. The CMC-degrading activity present in supernatant fluid was largely associated with sedimentable membranous fragment (molecular weight greater than 4 × 10^6), but partly also with a soluble protein fraction (molecular weight about 45,000) (Groleau and Forsberg, 1983). This pattern is somewhat similar to that found for *C. thermocellum* (Section 3.4.1b), and indicates a cellulase complex consisting of several components. Cellobiase is present in *B. succinogenes* (Forsberg and Groleau, 1982). This enzyme is associated with the cells. It is sensitive to air, but the activity is maintained in an atmosphere of nitrogen in the presence of dithiothreitol. However, the CMC-degrading activity is not oxygen sensitive. The cellobiase appears to be constitutive, whereas CMC-degrading activity is eight times higher in cells grown on cellulose than in cells grown on cellobiose or glucose (Groleau and Forsberg, 1981).

Butyrivibrio fibrisolvens, first described by Bryant and Small (1956), comprises strains of almost spindle-shaped bacteria with pointed ends. The strains differ substantially from one another. For instance, isolates from the rumen of cows (Clarke *et al.,* 1969) ferment a number of mono-, di-, and polysaccharides, including xylan, but not cellulose, whereas other strains obtained from sheep rumens ferment cellulose (van Gylswyk and Roche, 1970) in addition to xylan. The products of fermentations of cellulose and cellobiose are hydrogen, carbon dioxide, butyric acid (the major product), and lactic acid. Minor products are isobutyrate, isovalerate, and valerate (Bryant and Small, 1956; van Gylswyk and Roche, 1970). Some strains utilize acetate, in contrast to other strains, which form acetate. Shane *et al.* (1969) have classified *Butyrivibrio* isolates into two groups: (1) the acetate-utilizers and (2) the acetate-producers. Roche *et al.* (1973) found that propionate generally is inhibitory to the acetate-utilizing strains and stimulatory to the acetate-producing strains. In the fermentation of carbohydrates, group 1 produces a high amount of lactate but low levels of formate, whereas group 2 makes little lactate and more formate. The groups also differ in the response to CO_2 or bicarbonate in the medium. Bryant and Small (1956) found no requirement for bicarbonate or CO_2. However, van Gylswyk (1976) and Jarvis *et al.* (1978) found that bicarbonate added to the medium increases the cell yield and perhaps is essential for growth. With both groups [^{14}C]-

bicarbonate is incorporated almost exclusively into the carboxyl group of lactate and formate. In group 1 the specific radioactivity of lactate and formate were similar, which indicated a common precursor, presumably the carboxyl of pyruvate. The latter may be labeled in an exchange reaction between the carboxyl of pyruvate and $^{14}CO_2$ catalyzed by pyruvate-ferredoxin oxidoreductase (Thauer *et al.*, 1975). In group 2 the formate had twice as high a specific activity as the lactate, which may indicate a direct reduction of CO_2 to formate by formate dehydrogenase, as found in acetogenic anaerobic bacteria (Ljungdahl, 1983).

The cellulolytic system has not been investigated to any great extent in *B. fibrisolvens,* but it is a general consensus that this bacterium plays a substantial role in degrading fibers in the rumen. According to van Gylswyk and Roche (1970), some cellulolytic strains lose their ability to degrade cellulose after repeated transfers in a cellobiose medium.

Two cocci, *Ruminococcus flavefaciens,* isolated by Sijpesteijn (1951), and *Ruminococcus albus,* described by Hungate (1957), are prominent cellulose-degraders (van Gylswyk and Labuschagne, 1971) found in the rumen of cattle and sheep and in the cecum of other herbivores. Most isolated strains ferment cellulose and xylan and all ferment cellobiose. Fermentation of glucose and some other carbohydrates depends on the particular strain. The major products of fermentations of cellobiose are acetic acid, formic acid, ethanol, succinic acid, and hydrogen. The formation of succinate involves fixation of CO_2 (Hopgood and Walker, 1967).

The cellulase systems in both *Ruminococcus* species have been investigated. They appear to be similar. Results of early studies by Fusee and Leatherwood (1971) suggested that the *Ruminococcus* cellulase system is depressed by the presence of disaccharides such as cellobiose, sucrose, and lactose and also that cellobiose, if present, must first be used before cellulose is degraded. Inhibition of cellulose digestion by cellobiose, and to a lesser degree by glucose, was also observed by Smith *et al.* (1973). However, recently Hiltner and Dehority (1983) demonstrated that the cellulolytic ability of *R. albus, R. flavefaciens,* and *B. succinogenes* is constitutive and that cellulose degradation by the three bacteria occurs simultaneously with fermentation of cellobiose if this disaccharide is present. Pettipher and Latham (1979a) also suggest that the cellulase enzyme system as well as xylanase are constitutive enzymes in *R. flavefaciens.*

Pettipher and Latham (1979b) found that the cellulolytic activity and the xylanase in *R. flavefaciens* are both present in a large enzyme complex of molecular weight greater than 3×10^6 as well as in a lower molecular weight fraction (molecular weight about 89,000). The cellulase and the xylanase activities, although present in the same enzyme complex, clearly reside in different active sites. The products of the cellulolytic activity are cellobiose, cellotriose, and a smaller amount of glucose. Reducing agents (thiols) and Ca^{2+} or Mg^{2+} activate the cellulase. It is possible that the large enzyme complex is associated with a glycoprotein coat that surrounds the *R. flavefaciens* cells. Latham *et al.* (1978) have suggested that this coat, which contains rhamnose, glucose, and galactose, aids in the adhesion of the bacterial cells to the cellulosic substrate. An aryl β-glucosidase is present in *R. flavefaciens* (Pettipher and Latham, 1979b) and it may be involved in hydrolysis of cellobiose. However, this bacterium, like *C. thermocellum*, contains a cellobiose phosphorylase, which catalyzes the reversible phosphorylation of cellobiose to glucose-1-phosphate and glucose (W. A. Ayers, 1959).

The main product of cellulase degradation by *R. albus* is cellobiose (Leatherwood, 1965). This bacterium, like *C. thermocellum* and *R. flavefaciens*, appears to produce an affinity factor that binds cellulase or the bacterial cells to cellulose (Leatherwood, 1973). *Ruminococcus albus* also produces a high-molecular-weight cellulase complex (molecular weight arount 1.5×10^6) as well as low-molecular-weight cellulases (Yu and Hungate, 1979; T. M. Wood *et al.*, 1982). The high-molecular-weight complex can be released from the bacterial cells by phosphate buffer or water. This complex is formed when rumen fluid is present in the growth medium, whereas without the rumen fluid, low-molecular-weight cellulolytic enzymes are found in the medium (T. M. Wood *et al.*, 1982). The low-molecular-weight enzyme is also found in the medium late in the growth phase of the cells.

Hungate and Stack (1982) and Stack *et al.* (1983) have identified isobutyrate, butyrate, 2-methyl- and 3-methylbutyrate, phenylacetic acid, and 3-phenylpropanoic acid as factors that may influence the formation of the large cellulase complex of *R. albus*. The two phenyl compounds in particular stimulate the degradation of cellulose. Recently Stack and Hungate (1984) reported that 3-phenylpropanoic acid is a precursor of a new amino acid, β-ketophenylalanine. This amino acid is involved in the formation of a capsular structure that surrounds the cell wall and contains the cellulase complex. *Ruminococcus albus* grown in absence of 3-phenylpropanoic acid does not form the capsule and the cellulolytic enzyme is then of a low molecular weight. Clearly, several of the anaerobic cellulolytic bacteria in the rumen as well as *C. thermocellum* contain low-molecular-weight factors that may be involved with either the formation of the cellulase complex or in binding of such a complex to the cellulose.

Eubacterium (Cillobacterium) cellulosolvens (Bryant *et al.*, 1958; van Gylswyk and Hoffman, 1970; Prins *et al.*, 1972) is a cellulolytic, mainly lactate-producing, peritrichous rod found in the rumen of cattle and sheep. Some strains are almost homolactate-fermenting, whereas other strains produce in addition significant amounts of hydrogen, carbon dioxide, formate, butyrate, and valerate, as well as a small amount of succinate. Some strains utilize acetate and propionate. Most strains of *E. cellulosolvens* have a rather broad substrate spectrum and thus differ from *Ruminococcus* species, which have a rather narrow substrate spectrum based on cellulose and its hydrolytic products. The role of *E. cellulosolvens* in the degradation of cellulose in the rumen is not considered great, and studies of its cellulolytic enzyme system are lacking.

3.4.2. Aerobic Cellulolytic Bacteria

The presence in soil of aerobic cellulolytic bacteria has been recognized for a long time. For example, Kellerman and McBeth (1912) isolated *"Bacterium flavigena,"* now recognized as *Cellulomonas flavigena,* and reported that it ferments cellulose. Other cellulolytic aerobic bacteria are found among the genera *Bacillus, Cytophaga, Herpetosiphon, Pseudomonas, Serratia, Streptomyces, Sporocytophaga, Thermoactinomyces,* and *Thermomonospora.* Although many aerobic cellulolytic bacterial strains have been isolated, most of them have not been fully characterized and it appears that only recently has there been a surge of interest in these bacteria. This interest is related to the possibility that their cellulolytic enzyme systems may be of industrial use; in addition, the bacteria themselves appear to be excellent sources of cell protein (Humphrey *et al.,* 1977).

3.4.2a. Cellulomonas Species. Cellulomonas constitutes a genus of bacteria all of whose members are cellulose-degrading and non-spore-forming, and can grow either aerobically or anaerobically. A great deal of controversy has existed regarding whether the genus *Cellulomonas* contains one or more species. Although over the years several species of *Cellulomonas* have been described, Keddie (1974) considered these species to have a close resemblance, and consequently placed them all together in one species, *C. flavigena.* However, biochemical (Stackebrandt and Kandler, 1974, 1980a), serological (Branden and Thayer, 1976), and DNA–DNA homology studies (Stackebrandt and Kandler, 1979, 1980b; Stackebrandt *et al.,* 1982) revealed that there are at least eight distinct, validly described *Cellulomonas* species: *C. biazotea, C. flavigena, C. cellasea, C. fimi, C. gelida, C. cartae, C. uda,* and *C. turbata.* In addition, several unclassified *Cellulomonas* isolates are available and are being used in studies of cellulases. These isolates include strain IIbc, obtained in Cuba by Dr. M. Ibarra (Beguin *et al.,* 1977), strain CS1-1, found in soil (Choi *et al.,* 1978), strain DK, from sludge (Ramasamy *et al.,* 1981), as well as strains without designations from soil (Han and Srinivasan, 1968;

Stewart and Leatherwood, 1976), bagasse (Cabello *et al.,* 1981), and the gut of a manure worm (Chosson and Dupuy, 1983).

Cellulomonas strains degrade cellulose from various sources, including crystalline cellulose (cotton and Avicel) (Choi *et al.,* 1978; Haggett *et al.,* 1979; Kim and Wimpenny, 1981; Nakamura and Kitamura, 1983). In addition, it has been demonstrated with strain CS1-1 that hemicelluloses are hydrolyzed by *Cellulomonas* species, which contain β-glucosidase, β-xylosidase, β-mannosidase, and β-arabinosidase (Rickard and Laughlin, 1980; Ide *et al.,* 1983).

The cellulase enzyme system in *Cellulomonas* appears to be somewhat similar to that discussed earlier for *Clostridium thermocellum.* The main product from hydrolysis of cellulose by the extracellular cellulolytic enzymes apparently is cellobiose (Han and Srinivasan, 1968; Hagget *et al.,* 1979; Stoppok *et al.,* 1982), although in degradation of alkali-treated bagasse with a *Cellulomonas* mutant CS1-17 most of the released sugars were glucose (35%), cellobiose (26%), and xylose (14%) (Choudhury *et al.,* 1980).

The cellulolytic active enzymes produced by *Cellulomonas* species appear to be bound tightly to the cellulose substrate. Thus, Beguin *et al.* (1977), with *Cellulomonas* isolate IIbc, found that most of the cellulolytic activity during the growth phase was associated with the cellulose, and appeared in the culture fluid only in the stationary phase. The cellulose-bound activity remained on the cellulose even after extensive washing with buffers of different concentrations and pH. It was eluted from the cellulose with 8 M guanidine hydrochloride (Beguin and Eisen, 1978). Stoppok *et al.* (1982), working with *C. uda,* also observed that soluble cellulolytic activity appears relatively late in the growth. These authors, who did not determine cellulose-bound activity, detected only a small amount to be cell-bound. In contrast, Kim and Wimpenny (1981) suggest from studies with *C. flavigena* that the cellulolytic enzymes are partly cell-bound and partly free in the culture fluid.

The cellulolytic system in the *Cellulomonas* group of bacteria is clearly quite complex. Beguin and Eisen (1978) have purified three β-endoglucanases from isolate IIbc. One of these was from the culture fluid and two were from the cellulose-bound fraction. They maintain, though, that these glucanases do not represent the entire cellulolytic enzyme system. Thus, additional components are needed for digestion of cellulose; moreover, at least two more cellulases were discovered during polyacrylamide gel electrophoresis. In addition, Nakamura and Kitamura (1983) have purified a β-1,4-glucan-cellobiohydrolase from *C. uda.* The enzyme found in the culture fluid hydrolyzes crystalline cellulose with cellobiose as the product.

As has been mentioned, cellobiose appears to be the principal prod-

uct of the cellulolytic enzymes from *Cellulomonas*. Cell-bound β-glucos-
idases are present in *Cellulomonas* species (Choi *et al.*, 1978; Stoppok *et al.*, 1982). These enzymes may be responsible for the hydrolysis of cellobiose to glucose, which then is fermented. However, several species of
Cellulomonas are similar to *Clostridium thermocellum* (Alexander, 1968)
in having cellobiose phosphorylase, which cleaves cellobiose into glucose
and glucose-1-phosphate (Sato and Takahashi, 1967; Schimz *et al.*, 1983).
The latter authors rather convincingly demonstrated that, at least in *C.
uda* and *C. flavigena*, the metabolism of cellobiose occurs by cleavage
with the phosphorylase. Glucose and glucose-1-phosphate are metabolized via the Embden–Meyerhof–Parnas glycolytic pathway, as was ascertained by fermenting differently labeled ^{14}C-glucose substrates (Stackebrandt and Kandler, 1974, 1980a). However, the ^{14}C distribution in the
acidic products (acetate and lactate) indicated a rather substantial redistribution of the carbon not compatible with strict use of the glycolytic
pathway. This randomization was found to be caused by reversible
actions of aldolase, transaldolase, and transketolase.

   3.4.2b. Bacillus Species. The genus *Bacillus* comprises a large number of spore-forming, aerobic or facultative anaerobic species. Some of
them, including *B. brevis, B. firmus, B. licheniformis, B. pumilus, B. subtilis* (Knösel, 1971), *B. polymyxa* (Greaves, 1971; Fogarty and Griffin,
1973), and *B. cereus* (Thayer, 1978) possess cellulolytic activity. The *B.
cereus* strain RW1 was isolated together with *Serrata marcescens* strain
RW3 from the hindgut of termites (Thayer, 1978), where both strains
may contribute to the digestion of cellulose. Some *Bacillus* species also
contain enzymes for the hydrolysis of hemicelluloses (Inaoka and Soda,
1956; Emi and Yamamoti, 1972; Fogarty and Ward, 1973; Esteban *et al.*,
1982).

   The cellulolytic enzymes of the *Bacillus* species have hardly been
characterized. The species mentioned above are all able to hydrolyze carboxymethylcellulose, with cellobiose as the main product. Greaves
(1971), studying *B. polymyxa*, found that the cellulase is extracellular and
can be extracted from the mixture of cells and remaining cellulosic substrate with distilled water, similar to the cellulase of *Clostridium thermocellum* (Ljungdahl *et al.*, 1983). The formation of the *B. polymyxa* cellulase is dependent on the carbon source present in the growth medium.
Thus, the enzyme was produced when starch, glucose, cellulose, cellobiose, carboxymethylcellulose, xylan, maltose, and sucrose were added to
a basal salts-peptone medium. It was not formed in the basal salts-peptone medium alone or in the presence of arabinose (Fogarty and Griffin,
1973). Similarly, Thayer (1978) found that glucose and cellobiose stimulate enzyme formation by *B. cereus*, and Robson and Chambliss (1984),
working with a new isolate (DLG) related to *B. subtilis*, demonstrated

that the production of the carboxymethylcellulase is enhanced by the presence of sucrose, maltose, glucose, cellobiose, or lactose in the medium. The cellulase of isolate DLG did not degrade crystalline forms of cellulose. This isolate apparently also lacks β-glucosidase and it was proposed that the degradation of cellobiose may occur by a cellobiose phosphorylase.

In a recent report Sashihara et al. (1984) described an alkalophilic Bacillus sp. strain N-4 producing two carboxymethyl cellulases, which are stable at 75°C and are active at pH 5–10.9. The two enzymes, of molecular weights 50,000 and 58,000, where cloned into Escherichia coli with the vector pBR322. The enzymes were expressed and observed in the periplasmic space.

3.4.2c. Pseudomonas Species. The genus Pseudomonas is a large group of rod-shaped, non-spore-forming, Gram-negative, aerobic bacteria. The genus has recently been thoroughly described by Palleroni (1984). Some Pseudomonas species are cellulolytic (Ueda et al., 1952). Among them is a strain isolated from soil and characterized as 'Pseudomonas fluorescens var. cellulosa.' This strain, the cellulolytic enzyme systems of which have been studied to some extent, is not considered by Palleroni (1984) as a valid species of the genus Pseudomonas. Similarly bacteria placed in the genus 'Cellvibrio,' which all have cellulolytic enzymes, were suggested by Doudoroff and Palleroni (1974) to perhaps belong to the genus Pseudomonas. However, Palleroni (1984) did not include the 'Cellvibrio' species in the new description of the genus Pseudomonas. Thus, the classification of the cellulolytic 'Pseudomonas' and 'Cellvibrio' species remains to be clarified.

The cellulolytic enzyme system in 'P. fluorescens var. cellulosa' is quite complex. It consists of two electrophoretically separable extracellular enzymes, A and B, as well as of an enzyme C located in the periplasmic fraction (Yamane et al., 1970). The cellulolytic enzymes contain carbohydrates as major components, with galactose in enzyme A and glucose in enzymes B and C. The substrate spectra for enzymes A and B are very similar, and cellotetrose and cellulose molecules with higher DP, including swollen cellulose and Avicel, are hydrolyzed. Enzyme C hydrolyzes the same compounds and in addition cellotriose.

The enzymes A and B are not homogeneous. Separation of the extracellular enzymes according to molecular weight indicated the presence of at least five cellulolytic proteins (peaks I–V). Peak I, having the highest molecular weight (approximately 85,000), consists almost exclusively of enzyme B, and peak V (molecular weight 40,000) of enzyme A. Peaks II–IV are composed of mixtures of enzymes A and B (Yoshikawa et al., 1974). It was suggested that enzyme B (peak I) is the cellulolytic enzyme that is originally produced by the cells, and that it is converted extracellularly to the other cellulolytic proteins either by modification of the protein moiety or by changes in the carbohydrate content. This indicates some similarity with the observation by Eriksson and Petersson (1982) that two proteases from *Sporotrichum pulverulentum* activate the endoglucanases from the same fungus as well as other fungi.

Among '*Cellvibrio*' species named in the literature are '*C. viridus*,' '*C. flavescens*,' '*C. ochraceus*' (Winogradsky, 1929), '*C. fulvus*,' '*C. vulgaris*' (Stapp and Bortels, 1934; Gubsch, 1979), and '*C. gilvus*' (Hulcher and King, 1958). The cellulolytic enzyme system of '*C. gilvus*' has been studied rather extensively and the results have been summarized by King and Vessal (1969). This organism, which grows well on cellobiose, cellodtrins, and cellulose, but less well on glucose, has, like '*P. fluorescens* var. cellulosa,' an extracellular multicomponent cellulolytic system. This system is constitutive and contains at least four electrophoretically distinct endo-1,4-β-glucanases, which degrade cellulose to cellobiose and some cellotriose (Storvick *et al.*, 1963). Cellobiose, cellotriose, and longer cello-oligosaccharides are metabolized by '*C. gilvus*' and the cell yield per glucose equivalent increases with the DP, which indicates that the energy of the glucosidic bonds is preserved, and that cleavage of these bonds is by phosphorolysis (Schafer and King, 1965). However, '*C. gilvus*' apparently only contains a cellobiose phosphorylase, which has been purified (Sasaki *et al.*, 1983). This enzyme does not cleave cellotriose or higher cellodextrins. Therefore, it has been suggested by King and Vessal (1969) that cellotriose and higher cellodextrins are metabolized by transglucosylation to yield cellobiose:

$$\text{Cellodextrin } (G_n) + \text{glucose} \rightarrow \text{cellodextrin } (G_{n-1}) + \text{cellobiose} \quad (3)$$

and that cellobiose is cleaved by cellobiose phosphorylase:

$$\text{Cellobiose} + \text{phosphate} \rightarrow \text{glucose-1-P} + \text{glucose} \quad (4)$$

Cleavage of cellobiose catalyzed by cellobiose phosphorylase to glucose-1-phosphate and glucose opens up the possibility that the two glucose moieties of cellobiose are nonequivalent and may be metabolized differently. This has actually been observed for '*C. gilvus*' (Swisher *et al.*, 1964) and for *Cellulomonas fimi* (Sato and Takahashi, 1967).

A multicomponent cellulase system is also produced by '*Cellvibrio fulvus*' (Berg *et al.*, 1972a). In contrast to the '*C. gilvus*' cellulase, the '*C. fulvus*' enzyme is repressed by glucose. The location of the cellulase in '*C. fulvus*' seems to depend on the carbon source for growth (Berg, 1975). With glucose and cellobiose the carboxymethylcellulase activity is cell-bound during the active growth phase, whereas with cellulose as substrate it is cell-free early in the growth phase and also partly bound to the cel-lulose. Similarly, when 'Cellvibrio vulgaris' grows on cellulose (filter paper) it produces a truly extracellular cellulolytic enzyme (carboxyme-thylcellulase) (Oberkotter and Rosenberg, 1978).

    3.4.2d. Cytophaga Group. The genera Cytophaga, Herpetosiphon, and Sporocytophaga belong to the family Cytophagaceae (Leadbetter, 1974). This group of bacteria has been covered in excellent reviews by Stanier (1942) and, more recently, Christensen (1977). These bacteria are nonflagellated, nonfruiting, and include forms resembling blue-green algae. They exhibit a gliding or flexing motion and differ substantially from other, rigid bacteria. They inhabit soil, sea, and freshwater and degrade a wide range of rather resistant polysaccharides, including cellu-lose. The genus Cytophaga includes the following cellulolytic species: C. hutchinsonii, C. aurantiaca, C. rubra, C. tenuissima, and C. winogradskii, as well as five other named species that are considered synonymous with C. hutchinsonii. It also contains halophilic "marine" cellulolytic species, of which C. krzemieniewskae is best known (Christensen, 1977). The gen-era Herpetosiphon and Sporocytophaga each has one well-characterized cellulolytic species, H. geysericolus (Lewin, 1970) and S. myxococcoides (Stanier, 1942), respectively. Herpetosiphon geysericolus, obtained from a hot spring (Baja California), is remarkable in that it grows at 60–80°C and at pH values from 8 to 9.

    Electron microscopic studies (Berg et al., 1972b) revealed that S. myxococcoides grows in close contact with cellulose fibers and creeps on as well as into the fibers. The close contact seems necessary for the deg-radation to take place, although a cellulase active on carboxymethylcel-lulose can be recovered from the cell-free medium. However, this enzyme does not hydrolyze crystalline cellulose, such as Avicel. During growth on cellulose, S. myxococcoides produces a slimy polysaccharide. Char-pentier (1965), using 10-day-old cultures, found that the cell-free medium of these cultures contained a cellobiase and in addition cellulolytic enzymes that degrade cellulose fibers and a soluble cellulose "Novacel" to glucose. She also noticed the formation of several oligosaccharides, perhaps precursors of the slime, which she suggested were formed by tranglucosylations catalyzed by the cellobiase. The formation of the slime depended also on oxidative processes involving formation of, for exam-ple, uronic acids. Charpentier and Robic (1974) subsequently found in S. myxococcoides an exoglucanase, which yielded glucose with cellobiose and soluble as well as insoluble oligosaccharides as substrates. The enzyme, which was extracted from the cells using Triton X-100, had a molecular weight of 67,000. However, the S. myxococcoides cellulase sys-tem contains, in addition to the exoglucanase, several cellulases charac-terized as endoglucanases (Osmundsvåg and Goksøyr, 1975). Two of these obtained from the cell-free medium were purified to almost homogeneity. The molecular weights of these enzymes are 46,000 and 52,000. They hydrolyze carboxymethylcellulose, but are completely inactive on native cellulose, which seems to confirm the observations by Berg *et al.* (1972b). Osmundsvåg and Goksøyr (1975) noticed, in addition to cell-free cellulases, that 10% of the total cellulolytic activity is loosely associated with the cells and that another 10% is bound to the cell membrane and can be released using Triton X-100.

Species of *Cytophaga* (strain 39072, BDH Ltd., Poole, England, and strain ATCC 29474) contain cellulases active on carboxymethylcellulose (Marshall, 1973; W. T. H. Chang and Thayer, 1977). Two cellulases described by W. T. H. Chang and Thayer (1977) appear to differ from cellulases obtained from most other cellulolytic microorganisms in that they have very low molecular weights, 6250 and 8650, as determined by gel filtration. The cellulase with molecular weight 6250 seems to be located in the periplasmic space. It has a high saccharifying activity and is active against crystalline cellulose and can perhaps be characterized as an exoglucanase. The cellulase with molecular weight 8650, on the other hand, is membrane-bound and its action is typical for an endoglucanase.

3.4.2e. Actinomycetes Group. A few mycelium-producing and spore-forming bacteria have been investigated with regard to cellulolytic ability. Examples of such bacteria are found among the genera *Streptomyces, Thermoactinomyces,* and *Thermomonospora.* The characteristics of *Streptomyces* species are covered by Shirling and Gottlieb (1966) and those of the thermophilic genera by Henssen (1957). A detailed discussion of actinomycetes and their importance in soil, compost, and fodder has been given by Lacey (1973).

Daigneault-Sylvestre and Kluepfel (1979) described an agar plate assay for screening of cellulolytic *Streptomyces* species and for determining production and catabolic repression of cellulase. With this method *Streptomyces flavogriseus* strain IAF 45-CD was isolated by Ishaque and Kluepfel (1980). This strain, when grown on Avicel (crystalline cellulose) as substrate, produces an extracellular β-1,4-glucan glucanohydrolase, which was active also with filter paper and cotton as substrates. The enzyme, which was not purified, appears to consist of several components, as indicated by differences in pH and temperature optima for the hydrolysis of carboxymethylcellulose and filter paper: hydrolysis of carboxymethylcellulose occurs best at 50°C and pH 7.0, whereas the degradation of filter paper occurs best at 40°C and pH 5.5. The bacterium also contains cellobiase, which is associated with the mycelial fraction.

Recently Veiga *et al.* (1983) isolated 36 *Streptomyces* strains from marine sediments obtained from La Coruna Bay (Spain). Nineteen of these strains have cellulolytic activity, and all are able to grow in media

containing 3.5% NaCl. Three of the strains seem to be superior as cellulose-degraders to *Trichoderma viride,* which is generally thought of as a
most efficient cellulolytic fungus.

Thermoactinomyces species (Su and Paulavicius, 1975) and *Thermomonospora* species (Stutzenberger, 1971; Hägerdal *et al.,* 1978, 1980)
also seem superior as degraders of cellulose in comparison with *T. viride.*
The cellulolytic system in these bacteria is clearly extracellular. During
active growth of the bacteria a substantial part of the cellulase appears
bound to the substrate (cellulose) and is then released during the stationary phase (Su and Paulavicius, 1975; Hägerdal *et al.,* 1978). Hägerdal *et
al.* (1978) observed that the cellulase bound to cellulose could be
extracted using distilled water, similar to what was later found for the
cellulase system in *Clostridium thermocellum* (Ljungdahl *et al.,* 1983).
The cellulase system produced by a *Thermomonospora* species
[previously believed to be a *Thermoactinomyces* species (Hägerdal *et al.,*
1980; Ferchak *et al.,* 1980)] is active on carboxymethylcellulase as well as
on crystalline cellulose (Avicel) and it seems to be very complex. Thus,
Hägerdal *et al.* (1978) found with isoelectric focusing at least six protein
fractions with cellulolytic activity. Similarly, Stutzenberger (1972a) found
that the cellulase system of *Thermomonospora curvata* can be resolved
into two or more fractions.

The pH optimum for the *Thermoactinomyces* cellulase system is
between 5.5 and 6.5 and the temperature optimum is about 65°C (Su and
Paulavicius, 1975). In a *Thermomonospora* species the enzyme system
that is active on Avicel has optima at pH 7.0 and 65°C, whereas the carboxymethylcellulase activity has optima at pH 5.9 and 65°C (Hägerdal *et
al.,* 1980). A β-glucosidase from the same bacterium is most active at pH
6.5 and 55°C. This enzyme has been characterized (Hägerdal *et al.,* 1979),
and is tightly associated with the cells. The product of the cellulolytic
activity of *T. curvata* is cellobiose (Stutzenberger, 1972a), which then
may be further metabolized after hydrolysis with β-glucosidase to
glucose.

The *Thermomonospora* cellulolytic system seems to be repressed by
cellobiose and glucose (Stutzenberger, 1972b). However, mutants of a
Thermomonospora species (Meyer and Humphrey, 1982) and of *T. curvata* (Fennington *et al.,* 1984) have been obtained that are repression-
resistant.

4. Interaction among Microorganisms in Ecological Systems

In the previous section we discussed individual microorganisms and
their cellulolytic enzyme systems. That discussion emphasized the great

diversity of cellulolytic microorganisms, which are found in almost if not all habitats. Although the list of cellulolytic microorganisms presented is large, it is only partial, and in addition new species previously not described are being found in nature. For instance, enrichment cultures that hydrolyze cellulose (filter paper) at 85°C and pH 2–9 found in Icelandic hot springs are presently being characterized (Ljungdahl *et al.,* 1981a). Considering the large quantities of cellulose produced by photosynthesis and available in nature, perhaps it is not surprising that so many different cellulolytic microorganisms have developed. These microorganisms seem to be quite efficient, because apparently very little cellulose or its hydrolytic products accumulate in nature.

4.1. Cellulose in Soils

Soil would be a natural depository for cellulose. Analyses of carbohydrates in soils have been reviewed by Mehta *et al.* (1961) and V. C. Gupta (1967). Results of such analyses differ enormously. They depend on the kind of soil, whether the sample is from the surface or from deeper down in the soil, the season the sample is collected, weather conditions, etc. Furthermore, soils seem to bind carbohydrates tightly, which makes their extraction for analysis a problem. Nevertheless, such analyses have revealed that the carbohydrate content in soil is low. Organic matter can be as low as 1% in mineral soils and as much as 90% in organic soils (peat). Of the organic matter, 5–20% consists of carbohydrates. Of these (total carbohydrates), less than 1% constitutes free sugars,—mostly glucose, but also galactose, fructose, xylose, and arabinose. From 8 to 14% is cellulose. Other polysaccharides, which apparently make up the main fraction, are those containing uronic acid, amino sugars, galactose, mannose, arabinose, xylose, ribose, and rhamnose. Thus, the carbohydrate composition in soil is more diverse than its source, which one would assume is mostly cellulose and hemicellulose. This can be interpreted in several ways. First, the fact that only 5–20% of organic matter in the soil is carbohydrates compared with about 70% in biomass indicates that cellulose and hemicellulose are selectively metabolized over other organic matter (mostly lignin) of the biomass. Second, the fact that the carbohydrate composition in the soil is different than in biomass seems to indicate that cellulose and also hemicellulose are converted to other carbohydrates. That this occurs is well known. Many fungi and aerobic bacteria synthesize large amounts of extracellular polysaccharides from glucose (Berkeley *et al.,* 1979). Some of these polysaccharides have substantial numbers of uronic acid residues and bind metals very efficiently. Such polysaccharides have been considered for use in the recovery of trace metals from dilute solutions (Norberg and Persson, 1984). These polysaccharides may certainly serve as metal chelators in the soil and thus have substantial ecological impact. Third, the fact that only small amounts of cellulose and glucose are found in the soil indicates that these carbohydrates are rapidly metabolized.

Evidence supporting at least two of the above interpretations has been obtained with the use of [^{14}C-lignin]- and [^{14}C-cellulose]-lignocelluloses. Crawford *et al.* (1977) developed methods to prepare the [^{14}C]-lignocelluloses; they found that soil samples from a forest (Shenandoah National Park, Virginia) converted [^{14}C-cellulose]-lignocellulose to ^{14}CO$_2$ 4–10 times faster than [^{14}C-lignin]-lignocellulose. About 50% of the ^{14}C of the [^{14}C-cellulose]-lignocellulose was converted to CO$_2$ within 700 hr. Similar results were obtained by Maccubbin and Hodson (1980) and Benner *et al.* (1984) with sediments from salt marshes (Sapelo Island, Georgia) and by Aumen *et al.* (1983) using samples from a wetted Douglas fir log as inoculum. Aumen *et al.* (1983) also observed that supplementation of the incubation medium with a nitrogen source, either (NH$_4$)$_2$SO$_4$, KNO$_3$ or NH$_4$NO$_3$, enhanced the degradation of both lignin and cellulose, whereas supplementation with glucose repressed the degradation of the cellulose and also the lignin.

4.2. Cellulose Degradation in an Aerobic Environment

The ultimate product of microbial degradation of cellulose is CO$_2$, but in an anaerobic environment methane is also produced. The degradation of cellulose is initiated by the cellulolytic microorganisms, which thus can be regarded as "primary microorganisms" (fungi or bacteria). The initial products of cellulose degradation are cellobiose and glucose. These sugars are utilized by the primary microorganisms partly for cell material and partly for metabolism to yield energy for cell growth and maintenance. In the natural environment the primary microorganisms are associated with a very diverse population of microorganisms that are unable to hydrolyze cellulose but use cellobiose, glucose, and other free sugars for energy. These microorganisms depend on the cellulolytic primary microorganisms for a supply of free sugars. They can, therefore, be regarded as "secondary microorganisms." Although the secondary microorganisms depend on the primary microorganisms, they also aid them by removing free sugars, which normally are inhibitory for the degradation of cellulose. Thus, this "symbiosis" promotes cellulose degradation.

In an aerobic environment where oxygen is available, the association of primary and secondary microorganisms should be sufficient for the complete and efficient oxidation of cellulose to CO$_2$, which can then be regarded as a two-stage process. A two-stage process can also be envisioned in an anaerobic environment, but it will be necessary to introduce substages to convert cellulose to CO_2 and methane. This will be discussed below (Section 4.3).

An example of an aerobic enrichment culture containing a mixed population of primary and secondary microorganisms has been described by von Hofsten *et al.* (1971). This culture contains a *Sporocytophaga* sp. (presumably *S. myxococcoides*) and a noncellulolytic bacterium with characteristics of a *Flavobacterium* sp. This mixed culture, isolated from a sewage treatment plant, was serially transferred for several months and it was far superior in degrading and oxidizing cellulose than the *Sporocytophaga* sp. alone. The fact that the two bacteria were transferred together demonstrates their close relationship and the stability of the coculture. As is now evident, stable cocultures (consortia) among bacteria are quite common. Other examples of synergism between aerobic primary and secondary microorganisms have been described by Hulme and Shields (1972), Blanchette and Shaw (1978), and Shortle *et al.* (1978).

4.3. Cellulose Degradation in an Anaerobic Environment

Although it is commonly believed that most cellulose is degraded in an aerobic environment, it is clear that a substantial amount is also degraded under anaerobic conditions. Anaerobic habitats are the rumen and intestinal tracts of animals, sewage sludge digestors, soils, composts, and muds and sediments in fresh and seawater. Of these habitats, the rumen system is probably the most investigated; it has been discussed in several reviews (see Section 3.4.1c). In the rumen system cellulose is not completely converted to gaseous products; rather, it is fermented to acids such as acetate, propionate, and butyrate, which are used by the animal (Hungate, 1975; Wolin, 1979). In the sewage sludge digestor and in most natural anaerobic habitats, on the other hand, methane and CO_2 are the main products. The complete conversion of cellulose to methane and CO_2, the methane fermentation, has recently been reviewed (Hobson *et al.*, 1974; Mah *et al.*, 1977; Zeikus, 1977; Bryant, 1979; McInerney and Bryant, 1981a; Peck and Odom, 1981). There are also reviews giving special emphasis to methane production in thermophilic bacteria (Varel, 1983) and to the role of bacteria in detritus food chains of aquatic ecosystems (Fenchel and Jørgensen, 1977).

In soils, composts, muds, and sediments, anaerobic activity starts rather close to the surface, which means that aerobic conditions normally prevail only in a thin crust. This has been ascertained by measuring redox potentials (*Eh*) and levels of oxygen and by demonstration of anaerobic microbial communities from, for example, salt marshes (Atkinson and Hall, 1976; Oremland *et al.*, 1982), marine sediments (Jørgensen, 1977), and a meromictic lake (Winfrey and Zeikus, 1979). A discussion of *Eh* values, which should be below -200 mV for methanogenesis in natural environments, can be found in the review by Mah *et al.* (1977).

A simplistic view of the anaerobic degradation of cellulose is to consider it to be a complete oxidation of carbon to CO_2 concomitant with the formation of dihydrogen (H_2), as shown in the following reaction:

$$C_6H_{12}O_6 + 6H_2O \rightarrow 6CO_2 + 12H_2 \tag{5}$$

In the anaerobic environment very little H_2 escapes into the atmosphere; instead, H_2 is assimilated by bacteria and used for the reduction of, for example, CO_2, SO_4^{2-}, and NO_3^-. Carbon dioxide is reduced to methane by the methanogenic bacteria (Zeikus, 1977; Balch *et al.,* 1979; McInerney and Bryant, 1981b),

$$CO_2 + 4H_2 \rightarrow CH_4 + 2H_2O \tag{6}$$

and also most likely to acetate by acetogenic bacteria (H. G. Wood *et al.,* 1982; Ljungdahl, 1983),

$$2CO_2 + 4H_2 \rightarrow CH_3COOH + 2H_2O \tag{7}$$

Acetate, which is also formed directly from cellulose by fermentations carried out by the cellulolytic (primary) bacteria and from cellobiose or glucose by secondary bacteria, is considered an important precursor of methane in the anaerobic environment (van den Berg *et al.,* 1976; Strayer and Tiedje, 1978; Winfrey and Zeikus, 1979; Mah, 1981). The acetate is converted to equal amounts of methane and CO_2 by several methanogenic bacteria, including *Methanosarcina barkeri, Methanobacterium soehngenii,* and *Methanococcus mazei* (Mah, 1981), by the reaction

$$CH_3COOH \rightarrow CH_4 + CO_2 \tag{8}$$

When sulfate is present in the anoxic environment it seems to be preferentially reduced over CO_2. Thus, there is competition for H_2 between the sulfate-reducing and the methanogenic bacteria. This competition, which results in less formation of methane and more of CO_2 from cellulose when sulfate is present in the environment, has been evaluated to a considerable extent by Robinson and Tiedje (1984). Since sulfate is present in sea water, the oxidation of cellulose in connection with the reduction of sulfate to H_2S in marine environments is of great interest (Jørgensen, 1977; Madden *et al.,* 1981). Thus, the dissimilatory reduction of SO_4^{2-} to H_2S is a very important ecological process. It is also biochemically very interesting, as demonstrated in two recent reviews (Peck, 1984; Peck and Odom, 1984).

Nitrate is reduced to ammonia in an anaerobic environment by some sulfate-reducing bacteria (Peck, 1984), as well as by other anaerobic bacteria (Caskey and Tiedje, 1980). The reduction clearly has ecological implications and has been observed in soil (Buresh and Patrick, 1978; Caskey and Tiedje, 1980) and in marine sediments (Koike and Hattori, 1978; Sørensen, 1978). Denitrification, a process discussed at length by Payne (1981) and Knowles (1982), may also occur.

As discussed above, cellulose degradation in an anaerobic environment consists of two interrelated processes. The first is the oxidation of cellulose carbon to CO_2 and reduction of protons to H_2; the second is the oxidation of H_2 and the reduction of CO_2 to acetate and methane, sulfate to H_2S, or nitrate to ammonia. This implies that H_2 plays an essential role in the anaerobic degradation of cellulose. In the first step the bacteria are H_2-producing, whereas in the second step they are H_2-consuming. Thus, H_2 is transferred between the H_2-producing and the H_2-consuming bacteria. This process is now called interspecies H_2 transfer, and the concept (Bryant et al., 1967; Iannotti et al., 1973) has had an enormous impact on the understanding of energy generation in anaerobic microorganisms (Peck and Odom, 1984) and on fermentations using mixed cultures.

Knowing the physiology of bacteria in cultures converting cellulose to CO_2 and methane, one can realize that the conversion is carried out in stages and that each stage involves metabolic groups of bacteria. Schemes picturing these metabolic groups have been set up by Bryant (1979) and McInerney and Bryant (1981a), who considered conversion a three-stage process, and by Wiegel and Ljungdahl (1979), Peck and Odom (1981), and Carreira and Ljungdahl (1983), who have suggested a four-stage process.

Table III shows a scheme for the complete conversion of cellulose to methane and CO_2 based on the isolation of thermophilic anaerobic bacteria from soil and mud samples collected in Georgia and in Yellowstone National Park in Wyoming (Wiegel and Ljungdahl, 1979, 1981; Wiegel et al., 1979, 1981) and on mesophilic anaerobic bacteria isolated by Dilworth et al. (1980). The conversion is viewed as a two-stage process: (I) The first stage consists of hydrogen-producing bacteria and has three substages (A–C). (II) The second stage consists of hydrogen-consuming bacteria and has two substages (D and E). The scheme must be modified if sulfate, nitrate, or other oxidized compounds are present and are reduced by, for example, sulfate- or nitrate-reducing bacteria.

In the scheme, the primary cellulolytic bacterium can be represented by *Clostridium thermocellum*. It hydrolyzes cellulose to cellobiose and glucose by an extracellular cellulolytic enzyme system discussed in Sec-

Table III. Bacterial Groups Involved in the Complete Conversion of Cellulose to Methane and CO_2 in an Anaerobic Environment

I. Hydrogen-producing bacteria
  A. Primary (cellulolytic) bacteria
    Cellulose $\rightarrow$ cellobiose + glucose $\rightarrow$ lactate + ethanol + acetate + CO_2 + H_2
  B. Secondary bacteria
    1.   cellobiose + glucose $\rightarrow$ lactate + ethanol + acetate + CO_2 + H_2
    2.   cellobiose + glucose $\rightarrow$ ethanol   + CO_2
    3.   glucose $\rightarrow$ acetate
  C. Ancillary bacteria
    1.   lactate $\rightarrow$ acetate + CO_2 + H_2
    2.   ethanol $\rightarrow$ acetate   + H_2
II. Hydrogen-consuming bacteria
  D. Acetogenic bacteria
    1. CO_2 + H_2 $\rightarrow$ acetate
  E. Methanogenic bacteria
    1.   acetate $\rightarrow$ CO_2 + CH_4
    2. CO_2 + H_2 $\rightarrow$ CH_4

[a]Based on isolation of bacterial species by Wiegel and Ljungdahl (1979) and Dilworth et al. (1980).

tion 3.4.1b. *Clostridium thermocellum* ferments part of the sugars (it prefers cellobiose over glucose) to lactate, ethanol, acetate, CO_2, and H_2. However, if *C. thermocellum* is alone, glucose and cellobiose accumlate in the medium (Carreira and Ljungdahl, 1983), and eventually concentrations are reached at which the sugars inhibit the cellulolytic system. If a secondary bacterium capable of fermenting glucose and cellobiose is present, it consumes the sugars, which are fermented to products determined by the nature of the secondary bacterium.

In the scheme presented there are three examples of secondary bacteria. Number 1 may be represented by *Clostridium thermohydrosulfuricum* (Wiegel et al., 1979). It ferments cellobiose and glucose to the same products as does *C. thermocellum*. The fermentation of cellulose in cocultures of *C. thermocellum* and *C. thermohydrosulfuricum* is speeded up considerably in comparison with fermentation with *C. thermocellum* alone (Ng et al., 1981; Carreira and Ljungdahl, 1983), and there is no accumulation of free sugars in the medium. Secondary bacterium number 2 is *Thermoanaerobacter ethanolicus* (Wiegel and Ljungdahl, 1981). It ferments cellobiose and glucose with ethanol and CO_2 as main products. This bacterium in cocultures with *C. thermocellum* ferments cellulose with ethanol, 1.6 mole per mole utilized glucose equivalent, as the principal product. Secondary bacterium number 3 is *Clostridium thermoautotrophicum* (Wiegel et al., 1981). This bacterium is acetogenic (homo-acetate-fermenting) and ferments 1 mole glucose to 3 moles of acetate, which is the only product. It also has the ability to synthesize acetate from CO_2 and H_2, and therefore may also serve as a H_2-consuming bacterium

terium in stage II of the scheme in Table III. However, it should be pointed out that the function in ecological systems of acetogenic bacteria as H_2-consuming is controversial. Winter and Wolfe (1980) obtained results with a mixed culture containing the acetogen *Acetobacterium woodii* and *Methanobacterium* strain AZ growing on fructose that demonstrated *A. woodii* as being H_2-producing [reaction (9)] in the presence of the methanogen [reaction (6)]. The result is summarized in reaction (10):

$$C_6H_{12}O_6 + 2H_2O \rightarrow 2CH_3COOH + 2CO_2 + 4H_2 \qquad (9)$$
$$CO_2 + 4H_2 \rightarrow CH_4 + 2H_2O \qquad (6)$$
$$\text{Sum: } C_6H_{12}O_6 \rightarrow 2CH_3COOH + CH_4 + CO_2 \qquad (10)$$

On the other hand, *Lachnospira multiporus,* when growing alone on pectin, produces acetate, methanol, ethanol, formate, lactate, CO_2, and H_2, and when in coculture with *Eubacterium limosum,* an acetogenic bacterium, forms acetate, butyrate, and CO_2 as the significant products (Rode et al., 1981). These results demonstrate interspecies hydrogen transfer and that *E. limosum* is H_2-consuming.

In mixed cultures between a primary cellulolytic bacterium and a secondary bacterium the products obtained generally reflect the product pattern of the secondary bacterium. This was noticed in the example above with the coculture *C. thermocellum* and *T. ethanolicus,* which produced ethanol and CO_2 as the main products from cellulose. Similarly, Enebo (1949) demonstrated that cellulose was fermented faster with a coculture of primary and secondary bacteria and that the products reflected the pattern of the secondary bacterium. Other examples of primary and secondary bacteria in coculture are *Bacteroides succinogenes/ Treponema bryantii* (Stanton and Canale-Parola, 1980), *C. thermocellum/Clostridium thermosaccharolyticum* (Avgerinos and Wang, 1980), *C. thermocellum/Thermoanaerobium brockii* (Lamed and Zeikus, 1980), and *Cellulomonas uda/Rhodopseudomonas capsulata* (Odom and Wall, 1983). In all cases cellulose is degraded faster with the coculture than with the cellulolytic bacterium alone, and with a product pattern reflecting the secondary bacterium.

The scheme in Table III indicates that the ancillary bacteria are converting the acidic and alcoholic products of fermentations by the primary and secondary bacteria to acetate, CO_2, and H_2, which are the substrates for the methanogenic bacteria. An example of an ancillary bacterium is *Desulfotomaculum nigrificans,* a strain of which was isolated with other thermophilic bacteria from soil samples (Wiegel and Ljungdahl, 1979). This bacterium, which is a sulfate-reducer, grows on lactate and ethanol in the absence of sulfate, with the formation of acetate, CO_2, and H_2. It has been successfully cocultured with *Methanobacterium thermoautotro-*

phicum and *Clostridium thermohydrosulfuricum*. The fermentation rate of glucose or cellobiose by *C. thermohydrosulfuricum* is increased in the coculture. The explanation is that *D. nigrificans* consumes lactate with formation of acetate, CO_2, and H_2 and that *M. thermoautotrophicum* consumes CO_2 and H_2 to form methane. Thus, lactate and H_2, which are products in the fermentation of glucose by *C. thermohydrosulfuricum,* are removed from the fermentation medium. Both lactate and H_2 are inhibitors of the fermentation of glucose by *C. thermohydrosulfuricum* and H_2 is inhibitory for *C. cellobioparum* (Chung, 1976).

Evidence for a function of sulfate-reducers as ancillary H_2-producing bacteria has also been obtained by McInerney and Bryant (1981b), who studied fermentations of lactate in the absence of sulfate with cocultures of sulfate-reducers and methanogens, and by Laube and Martin (1981), who investigated cellulose fermentations with tricultures containing *Acetivibrio cellulolyticus, Desulfovibrio* sp., and *Methanosarcina barkeri.*

In the example above, *D. nigrificans* is a H_2-producing bacterium. When sulfate is present, this organism will function instead as a H_2-consuming bacterium. The hydrogen produced by the primary and secondary bacteria during cellulose fermentations is then used by *D. nigrificans* for the reduction of sulfate to H_2S. The scheme in Table III may then be altered to involve only the primary and secondary bacteria as H_2-producing, whereas the ancillary bacteria become H_2-consuming. The important methanogenic bacteria under such circumstances will be those that convert acetate to methane and CO_2, whereas the H_2-consuming methanogens will have less significance. This seems to agree well with findings that acetate is the main source of methane in the natural environment (Mah, 1981).

The influence of *Desulfovibrio vulgaris* in the presence of sulfate on the fermentation of cellulose by an anaerobic cellulolytic bacterium have been evaluated by Bharati *et al.* (1982). The cellulolytic bacterium converts cellulose to formate, acetate, lactate, ethanol, CO_2, and H_2. In the presence of *D. vulgaris* the products are sulfide and acetate. These results demonstrate the role of the sulfate-reducing bacterium as H_2-consuming. Bharati *et al.* (1982) carried their mixed culture studies further by adding a phototrophic bacterium *(Chromatium vinosum)* to the coculture of the cellulolytic and sulfate-reducing bacteria. In this triculture, practically all cellulose that was fermented was converted to cell protein.

5. Conclusion

Cellulose is degraded in nature by a large number of different microorganisms—eukaryotes and prokaryotes, aerobes and anaerobes, mesophiles and thermophiles. Although the cellulolytic enzyme systems in many of the microorganisms have been and are under investigation, much needs to be added to our knowledge of these systems. Cellulolytic enzyme systems appear almost as diverse as there are types of cellulolytic microorganisms. One common property of these systems is that they contain several proteins with cellulolytic activity and that these proteins must act in combination to degrade cellulose efficiently.

Although it may be that most cellulose in nature is converted by aerobic microorganisms to CO_2, anaerobic processes have considerable ecological importance. We believe that these have been emphasized in our discussion. Even so, it has not been possible or practical to include many aspects of cellulose degradation in anaerobic systems, or even of aerobic systems for that matter. These systems are complex and involve several types of microorganisms that interact. These interactions are generally symbiotic and lead to efficient degradation of cellulose.

Two areas of neglect are clearly outstanding. Many insects, for example, termites, feed on cellulose, which is degraded by anaerobic microorganisms in the gut of the insect before it can be used as a nutritional source. Aspects of microorganisms in insects are covered in recent reviews by O'Brien and Slaytor (1982), Breznak (1982), and K.-P. Chang *et al.* (1984). The second area involves phytopathogenic microorganisms, mostly fungi. Here the reader is referred to a review by Bateman (1976).

ACKNOWLEDGMENTS. Investigations of anaerobic cellulolytic bacterial systems are supported by project DE-AS09-79ER10499 from the U.S. Department of Energy for work at the University of Georgia. L. G. Ljungdahl is also grateful for support provided by the Swedish Board of Energy Conservation during a year as guest researcher at the Swedish Forest Products Research Laboratory. The authors would also like to thank many individuals for reprints, preprints, and comments during work on the manuscript.

References

Ait, N., Creuzet, N., and Cattaneo, J., 1982, Properties of β-glucosidase purified from *Clostridium thermocellum, J. Gen. Microbiol.* **128**:569–577.

Akin, D. E., 1985, Chemical and biological structure in plants as related to microbial degradation of forage cell walls, in: *The Proceedings of VIth International Symposium on Ruminent Physiology,* Banff, Canada (in press).

Akin, D. E., and Barton II, F. E., 1983, Forage ultrastructure and the digestion of plant cell walls by rumen microorganisms, in: *Wood and Agricultural Residues* (E. J. Soltes, ed.), pp. 33–57, Academic Press, New York.

Akin, D. E., Gordon, G. L. R., and Hogan, J. P., 1983, Rumen bacterial and fungal degradation of *Digitaria pentzii* grown with or without sulfur, *Appl. Environ. Microbiol.* **46**:738–748.

Alexander, J. K., 1968, Purification and specificity of cellobiose phosphorylase from *Clostridium thermocellum*, *J. Biol. Chem.* **243**:2899–2904.

Allcock, E. R., and Woods, D. R., 1981, Carboxymethyl cellulase and cellobiase production by *Clostridium acetobutylicum*, in an industrial fermentation medium, *Appl. Environ. Microbiol.* **41**:539–541.

Ander, P., and Eriksson, K.-E., 1976a, Influence of carbohydrates on lignin degradation by the white-rot fungus *Sporotrichum pulverulentum*, *Svensk. Papperstidn.* **78**:643–652.

Ander, P., and Eriksson, K.-E., 1976b, The importance of phenol oxidase activity in lignin degradation by the white-rot fungus *Sporotrichum pulverulentum*, *Arch. Microbiol.* **109**:1–8.

Ander, P., and Eriksson, K.-E., 1977, Selective degradation of wood components by white-rot fungi, *Physiol. Plant.* **41**:239–248.

Araujo, E. F., Barros, E. G., Caldas, R. A., and Silva, D. O., 1983, Beta-glucosidase activity of a thermophilic cellulolytic fungus, *Humicola* sp., *Biotechnol. Lett.* **5**:781–784.

Atalla, R. H., 1979, Conformational effects in the hydrolysis of cellulose, *Adv. Chem. Series* **181**:55–69.

Atalla, R. H., 1983, The structure of cellulose: Recent developments, in: *Wood and Agricultural Residues* (E. J. Soltes, ed.), pp. 59–77, Academic Press, New York.

Atalla, R. H., and VanderHart, D. L., 1984, Native cellulose: A composite of two distinct crystalline forms, *Science* **223**:283–285.

Atkinson, L. P., and Hall, J. R., 1976, Methane distribution and production in the Georgia salt marsh, *Estuarine Coastal Mar. Sci.* **4**:677–686.

Aumen, N. G., Bottomley, P. J., Ward, G. M., and Gregory, S. V., 1983, Microbial decomposition of wood in streams: Distribution of microflora and factors affecting (^{14}C)lignocellulose mineralization, *Appl. Environ. Microbiol.* **46**:1409–1416.

Avgerinos, G. C., and Wang, D. I. C., 1980, Direct microbiological conversion of cellulosics to ethanol, *Annu. Rep. Ferment. Proc.* **4**:165–191.

Ayers, A. R., Ayers, S. B., and Eriksson, K.-E., 1978, Cellobiose oxidase, purification and partial characterization of a hemeprotein from *Sporotrichum pulverulentum*, *Eur. J. Biochem.* **90**:171–181.

Ayers, W. A., 1959, Phosphorolysis and synthesis of cellobiose by cell extracts from *Ruminococcus flavefaciens*, *J. Biol. Chem.* **234**:2819–2822.

Balch, W. E., Fox, G. E., Mangram, L. J., Woese, C. R., and Wolfe, R. S., 1979, Methanogens: Reevaluation of a unique biological group, *Microbiol. Rev.* **43**:260–296.

Bassham, J. A., 1975, General considerations, *Biotechnol. Bioeng. Symp.* **5**:9–19.

Bateman, D. F., 1976, Plant cell wall hydrolysis by pathogens, in: *Biochemical Aspects of Plant Parasite Relationships* (J. Friend and D. R. Threlfall, eds.), pp. 79–103, Academic Press, London.

Bauchop, T., 1981, The anaerobic fungi in rumen fiber digestion, *Agric. Environ.* **6**:339–348.

Bauchop, T., and Mountfort, D. O., 1981, Cellulose fermentation by a rumen anaerobic fungus in both the absence and the presence of rumen methanogens, *Appl. Environ. Microbiol.* **42**:1103–1110.

Bavendamm, W., 1928, Ueber das Vorkommen und den Nachweis von Oxydasen bei holzzerstorenden Pilzen, *Z. Pflanzenkrankh.* **38**:257–320.

Bayer, E. A., Kenig, R., and Lamed, R., 1983, Studies on the adherence of *Clostridium thermocellum* to cellulose, *J. Bacteriol.* **156**:818–827.

Beguin, P., and Eisen, H., 1978, Purification and partial characterization of three extracellular cellulases from *Cellulomonas* sp., *Eur. J. Biochem.* **87**:525–531.

Beguin, P., Eisen, H., and Roupas, A., 1977, Free and cellulose-bound cellulases in a *Cellulomonas* sp., *J. Gen. Microbiol.* **101**:191–196.

Benner, R., Maccubbin, A. E., and Hodson, R. E., 1984, Preparation, characterization, and microbial degradation of specifically radiolabeled [^{14}C] lignocelluloses from marine and freshwater macrophytes, *Appl. Environ. Microbiol.* **47**:381–389.

Berg, B., 1975, Cellulase in *Cellvibrio fulvus, Can. J. Microbiol.* **21**:51–57.

Berg, B., 1978, Cellulose degradation and cellulase formation by *Phialophora malorum, Arch. Microbiol.* **118**:61–65.

Berg, B., von Hofsten, B., and Pettersson, L. G., 1972a, Growth and cellulase formation by *Cellvibrio fulvus, J. Appl. Bacteriol.* **35**:201–214.

Berg, B., von Hofsten, B., and Pettersson, G., 1972b, Electron microscopic observations on the degradation of cellulose fibres by *Cellvibrio fulvus* and *Sporocytophaga myxococcoides, J. Appl. Bacteriol.* **35**:215–219.

Berghem, L. E. R., Pettersson, L. G., and Axiö-Fredriksson, U.-B., 1976, The mechanism of enzymatic cellulose degradation. Purification and some properties of two different 1,4-β-glucan glucanohydrolases from *Trichoderma viride, Eur. J. Biochem.* **61**:621–630.

Berkeley, R. C. W., Gooday, G. W., and Ellwood, D. C. (eds.), 1979, *Microbial Polysaccharides and Polysaccharases,* Academic Press, London.

Berner, K. E., and Chapman, E. S., 1977, The cellulolytic activity of six oomycetes, *Mycologia* **69**:1232–1236.

Bharati, P. A. L., Baulaigue, R., and Matheron, R., 1982, Degradation of cellulose by mixed cultures of fermentative bacteria and anaerobic sulfur bacteria, *Zentralbl. Bakteriol. Hyg. I Abt. Orig. C* **3**:466–474.

Bikales, N. M., and Segal, L. (eds.), 1971, *Cellulose and Cellulose Derivatives: High Polymers,* Vol. V, Parts IV and V, Wiley-Interscience, New York.

Blackwell, J., 1982, The macromolecular organization of cellulose and chitin, in: *Cellulose and Other Natural Polymer Systems* (R. M. Brown, Jr., ed.), pp. 403–428, Plenum Press, New York.

Blanchette, R. A., 1980a, Wood decomposition by *Phellinus (Fomes) pini:* A scanning electron microscopy study, *Can. J. Bot.* **58**:1496–1503.

Blanchette, R. A., 1980b, Wood decay: A submicroscopic view, *J. For. Res.* **78**:734–737.

Blanchette, R. A., 1982, *Phellinus (Fomes) pini* decay associated with sweetgum rust in sapwood of jack pine, *Can. J. For. Res.* **12**:304–310.

Blanchette, R. A., and Shaw, C. G., 1978, Associations among bacteria, yeasts and basidiomycetes during wood decay, *Phytopathology* **68**:631–637.

Blanchette, R. A., Shaw, C. G., and Cohen, A. L., 1978, A SEM study of the effects of bacteria and yeasts on wood decay by brown- and white-rot fungi, *Scanning Electron Microsc.* **11**:61–67.

Branden, A. R., and Thayer, D. W., 1976, Serological study of *Cellulomonas, Int. J. Syst. Bacteriol.* **26**:123–126.

Breznak, J. A., 1982, Intestinal microbiota of termites and other xylophagous insects, *Annu. Rev. Microbiol.* **36**:323–343.

Brown, Jr., R. M., 1981, Integration of biochemical and visual approaches to the study of cellulose biosynthesis and degradation, in: *The Ekman Days,* International Symposium on Wood and Pulping Chemistry, Stockholm, Vol. 3, pp. 3–15.

Brown, Jr., R. M. (ed.), 1982, *Cellulose and Other Natural Polymer Systems: Biogenesis, Structure and Degradation,* Plenum Press, New York.

Bryant, M. P., 1973, Nutritional requirements of the predominant rumen cellulolytic bacteria, *Fed. Proc.* **32**:1809–1813.

Bryant, M. P., 1979, Microbial methane production—theoretical aspects, *J. Anim. Sci.* **48**:193–201.

Bryant, M. P., and Doetsch, R. N., 1954, A study of actively cellulolytic rod-shaped bacteria of the bovine rumen, *J. Dairy Sci.* **37**:1176–1183.

Bryant, M. P., and Robinson, I. M., 1961, An improved nonselective culture medium for ruminal bacteria and its use in determining diurnal variation in numbers of bacteria in the rumen, *J. Dairy Sci.* **41**:1446–1456.

Bryant, M. P., and Small, N., 1956, The anaerobic monotrichous butyric acid-producing curved rod-shaped bacteria of the rumen, *J. Bacteriol.* **76**:16–21.

Bryant, M. P., Small, N., Bouma, C., and Robinson, I. M., 1958, Characteristics of ruminal anaerobic cellulolytic cocci and *Cillobacterium cellulosolvens* n. sp., *J. Bacteriol.* **76**:529–537.

Bryant, M. P., Wolin, E. A., Wolin, M. J., and Wolfe, R. S., 1967, *Methanobacillus omelianski;* A symbiotic association of two species of bacteria, *Arch. Mikrobiol.* **59**:20–31.

Buresh, R. J., and Patrick, W. H., 1978, Nitrate reduction to ammonium in anaerobic soil. *Soil Sci. Soc. Am. J.* **42**:913–918.

Cabello, A., Conde, J., and Otero, M. A., 1981, Prediction of the degradability of sugarcane cellulosic residues by indirect methods, *Biotechnol. Bioeng.* **23**:2737–2745.

Carpita, N. C., 1982, Cellulose synthesis in detached cotton fibers, in: *Cellulose and Other Natural Polymer Systems: Biogenesis, Structure and Degradation* (R. M. Brown, Jr., ed.), pp. 225–242, Plenum Press, New York.

Carreira, L., and Ljungdahl, L. G., 1983, Production of ethanol from biomass using anaerobic thermophilic bacteria, in: *Liquid Fuel Developments* (D. Wise, ed.), pp. 1–29, CRC Press, Boca Raton, Florida.

Caskey, W. H., and Tiedje, J. M., 1980, The reduction of nitrate to ammonium by a *Clostridium* sp. isolated from soil, *J. Gen. Microbiol.* **119**:217–223.

Cato, E. P., Moore, W. E. C., and Bryant, M. P., 1978, Designation of neotype strains for *Bacteroides amylophilus* Hamlin and Hungate 1956 and *Bacteroides succinogens* Hungate 1950, *Int. J. Syst. Bacteriol.* **28**:491–495.

Caulfield, D. F., and Moore, W. E., 1974, Effect of varying crystallinity of cellulose on enzymic hydrolysis, *Wood Sci.* **6**:375–379.

Chahal, D. S., and Hawksworth, D. L., 1976, *Chaetomium cellulolyticum,* a new thermotolerant and cellulolytic *Chaetomium,* I. Isolation, description and growth rate, *Mycologia* **68**:600–610.

Chang, K.-P., Dasch, G. A., and Weiss, E., 1984, Endosymbionts of fungi and invertebrates other than arthropods, in: *Bergey's Manual of Systematic Bacteriology* (N. A. Krieg and J. G. Holt, eds.), Vol. 1, pp. 833–836, Williams and Wilkins, Baltimore.

Chang, W. T. H., and Thayer, D. W., 1977, The cellulase system of a *Cytophaga* species, *Can. J. Microbiol.* **23**:1285–1292.

Chapman, E. S., Evans, E., Jacobelli, M. C., and Logan, A. A., 1975, The cellulolytic and amylolytic activity of *Papulaspora thermophila, Mycologia* **67**:608–615.

Charpentier, M., 1965, Étude de l'activité cellulolytique de *Sporocytophaga myxococcoides, Ann. Inst. Pasteur* **109**:771–797.

Charpentier, M., and Robic, D., 1974, Degradation de la cellulose par un micro-organisme du sol: *Sporocytophaga myxococcoides:* caracterisation d'une exoglucanase, *C. R. Acad. Sci. Paris* **279**:863–866.

Choi, W. Y., Haggett, K. D., and Dunn, N. W., 1978, Isolation of a cotton wool degrading strain of *Cellulomonas;* Mutants with altered ability to degrade cotton wool, *Aust. J. Biol. Sci.* **31**:553–564.

Chosson, J., and Dupuy, P., 1983, Improvement of the cellulolytic activity of a natural population of aerobic bacteria, *Eur. J. Appl. Microbiol. Biotechnol.* **18**:163–167.

Choudhury, N., Gray, P. P., and Dunn, N. W., 1980, Reducing sugar accumulation from alkali pretreated sugar cane bagasse using *Cellulomonas, Eur. J. Appl. Microbiol. Biotechnol.* **11**:50–54.

Christensen, P. J., 1977, The history, biology, and taxonomy of the *Cytophaga* group, *Can. J. Microbiol.* **23**:1599–1653.

Chu, S. C., and Jeffrey, G. A., 1968, The refinement of the crystal of β-D-glucose and cello-biose, *Acta Cryst. B* **24**:830–838.

Chung, K.-T., 1976, Inhibitory effects of H_2 on growth of *Clostridium cellobioparum, Appl. Environ. Microbiol.* **31**:342–348.

Clarke, R. T. J., Bailey, R. W., and Gaillard, B. D. E., 1969, Growth of rumen bacteria on plant cell wall polysaccharides, *J. Gen. Microbiol.* **56**:79–86.

Clermont, S., Charpentier, M., and Percheron, F., 1970, Polysaccharidases de *Sporocytophaga myxococcoides* β-mannanase, cellulase et xylanase, *Bull. Soc. Chim. Biol.* **52**:1481–1495.

Cobb, T. L., 1982, The nonenzymatic decomposition of cellulose by the brown-rot fungus *Gloeophyllum trabeum*, Thesis, Michigan Tech University.

Colvin, J. R., Sowden, L. C., Patel, G. B., and Khan, A. W., 1982, The ultrastructure of *Acetivibrio cellulolyticus*, a recently isolated cellulolytic anaerobe, *Curr. Microbiol.* **7**:13–17.

Coudray, M. R., Canevascini, G., and Meier, H., 1982, Characterization of a cellobiose dehydrogenase in the cellulolytic fungus *Sporotrichum (Chrysosporium) thermophile, Biochem. J.* **203**:277–284.

Cowling, E. B., 1961, Comparative biochemistry of the decay of sweetgum by white-rot and brown-rot fungi, *U. S. Dept. Agric. Tech. Bull.* **1258**:1–79.

Cowling, E. B., 1963, Structural features of cellulose that influence its susceptibility to enzymatic hydrolysis, in: *Advances in Enzymic Hydrolysis of Cellulose and Related Materials* (E. T. Reese, ed.), pp. 1–32, Macmillan, New York.

Crawford, D. L., Crawford, R. L., and Pometto III, A. I., 1977, Preparation of specifically labeled ^{14}C-(lignin)- and ^{14}C-(cellulose)-lignocelluloses and their decomposition by the microflora of soil, *Appl. Environ. Microbiol.* **33**:1247–1251.

Creuzet, N., and Frixon, C., 1983, Purification and characterization of an endoglucanase from a newly isolated thermophilic bacterium, *Biochimie* **65**:149–156.

Daigneault-Sylvestre, N., and Kluepfel, D., 1979, Method for rapid screening of cellulolytic streptomycetes and their mutants, *Can. J. Microbiol.* **25**:858–860.

Dekker, R. F. H., 1980, Induction and characterization of a cellobiose dehydrogenase produced by a species of *Monilia, J. Gen. Microbiol.* **120**:309–316.

Deshpande, V., Eriksson, K.-E., and Pettersson, B., 1978, Production, purification and partial characterization for 1,4-β-glucosidase enzymes from *Sporotrichum pulverulentum, Eur. J. Biochem.* **90**:191–198.

Deshpande, V., Mishra, C., Ghadge, G. D., Seeta, R., and Rao, M., 1983, Separation and recovery of *Penicillium funiculosum* cellulase and glucose from cellulosic hydrolysates using polyacrylate gel, *Biotechnol. Lett.* **5**:391–394.

Desrochers, M., Jurasek, L., and Paice, M. G., 1981, High production of β-glucosidase in *Schizophyllum commune:* Isolation of the enzyme and effect of the culture filtrate on cellulose hydrolysis, *Appl. Environ. Microbiol.* **41**:222–228.

Dilworth, G., Wiegel, J., Ljungdahl, L. G., and Peck, Jr., H. D., 1980, Reconstitution of mesophilic microbial associations which ferment cellulose to various products, in: *Proc. Colloque Cellulolyse Microbienne,* Marseilles Centre National de la Rescherche Scientifique, pp. 111–126.

Doudoroff, M., and Palleroni, N. J., 1974, Pseudomonas, in: *Bergey's Manual of Determinative Bacteriology,* 8th ed. (R. E. Buchanan and N. E. Ribbons, eds.), pp. 217–249, Williams and Wilkins, Baltimore.

Duong, T.-V. C., Johnson, E. A., and Demain, A. L., 1983, Thermophilic, anaerobic and cellulolytic bacteria, in: *Topics in Enzyme and Fermentation Technology 7* (A. Wiseman, ed.), pp. 156–195, John Wiley, New York.

Eberhart, B. M., Beck, R. S., and Goolsby, K. M., 1977, Cellulase of *Neurospora crassa, J. Bacteriol.* **130**:181–186.

Ehhalt, D. H., 1976, The atmospheric cycle of methane, in: *Microbial Production and Uti-*

lization of Gases (H. G. Schlegel, G. Gottschalk, and N. Pfennig, eds.), pp. 13–22, E. Goltze KG, Göttingen.

Ellefsen, Ø., and Tønnesen, B. A., 1971, Polymeric forms, in: *Cellulose and Cellulose Derivatives* (N. M. Bikales and L. Segal, eds.), Vol. V, Part IV, pp. 151–180, Wiley, New York.

Emi, S., and Yamamoto, T., 1972, Purification and properties of several galactanases of *Bacillus subtilis* var. *amylosacchariticus, Agric. Biol. Chem.* **36:**1945–1954.

Enebo, L., 1949, Symbiosis in thermophilic cellulose fermentation, *Nature* 163:805.

Enebo, L., 1951, On three bacteria connected with thermophilic cellulose fermentation, *Physiol. Plant.* **4:**652–666.

Erdtman, H., 1939, Phenolic constituents of the pine heartwood, their physiological importance and their retarding action upon the normal digestion of pine heartwood according to the sulfite process, *Justus Liebig's Ann. Chem.* **539:**116–127.

Eriksson, K.-E., 1981a, Cellulases of fungi, in: *Trends in the Biology of Fermentations for Fuels and Chemicals* (A. Hollaender, ed.), pp. 19–32, Plenum Press, New York.

Eriksson, K.-E., 1981b, Fungal degradation of wood components, *Pure Appl. Chem.* **53:**33–43.

Eriksson, K.-E., and Hamp, S. G., 1978, Regulation of endo-1,4-β-glucanase production in *Sporotrichum pulverulentum, Eur. J. Biochem.* **90:**183–190.

Eriksson, K.-E., and Johnsrud, C., 1982, Mineralization of carbon, in: *Experimental Microbial Ecology* (A. Hollaender, ed.), pp. 134–153, Plenum Press, New York.

Eriksson, K.-E., and Pettersson, B., 1982, Purification and partial characterization of two acidic proteases from the white-rot fungus *Sporotrichum pulverulentum, Eur. J. Biochem.* **124:**635–642.

Eriksson, K.-E., and Wood, T. M., 1984, Biodegradation of cellulose, in: *Biosynthesis and Biodegradation of Wood Components* (T. Higuchi, ed.), pp. 469–503, Academic Press, New York.

Eriksson, K.-E., Grünewald, A., and Vallander, L., 1980a, Studies of growth conditions in wood for three white-rot fungi and their cellulase-less mutants, *Biotechnol. Bioeng.* **22:**363–376.

Eriksson, K.-E., Grünewald, A., Nilsson, T., and Vallander, L., 1980b, A scanning electron microscopy study of the growth and attack on wood by three white-rot fungi and their cellulase-less mutants, *Holzforschung* **34:**207–213.

Esteban, R., Villanueva, J. R., and Villa, T. G., 1982, β-D-Xylanases of *Bacillus circulans* WL-12, *Can. J. Microbiol.* **28:**733–739.

Fähnrich, P., and Irrang, K., 1982, Conversion of cellulose to sugars and cellobionic acid by the extracellular enzyme system of *Chaetomium cellulolyticum, Biotechnol. Lett.* **4:**775–780.

Fan, L. T., Lee, Y.-H., and Beardmore, D. H., 1980, Mechanism of the enzymatic hydrolysis of cellulose: Effects of major structural features of cellulose on enzymatic hydrolysis, *Biotechnol. Bioeng.* **22:**177–199.

Fenchel, T. M., and Jørgensen, B. B., 1977, Detritus food chains of aquatic ecosystems: The role of bacteria, in *Advances in Microbial Ecology*, Vol. 1 (M. Alexander, ed.), pp. 1–58, Plenum Press, New York.

Fennington, G., Neubauer, D., and Stutzenberger, F., 1984, Cellulase biosynthesis in a catabolite repression-resistant mutant of *Thermomonospora curvata, Appl. Environ. Microbiol.* **47:**201–204.

Ferchak, J. D., Hägerdal, B., and Pye, E. K., 1980, Saccharification of cellulose by the cellulolytic enzyme system of *Thermomonospora* sp. II. Hydrolysis of cellulosic substrates, *Biotechnol. Bioeng.* **22:**1527–1542.

Fergus, C. L., 1969, The cellulolytic activity of thermophilic fungi and actinomycetes, *Mycologia* **61:**120–129.

Fogarty, W. M., and Griffin, P. J., 1973, Some preliminary observations on the production and properties of a cellulolytic enzyme elaborated by *Bacillus polymyxa*, *Biochem. Soc. Trans.* **1**:1297–1298.

Fogarty, W. M., and Ward, O. P., 1973, A preliminary study on the production, purification and properties of a xylan-degrading enzyme from a *Bacillus* sp. isolated from water-stored sitka spruce *(Picea sitchensis)*, *Biochem. Soc. Trans.* **1**:260–262.

Folan, M. A., and Coughlan, M. P., 1981, Cellulase activity of colour "variants" of *Talaromyces emersonii*, *Int. J. Biochem.* **13**:243–245.

Forsberg, C. W., and Groleau, D., 1982, Stability of the endo-β-1,4-glucanase and β-1,4-glucosidase from *Bacteroides succinogenes*, *Can. J. Microbiol.* **28**:144–148.

Fusee, M. C., and Leatherwood, J. M., 1971, Regulation of cellulase from *Ruminococcus*, *Can. J. Microbiol.* **18**:347–353.

Gardner, K. H., and Blackwell, J., 1974, The structure of native cellulose, *Biopolymers* **13**:1975–2001.

Garg, S. K., and Neelakantan, S., 1982, Studies on the properties of cellulase enzyme from *Aspergillus terreus* GN1, *Biotechnol. Bioeng.* **24**:737–742.

Garrett, S. D., 1963, *Soil Fungi and Soil Fertility*, Pergamon Press, Oxford.

Giallo, J., Gaudin, C., Belaich, J. P., Petitdemange, E., and Caillet-Mangin, F., 1983, Metabolism of glucose and cellobiose by cellulolytic mesophilic *Clostridium* sp. strain H10, *Appl. Environ. Microbiol.* **45**:843–849.

Greaves, H., 1971, The effect of substrate availability on cellulolytic enzyme production by selected wood-rotting microorganisms, *Aust. J. Biol. Sci.* **24**:1169–1180.

Groleau, D., and Forsberg, C. W., 1981, Cellulolytic activity of the rumen bacterium *Bacteroides succinogenes*, *Can. J. Microbiol.* **27**:517–530.

Groleau, D., and Forsberg, C. W., 1983, Partial characterization of the extracellular carboxymethylcellulase activity produced by the rumen bacterium *Bacteroides succinogenes*, *Can. J. Microbiol.* **29**:504–517.

Gubsch, G., 1979, Investigations on the effects of temperature on the degradation of cellulose by bacteria from aquatic environments in different climatic zones, *Acta Hydrochim. Hydrobiol.* **7**:307–316.

Gupta, D. P., and Heale, J. B., 1971, Induction of cellulase (C_x) in *Verticillium albo-atrum*, *J. Gen. Microbiol.* **63**:163–173.

Gupta, V. C., 1967, Carbohydrates, in: *Soil Biochemistry*, Vol. 1 (A. D. McLaren and G. H. Peterson, eds.), pp. 91–118, Marcel Dekker, New York.

Hägerdal, B. G. R., Ferchak, J. D., and Pye, E. K., 1978, Cellulolytic enzyme system of *Thermoactinomyces* sp. grown on microcrystalline cellulose, *Appl. Environ. Microbiol.* **36**:606–612.

Hägerdal, B., Harris, H., and Pye, E. K., 1979, Association of β-glucosidase with intact cells of *Thermoactinomyces*, *Biotechnol. Bioeng.* **21**:345–355.

Hägerdal, B., Ferchak, J. D., and Pye, E. K., 1980, Saccharification of cellulose by the cellulolytic enzyme system of *Thermomonospora* sp. I. Stability of cellulolytic activities with respect to time, temperature, and pH, *Biotechnol. Bioeng.* **22**:1515–1526.

Haggett, K. D., Gray, P. P., and Dunn, N. W., 1979, Crystalline cellulose degradation by a strain of *Cellulomonas* and its mutant derivatives, *Eur. J. Appl. Microbiol.* **8**:183–190.

Haigler, C. H., and Benziman, M., 1982, Biogenesis of cellulose I microfibrils occurs by cell-directed self-assembly in *Acetobacter xylinum*, in: *Cellulose and Other Natural Polymer Systems: Biogenesis, Structure and Degradation* (R. M. Brown, Jr., ed.), pp. 273–297, Plenum Press, New York.

Halliwell, G., 1965, Hydrolysis of fibrous cotton and reprecipitated cellulose by cellulolytic enzymes from soil microorganisms, *Biochem. J.* **95**:270–281.

Halliwell, G., and Bryant, M. P., 1963, The cellulolytic activity of pure strains of bacteria from the rumen of cattle, *J. Gen. Microbiol.* **32**:441–448.

Halliwell, G., and Vincent, R., 1981, The action on cellulose and its derivatives of a purified 1,4-β-glucanase from *Trichoderma koningii, Biochem. J.* **199**:409–417.

Ham, J. T., and Williams, D. G., 1970, The crystal and molecular structure of methyl β-cellobioside-methanol, *Acta Cryst.* B **26**:1373–1383.

Han, Y. W., and Srinivasan, V. R., 1968, Isolation and characterization of a cellulose-utilizing bacterium, *Appl. Microbiol.* **16**:1140–1145.

Harrer, W., Kubicek, C. P., Rohr, M., Wurth, H., and Marihart, J., 1983, The effect of carboxymethyl cellulose addition on extracellular enzyme formation in *Trichoderma pseudokoningii, Eur. J. Appl Microbiol. Biotechnol.* **17**:339–343.

Heale, J. B., and Gupta, D. P., 1971, The utilization of cellobiose by *Verticillium alboatrum, J. Gen. Microbiol.* **63**:175–181.

Heath, I. B., Bauchop, T., and Skipp, R. A., 1983, Assignment of the rumen anaerobe *Neocallimastix frontalis* to the *Spizellomycetales (Chytridiomycetes)* on the basis of its polyflagellate zoospore ultrastructure, *Can. J. Bot.* **61**:295–307.

Henssen, A., 1957, Beitrage zur Morphologie und Systematik der thermophilen *Actinomyceten, Arch. Mikrobiol.* **26**:373–414.

Highley, T. L., 1973, Influence of carbon source on cellulase activity of white-rot and brown-rot fungi, *Wood Fiber* **5**:50–58.

Highley, T. L., 1975a, Can wood-rot fungi degrade cellulose without other wood constituents? *For. Prod. J.* **25**:38–39.

Highley, T. L., 1975b, Properties of cellulases of two brown-rot fungi and two white-rot fungi, *Wood Fiber* **6**:275–281.

Highley, T. L., 1977, Requirements for cellulose degradation by a brown rot fungus, *Mater. Org. (Berl.)* **12**:25–36.

Highley, T. L., 1980, Cellulose degradation by cellulose-clearing and non-cellulose clearing brown-rot fungi, *Appl. Environ. Microbiol.* **40**:1145–1147.

Highley, T. L., Wolter, K. E., and Evans, F. J., 1981, Polysaccharide-degrading complex produced in wood and liquid media by the brown-rot fungus *Poria placenta, Wood Fiber* **13**:265–274.

Hiltner, P., and Dehority, B. A., 1983, Effect of soluble carbohydrates on digestion of cellulose by pure cultures of rumen bacteria, *Appl. Environ. Microbiol.* **46**:642–648.

Hiroi, T., and Eriksson, K.-E., 1976, Microbiological degradation of lignin. I. Influence of cellulose on the degradation of lignins by the white-rot fungus *Pleurotus ostreatus, Svensk. Papperstidn.* **79**:157–161.

Hobson, P. N., 1971, Rumen micro-organisms, in: *Progress in Industrial Microbiology* (D. J. D. Hockenhull, ed.), pp. 41–77, J. & A. Churchill, London.

Hobson, P. N., Bousfield, S., and Summers, R., 1974, Anaerobic digestion of organic matter, in: *CRC Critical Reviews in Environmental Control,* pp. 131–191, CRC Press, Cleveland.

Hopgood, M. F., and Walker, D. J., 1967, Succinic acid production by rumen bacteria. II. Radioisotope studies on succinate production by *Ruminococcus flavefaciens, Aust. J. Biol. Sci.* **20**:165–182.

Hulcher, F. H., and King, K. W., 1958, Disaccharide preference of an aerobic cellulolytic bacterium, *Cellvibrio gilvus* n. sp., *J. Bacteriol.* **76**:565–570.

Hulme, M. A., and Shields, J. K., 1972, Interaction between fungi in wood blocks, *Can. J. Bot.* **50**:1421–1427.

Humphrey, A. E., Moreira, A., Armiger, W., and Zabriskie, D., 1977, Production of single cell protein from cellulose wastes, *Biotechnol. Bioeng. Symp.* **7**:45–64.

Hungate, R. E., 1944, Studies on cellulose fermentation I. The culture and physiology of an anaerobic cellulose-digesting bacterium, *J. Bacteriol.* **48**:499–513.

Hungate, R. E., 1946, Studies on cellulose fermentation. II. An anaerobic cellulose decomposing *Actinomycete, Micromonospora propionici* N. sp., *J. Bacteriol.* **51**:51–56.

Hungate, R. E., 1950, The anaerobic mesophilic cellulolytic bacteria, *Bacteriol. Rev.* **14:**1–50.

Hungate, R. E., 1957, Microorganisms in the rumen of cattle fed a constant ration, *Can. J. Microbiol.* **3:**289–311.

Hungate, R. E., 1966, *The Rumen and Its Microbes,* Academic Press, New York.

Hungate, R. E., 1969a, A roll tube method for cultivation of strict anaerobes, in: *Methods in Microbiology,* Vol. 3B (J. R. Norris and D. W. Ribbons, eds.), pp. 117–132, Academic Press, New York.

Hungate, R. E., 1969b, Interrelationships in the rumen microbiota, in: *Physiology of Digestion and Metabolism in the Ruminant* (A. T. Phillipson, ed.), pp. 292–305, Oriel Press, Newcastle-upon-Tyne, England.

Hungate, R. E., 1975, The rumen microbial ecosystem, *Annu. Rev. Ecol. Syst.* **6:**39–66.

Hungate, R. E., and Stack, R. J., 1982, Phenylpropanoic acid: Growth factor for *Ruminococcus albus, Appl. Environ. Microbiol.* **44:**79–83.

Hurst, P. L., Sullivan, P. A., and Shepherd, M. G., 1978, Substrate specificity and mode of action of a cellulase from *Aspergillus niger, Biochem. J.* **169:**389–395.

Hutterman, A., and Noelle, A., 1982, Characterization and regulation of cellobiose dehydrogenase in *Fomes annosus, Holzforschung* **36:**283–286.

Iannotti, E. L., Kafkewitz, D., Wolin, M. J., and Bryant, M. P., 1973, Glucose fermentation products of *Ruminococcus albus* grown in continuous culture with *Vibrio succinogenes:* Changes caused by interspecies transfer of H_2, *J. Bacteriol.* **114:**1231–1240.

Ide, J. A., Daly, J. M., and Rickard, P. A. D., 1983, Production of glycosidase activity by *Cellulomonas* during growth on various carbohydrate substrates, *Appl. Microbiol. Biotechnol.* **18:**100–102.

Ikeda, R., Yamamoto, T., and Funatsu, M., 1973, Chemical and enzymatic properties of acid-cellulase produced by *Aspergillus niger, Agric. Biol. Chem.* **37:**1169–1175.

Inaoka, M., and Soda, H., 1956, Crystalline xylanase, *Nature* **178:**202–203.

Ishaque, M., and Kluepfel, D., 1980, Cellulase complex of a mesophilic *Streptomyces* strain, *Can. J. Microbiol.* **26:**183–189.

Iwasaki, T., Hayashi, K., and Funatsu, M., 1964, Purification and characterization of two types of cellulase from *Trichoderma koningi, J. Biochem.* **55:**209–212.

Jain, M. K., Kapoor, K. K., and Mishra, M. M., 1979, Cellulase activity, degradation of cellulose and lignin, and humus formation by thermophilic fungi, *Trans. Br. Mycol. Soc.* **73:**85–89.

Jarvis, B. D. W., Henderson, C., and Asmundson, R. V., 1978, The role of carbonate in the metabolism of glucose by *Butyrivibrio fibrisolvens, J. Gen. Microbiol.* **105:**287–295.

Jeffries, T. W., Choi, S., and Kirk, T. K., 1981, Nutritional regulation of lignin degradation by *Phanerochaete chrysosporium, Appl. Environ. Microbiol.* **42:**290–296.

Joglekar, A. V., Srinivasan, M. C., Manchanda, A. C., Jogdand, V. V., and Karanth, N. G., 1983, Studies on cellulase production by *Penicillium funiculosum* strain in an instrumented fermenter, *Enzyme Microb. Technol.* **5:**22–24.

Johnson, E. A., Sakajoh, M., Halliwell, G., Madia, A., and Demain, A. L., 1982, Saccharification of complex cellulosic substrates by the cellulase system from *Clostridium thermocellum, Appl. Environ. Microbiol.* **43:**1125–1132.

Jørgensen, B. B., 1977, The sulfur cycle of a coastal marine sediment (Limfjorden, Denmark), *Limnol. Oceanogr.* **22:**814–832.

Käärik, A. A., 1974a, Succession of microorganisms during wood decay, in: *Biological Transformation of Wood by Microorganisms* (W. Liese, ed.), pp. 39–51, Springer-Verlag, New York.

Käärik, A. A., 1974b, Decomposition of wood, in: *Biology of Plant Litter Decomposition* (C. H. Dickinson and G. J. F. Pugh, eds.), Vol. 1, pp. 129–174, Academic Press, London.

Kanda, T., Noda, I., Wakabayashi, K., and Nisizawa, K., 1983, Transglycosylation activities of exo- and endo-type cellulases from *Irpex lacteus (Polyporus tulipiferae), J. Biochem.* **93:**787–794.

Keddie, R. M., 1974, Cellulomonas, in: *Bergey's Manual of Determinative Bacteriology,* 8th ed. (R. E. Buchanan and N. E. Gibbons, eds.), pp. 629–631, Williams and Wilkins, Baltimore.

Kelleher, T. J., Jr., 1981, The lignocellulolytic activity of *Phanerochaete chrysosporium* Burds: Regulation and application, Thesis, Department of Biochemistry and Microbiology, Rutgers University, New Brunswick, New Jersey.

Kellerman, K. F., and McBeth, I. G., 1912, The fermentation of cellulose, *Zentralbl. Bakteriol. Parasitenkd. Infektionskr. Hyg. Abt. II* **34:**485–494.

Khan, A. W., 1980, Cellulolytic enzyme system of *Acetivibrio cellulolyticus,* a newly isolated anaerobe, *J. Gen. Microbiol.* **121:**499–502.

Khan, A. W., Saddler, J. N., Patel, G. B., Colvin, J. R., and Martin, S. M., 1980, Degradation of cellulose by a newly isolated mesophilic anaerobe, Bacteriodaceae family, *FEMS Microbiol. Lett.* **7:**47–50.

Kim, B. H., and Wimpenny, J. W. T., 1981, Growth and cellulolytic activity of *Cellulomonas flavigena, Can. J. Microbiol.* **27:**1260–1266.

King, K. W., and Vessal, M. I., 1969, Enzymes of the cellulase complex, *Adv. Chem. Series* **95:**7–25.

Kirk, T. K., 1973, Polysaccharide integrity as related to the degradation of lignin in wood by white-rot fungi, *Phytopathology* **63:**1504–1507.

Kirk, T. K., and Highley, T. L., 1973, Quantitative changes in structural components of conifer woods during decay by white- and brown-rot fungi, *Phytopathology* **63:**1338–1342.

Kirk, T. L., Connors, W. J., and Zeikus, J. G., 1976, Requirement for a growth substrate during lignin decomposition by two wood-rotting fungi, *Appl. Environ. Microbiol.* **32:**192–194.

Knösel, D., 1971, Continued investigation for pectolytic and cellulolytic activity of different *Bacillus* species, *Zentralbl. Bakteriol. Parasitenkd. Infektionskr. Hyg. Abt. II* **126:**604–609.

Knowles, R., 1982, Dentrification, *Microbiol. Rev.* **46:**43–70.

Koenigs, J. W., 1974a, Production of hydrogen peroxide by wood-rotting fungi in wood and its correlation with weight loss, depolymerization, and pH changes, *Arch. Microbiol.* **99:**129–145.

Koenigs, J. W., 1974b, Hydrogen peroxide and iron: A proposed system for degradation of wood by brown-rot basidomycetes, *Wood Fiber* **6:**66–80.

Koevenig, J. L., and Liu, E. H., 1981, Carboxymethyl cellulase activity in the myxomycete *Physarum polycephalum, Mycologia* **73:**1085–1091.

Koike, I., and Hattori, A., 1978, Denitrification and ammonia formation in anaerobic coastal sediments, *Appl. Environ. Microbiol.* **35:**278–282.

Kolpak, F. J., and Blackwell, J., 1976, Determination of the structure of cellulose II, *Macromolecules* **9:**273–278.

Kozlik, I., and Schanel, L., 1974, Changes of the atmosphere during wood decay by fungi under conditions of stopped gas diffusion, *Drevarsky Vyskum* **19:**169–179.

Lacey, J., 1973, Actinomycetes in soils, composts and fodders, in: *Actinomycetales: Characteristics and Practical Importance* (G. Sykes and F. A. Skinner, eds.), pp. 231–251, Academic Press, London.

Lamed, R., and Zeikus, J. G., 1980, Ethanol production by thermophilic bacteria: Relationship between fermentation product yields of and catabolic enzyme activities in *Clostridium thermocellum* and *Thermoanaerobium brockii, J. Bacteriol.* **144:**569–578.

Lamed, R., Setter, E., and Bayer, E. A., 1983, Characterization of a cellulose-binding, cellulase-containing complex in *Clostridium thermocellum, J. Bacteriol.* **156**:828–836.

Latham, M. J., Brooker, B. E., Pettipayerher, G. L., and Harris, P. J., 1978, *Ruminococcus flavefaciens* cell coat and adhesion to cotton cellulose and to cell walls in leaves of perennial ryegrass *(Lolium perenne), Appl. Environ. Microbiol.* **35**:156–165.

Laube, V. M., and Martin, S. M., 1981, Conversion of cellulose to methane and carbon dioxide by triculture of *Acetivibrio cellulolyticus, Desulfovibrio sp.* and *Methanosarcina barkeri, Appl. Environ. Microbiol.* **42**:413–420.

Leadbetter, E. R., 1974, Cytophagaceae, in: *Bergey's Manual of Determinative Bacteriology,* 8th ed. (R. E. Buchanan and N. E. Gibbons, eds.), pp. 99–100, Williams and Wilkins, Baltimore.

Leatherwood, J. M., 1965, Cellulase from *Ruminococcus albus* and mixed rumen microorganisms, *Appl. Microbiol.* **13**:771–775.

Leatherwood, J. M., 1973, Cellulose degradation by *Ruminococcus, Fed. Proc.* **32**:1814–1818.

Lee, B. H., and Blackburn, T. H., 1975, Cellulase production by a thermophilic *Clostridium* species, *Appl. Microbiol.* **30**:346–353.

Leschine, S. B., and Canale-Parola, E., 1983, Mesophilic cellulolytic clostridia from freshwater environments, *Appl. Environ. Microbiol.* **46**:728–737.

Lewin, R. A., 1970, New *Herpetosiphon* species *(Flexibacterales), Can J. Microbiol.* **16**:517–520.

Liese, W., 1970, Ultrastructural aspects of woody tissue disintegration, *Annu. Rev. Phytopathol.* **8**:231–258.

Lieth, H., 1973, Primary production: Terrestrial ecosystem, *Hum. Eco.* **1**:303–331.

Ljungdahl, L. G., 1983, Formation of acetate using homoacetate fermenting bacteria, in: *Organic Chemicals from Biomass* (D. L. Wise, ed.), pp. 219–248, Benjamin/Cummings, Menlo Park, California.

Ljungdahl, L. G., Bryant, F., Careira, L., Saiki, T., and Wiegel, J., 1981a, Some aspects of thermophilic and extreme thermophilic anaerobic microorganisms, in: *Trends in the Biology of Fermentations for Fuels and Chemicals* (A. Hollaender, ed.), pp. 397–419, Plenum Press, New York.

Ljungdahl, L. G., Careira, L., and Wiegel, J., 1981b, Production of ethanol from carbohydrates using anaerobic thermophilic bacteria, in: *The Ekman Days,* International Symposium on Wood and Pulping Chemistry, Stockholm, Vol. 4, pp. 23–28.

Ljungdahl, L. G., Pettersson, B., Eriksson, K.-E., and Wiegel, J., 1983, A yellow affinity substance involved in the cellulolytic system of *Clostridium thermocellum, Curr. Microbiol.* **9**:195–199.

Lynch, J. M., Slater, J. H., Bennett, J. A., and Harper, S. H. T., 1981, Cellulase activities of some aerobic micro-organisms isolated from soil, *J. Gen. Microbiol.* **127**:231–236.

Maccubbin, A. E., and Hodson, R. E., 1980, Mineralization of detrital lignocelluloses by salt marsh sediment microflora, *Appl. Environ. Microbiol.* **40**:735–740.

MacKenzie, C. R., and Bilous, D., 1982, Location and kinetic properties of the cellulase system of *Acetivibrio cellulolyticus, Can. J. Microbiol.* **28**:1158–1164.

Madan, M., and Bisaria, R., 1983, Cellulolytic enzymes from an edible mushroom, *Pleurotus sajor-caju, Biotechnol. Lett.* **5**:601–604.

Madden, R. H., 1983, Isolation and characterization of *Clostridium stercorarium* sp. nov., cellulolytic thermophile, *Int. J. Syst. Bacteriol.* **33**:837–840.

Madden, R. H., Bryder, M. J., and Poole, N. J., 1981, The cellulolytic community of an anaerobic estuarine sediment, in *Proceedings of the International Conference on Energy from Biomass* (P. Chartier and D. O. Hall, eds.), pp. 366–371, Applied Science, London.

Madden, R. H., Bryder, M. J., and Poole, N. J., 1982, Isolation and characterization of an anaerobic, cellulolytic bacterium, *Clostridium papyrosolvens* sp. nov., *Int. J. Syst. Bacteriol.* **32**:87–91.

Mah, R. A., 1981, The methanogenic bacteria, their ecology and physiology, in: *Trends in the Biology of Fermentations for Fuels and Chemicals* (A. Hollaender, ed.), pp. 357–374, Plenum Press, New York.

Mah, R. A., Ward, D. M., Baresi, L., and Glass, T. L., 1977, Biogenesis of methane, *Annu. Rev. Microbiol.* **31**:309–341.

Maluszynska, G. M., and Janota-Bassalik, L., 1974, A cellulolytic rumen bacterium, *Micromonospora ruminantium* sp. nov., *J. Gen. Microbiol.* **82**:57–65.

Mandels, M., and Reese, E. T., 1960, Induction of cellulase in fungi by cellobiose, *J. Bacteriol.* **79**:816–826.

Manning, K., and Wood, D. A., 1983, Production and regulation of extracellular endocellulase by *Agaricus bisporus*, *J. Gen. Microbiol.* **129**:1839–1847.

Margaritis, A., and Merchant, R., 1983, Xylanase, CM-cellulase and avicelase production by the thermophilic fungus *Sporotrichum thermophile*, *Biotechnol. Lett.* **5**:265–270.

Marshall, J. J., 1973, Nature of the binding of a β-1,4-glucan hydrolase to ion exchangers, *J. Chromatogr.* **76**:257–260.

Marx-Figini, M., 1982, The control of molecular weight and molecular-weight distribution in the biogenesis of cellulose, in: *Cellulose and Other Natural Polymer Systems: Biogenesis, Structure and Degradation* (R. M. Brown, Jr., ed.), pp. 243–271, Plenum Press, New York.

Marx-Figini, M., and Schulz, G. V., 1966, Zur Biosynthese der Cellulose, *Naturwissenschaften* **53**:466–474.

McBee, R. H., 1950, The anaerobic thermophilic cellulolytic bacteria, *Bacteriol. Rev.* **14**:51–63.

McBee, R. H., 1954, The characteristics of *Clostridium thermocellum*, *J. Bacteriol.* **67**:505–506.

McHale, A., and Coughlan, M. P., 1981a, A convenient zymogram stain for cellulases, *Biochem. J.* **199**:267–268.

McHale, A., and Coughlan, M. P., 1981b, The cellulolytic system of *Talaromyces emersonii*. Identification of the various components produced during growth on cellulosic media, *Biochim. Biophys. Acta* **661**:145–151.

McInerney, M. J., and Bryant, M. P., 1981a, Review of fermentation fundamentals, in: *Fuel Gas Production from Biomass* (D. L. Wise, ed.), pp. 19–46, CRC Press, Boca Raton, Florida.

McInerney, M. J., and Bryant, M. P., 1981b, Anaerobic degradation of lactate by syntrophic associations of *Methanosarcina barkeri* and *Desulfovibrio* species and effect of H_2 on acetate degradation, *Appl. Environ. Microbiol.* **41**:346–354.

Mehta, N. C., Dubach, P., and Deuel, H., 1961, Carbohydrates in the soil, *Adv. Carbohyd. Chem.* **16**:335–355.

Meyer, H. P., and Humphrey, A. E., 1982, Cellulase production by a wild and a new mutant strains of *Thermomonospora* sp., *Biotechnol. Bioeng.* **24**:1901–1904.

Miller, T. L., and Wolin, M. J., 1974, A serum bottle modification of the Hungate technique for cultivating obligate anaerobes, *Appl. Microbiol.* **27**:985–987.

Mishra, M. M., Yadav, K. S., and Kapoor, K. K., 1981, Degradation of lignocellulose by mixed cultures of cellulolytic fungi and their competitive ability, *Zentralbl. Bakteriol. Parasitenkd. Infektionskr. Hyg. Abt.* **136**:603–608.

Miyoshi, H., 1978, Characterization of anaerobic cellulolytic bacteria isolated from marine environments, *Bull. Jpn. Soc. Sci. Fish.* **44**:197–202.

Moloney, A. P., Considine, P. J., and Coughlan, M. P., 1983, Cellulose hydrolysis by the cellulases produced by *Talaromyces emersonii* when grown on different inducing substrates, *Biotechnol. Bioeng.* **25**:1169–1173.

Montenecourt, B. S., Nhlapo, S. D., Trimino-Vazquez, H., Cuskey, S., Schamhart, D. H. J., and Eveleigh, D. E., 1981, Regulatory controls in relation to over-production of fungal cellulases, in: *Trends in the Biology of Fermentations for Fuels and Chemicals* (A. Hollaender, ed.), pp. 33–53, Plenum Press, New York.

Montgomery, L., and Macy, J. M., 1982, Characterization of rat cecum cellulolytic bacteria, *Appl. Environ. Microbiol.* **44:**1435–1443.

Moo-Young, M., Chahal, D. S., and Vlach, D., 1978, Single cell protein from various chemically pretreated wood substrates using *Chaetomium cellulolyticum, Biotechnol. Bioeng.* **20:**107–118.

Mountfort, D. O., and Asher, R. A., 1983, Role of catabolite regulatory mechanics in control of carbohydrate utilization by the rumen anaerobic fungus *Neocallimastix frontalis, Appl. Environ. Microbiol.* **46:**1331–1338.

Mueller, S. C., 1982, Cellulose-microfibril assembly and orientation in higher plant cells with particular reference to seedlings of *Zea mays,* in: *Cellulose and Other Natural Polymer Systems: Biosynthesis, Structure and Degradation* (R. M. Brown, Jr., ed.), pp. 87–103, Plenum Press, New York.

Nakamura, K., and Kitamura, K., 1983, Purification and some properties of a cellulase active on crystalline cellulose from *Cellulomonas uda, J. Ferment. Technol.* **61:**379–382.

Ng, T. K., and Zeikus, J. G., 1981a, Comparison of extracellular cellulase activities of *Clostridium thermocellum* LQRI and *Trichoderma reesei* QM9414, *Appl. Environ. Microbiol.* **42:**231–240.

Ng, T. K., and Zeikus, J. G., 1981b, Purification and characterization of an endoglucanase (1,4-β-D-glucan glucanohydrolase) from *Clostridium thermocellum, Biochem. J.* **199:**341–350.

Ng, T. K., Ben-Bassat, A., and Zeikus, J. G., 1981, Ethanol production by thermophilic bacteria: Fermentation of cellulosic substrates by coculture of *Clostridium thermocellum* and *Clostridium thermohydrosulfuricum, Appl. Environ. Microbiol.* **41:**1337–1343.

Nilsson, T., 1974a, Comparative study of the cellulolytic activity of white-rot and brown-rot fungi, *Mater. Organ. (Berl.)* **9:**173–198.

Nilsson, T., 1974b, Formation of soft rot cavities in various cellulose fibers by *Humicola alopsallonella* Meyer and Moore, *Studia Forestalia Suecia,* No. 112.

Norberg, A. B., and Persson, H., 1984, Accumulation of heavy-metal ions by *Zoogloea ramigera, Biotechnol. Bioeng.* **26:**239–246.

Norkrans, B., 1950, Influence of cellulolytic enzymes from hymenomycetes on cellulose preparations of different crystallinity, *Physiol. Plant.* **3:**75–78.

Oberkotter, L. V., and Rosenberg, F. A., 1978, Extracellular endo-β-1,4-glucanase in *Cellvibrio vulgaris, Appl. Environ. Microbiol.* **36:**205–209.

O'Brien, R. W., and Slaytor, M., 1982, Role of microorganisms in the metabolism of termites, *Aust. J. Biol. Sci.* **35:**239–262.

Odom, J. M., and Wall, J. D., 1983, Photoproduction of H_2 from cellulose by an anaerobic bacterial coculture, *Appl. Environ. Microbiol.* **45:**1300–1305.

Ohmine, K., Ooshima, H., and Harano, Y., 1983, Kinetic study on enzymatic hydrolysis of cellulose by cellulase from *Trichoderma viride, Biotechnol. Bioeng.* **25:**2041–2053.

Oremland, R. S., Marsh, L. M., and Polcin, S., 1982, Methane production and simultaneous sulphate reduction in anoxic, salt marsh sediments, *Nature* **296:**143–145.

Orpin, C. G., 1975, Studies on the rumen flagellate *Neocallimastix frontalis, J. Gen. Microbiol.* **91:**249–262.

Orpin, C. G., 1977a, The rumen flagellate *Piromonas communis:* Its life-history and invasion of plant material in the rumen, *J. Gen. Microbiol.* **99:**107–117.

Orpin, C. G., 1977b, On the induction of zoosporogenesis in the rumen phycomycetes *Neo-*

callimastix frontalis, Piromonas communis and *Sphaeromonas communis, J. Gen. Microbiol.* **101**:181–189.

Orpin, C. G., 1977c, Invasion of plant tissue in the rumen by the flagellate *Neocallimastix frontalis, J. Gen. Microbiol.* **98**:423–430.

Orpin, C. G., 1981, Isolation of cellulolytic phycomycete fungi from the caecum of the horse, *J. Gen. Microbiol.* **123**:287–296.

Orpin, C. G., and Letcher, A. J., 1979, Utilization of cellulose, starch, xylan, and other hemicelluloses for growth by the rumen phycomycete *Neocallimastix frontalis, Curr. Microbiol.* **3**:121–124.

Osmundsvåg, K., and Goksøyr, J., 1975, Cellulases from *Sporocytophaga myxococcoides:* Purification and properties, *Eur. J. Biochem.* **57**:405–409.

Otjen, L., and Blanchette, R. A., 1982, Patterns of decay caused by *Inonotus dryophilus* (Aphyllophorales: Hymenochaetaceae), a white-pocket rot fungus of oaks, *Can. J. Bot.* **60**:2770–2779.

Palleroni, N. J., 1984, Pseudomonas, in: *Bergey's Manual of Systematic Bacteriology,* Vol. 1 (N. R. Krieg and J. G. Holt, eds.), pp. 141–199, Williams and Wilkins, Baltimore.

Patel, G. B., and Breuil, C., 1981, Accumulation of an iodophilic polysaccharide during growth of *Acetivibrio cellulolyticus* on cellobiose, *Arch. Microbiol.* **129**:265–267.

Patel, G. B., Khan, A. W., Agnew, B. J., and Colvin, J. R., 1980, Isolation and characterization of an anaerobic cellulolytic microorganism, *Acetivibrio cellulolyticus* gen. nov., sp. nov., *Int. J. Syst. Bacteriol.* **30**:179–185.

Payne, W. J., 1981, *Denitrification,* Wiley, New York.

Peck Jr., H. D., 1984, Physiological diversity of the sulfate bacteria, in: *Microbial Chemoautotrophy* (W. R. Strohl and O. H. Tuovinen, eds.), pp. 309–335, Ohio State University Press, Columbus.

Peck Jr., H. D., and Odom, M., 1981, Anaerobic fermentations of cellulose to methane, in: *Trends in the Biology of Fermentations for Fuels and Chemicals* (A. Hollaender, ed.), pp. 375–395, Plenum Press, New York.

Peck, Jr., H. D., and Odom, J. M., 1984, Hydrogen cycling in *Desulfovibrio:* A new mechanism for energy coupling in anaerobic microorganisms, in: *Microbial Mats: Stromatolites,* pp. 215–243, Alan R. Liss, New York.

Petre, J., Longin, R., and Millet, J., 1981, Purification and properties of endo-β-1,4-glucanase from *Clostridium thermocellum, Biochimie* **7**:629–639.

Pettipher, G. L., and Latham, M. J., 1979a, Production of enzymes degrading plant cell walls and fermentation of cellobiose by *Ruminococcus flavefaciens* in batch and continuous culture, *J. Gen. Microbiol.* **110**:29–38.

Pettipher, G. L., and Latham, M. J., 1979b, Characteristics of enzymes produced by *Ruminococcus flavefaciens* which degrade plant cell walls, *J. Gen. Microbiol.* **110**:21–27.

Prins, R. A., van Vught, F., Hungate, R. E., and van Vorstenbosch, C. J. A. H. V., 1972, A comparison of strains of *Eubacterium cellulosolvens* from the rumen, *Antonie Leeuwenhoek J. Microbiol.* **38**:153–161.

Ramasamy, K., Meyers, M., Bevers, J., and Verachtert, H., 1981, Isolation and characterization of cellulolytic bacteria from activated sludge, *J. Appl. Bacteriol.* **51**:475–481.

Rao, M., Deshpande, V., Keskar, S., and Srinivasan, M. C., 1983a, Cellulase and ethanol production from cellulose by *Neurospora crassa, Enzyme Microb. Technol.* **5**:133–136.

Rao, M. N. A., Mithal, B. M., Thakur, R. N., and Sastry, K. S. M., 1983b, Productions of cellulase from *Pestalotiopsis versicolor, Biotechnol. Bioeng.* **25**:2395–2398.

Reid, I. D., 1983a, Effects of nitrogen sources on cellulose and synthetic lignin degradation by *Phanerochaete chrysosporium, Appl. Environ. Microbiol.* **45**:838–842.

Reid, I. D., 1983b, Effects of nitrogen supplements on degradation of aspen wood lignin and carbohydrate components by *Phanerochaete chrysosporium, Appl Environ. Microbiol.* **45**:830–837.

Rennerfelt, E., 1944, Investigations on the toxicity to rot fungi of the phenolic components of pine heartwood, *Medd. Skogsförsöksanstalten* **33**:331–364.

Rho, D., Desrochers, M., Jurasek, L., Driguez, H., and DeFaye, J., 1982, Induction of cellulase in *Schizophyllum commune*: Thiocellobiose as a new inducer, *J. Bacteriol.* **249**:47–53.

Rickard, P. A. D., and Laughlin, T. A., 1980, Detection and assay of xylanolytic enzymes in a *Cellulomonas* isolate, *Biotechnol. Lett.* **2**:363–368.

Robinson, J. A., and Tiedje, J. M., 1984, Competition between sulfate-reducing and methanogenic bacteria for H_2 under resting and growing conditions, *Arch. Microbiol.* **137**:26–32.

Robson, L. M., and Chambliss, G. H., 1984, Characterization of the cellulolytic activity of a *Bacillus* isolate, *Appl. Environ. Microbiol* **47**:1039–1046.

Roche, C. H., Albertyn, H., van Gylswyk, N. O., and Kistner, A., 1973, The growth response of cellulolytic acetate-utilizing and acetate producing *Butyrivibrios* to volatile fatty acids and other nutrients, *J. Gen. Microbiol.* **78**:253–260.

Rode, L. M., Genthner, B. R. S., and Bryant, M. P., 1981, Syntrophic association by cocultures of the methanol- and CO_2-H_2-utilizing species *Eubacterium limosum* and pectin-fermenting *Lachnospira multiparus* during growth in a pectin medium, *Appl. Environ. Microbiol.* **42**:20–22.

Rogers, C. J., Coleman, E., Spino, D. F., and Purcell, T. C., 1972, Production of fungal protein from cellulose and waste cellulosics, *Environ. Sci. Technol.* **6**:715–718.

Romanelli, R. A., Houston, C. W., and Barrett, S. M., 1975, Studies on thermophilic cellulolytic fungi, *Appl. Microbiol.* **30**:276–281.

Rosenberg, S. L., 1978, Cellulose and lignocellulose degradation by thermophilic and thermotolerant fungi, *Mycologia* **70**:1–13.

Rudman, P., and DaCosta, E. W., 1958, The causes of natural durability in timber. The role of toxic extractives in the resistance of silvertop ash *(Eucalyptus sieberiana)* to decay, Australia Commonwealth Scientific and Industrial Research Organization, Division of Forest Products and Technology, Paper 1,8.

Ruel, K., Barnoud, F., and Eriksson, K.-E., 1981, Micromorphological and ultrastructural aspects of spruce wood degradation by wild-type *Sporotrichum pulverulentum* and its cellulase-less mutant Cel 44, *Holzforschung* **35**:157–171.

Ryu, D. D. Y., and Mandels, M., 1980, Cellulase: Biosynthesis and applications, *Enzyme Microbial. Technol.* **2**:91–102.

Saddler, J. N., and Khan, A. W., 1979, Cellulose degradation by a new isolate from sewage sludge, a member of the Bacteroidaceae family, *Can. J. Microbiol.* **25**:1427–1431.

Saddler, J. N., and Khan, A. W., 1980, Cellulase production by *Acetivibrio cellulolyticus*, *Can. J. Microbiol.* **26**:760–765.

Saddler, J. N., and Khan, A. W., 1981, Cellulolytic enzyme system of *Acetivibrio cellulolyticus, Can. J. Microbiol.* **27**:288–294.

Saddler, J. N., Khan, A. W., and Martin, S. M., 1980, Regulation of cellulase synthesis in *Acetivibrio cellulolyticus, Microbios* **28**:97–106.

Sarkanen, K. V., and Hergert, H. L., 1971, Classification and distribution, in: *Lignins: Occurrence, Formation, Structure and Reactions* (K. V. Sarkanen, and C. H. Ludwig, eds), p. 43–94, Wiley-Interscience, New York.

Sarko, A., and Muggli, R. 1974, Packing analysis of carbohydrates and polysaccharides, IV. *Valonia* cellulose and cellulose II, *Macromolecules* **7**:486–494.

Sasaki, T., Tanaka, T., Nakagawa, S., and Kainuma, K., 1983, Purification and properties of *Cellvibrio gilvus* cellobiose phosphorylase, *Biochem. J.* **209**:803–807.

Sashihara, N., Kudo, T., and Horikoshi, K., 1984, Molecular cloning and expression of cellulase genes of alkalophilic *Bacillus* sp. strain N-4 in *Escherichia coli, J. Bacteriol.* **158**:503–506.

Sato, M., and Takahashi, H., 1967, Fermentation of ^{14}C-labeled cellobiose by *Cellulomonas fimi*, *Agric. Biol. Chem.* **31**:470–474.

Schafer, M. L., and King, K. W., 1965, Utilization of cellulose oligosaccharides by *Cellvibrio gilvus*, *J. Bacteriol.* **89**:113–116.

Schimz, K.-L., Broll, B., and John, B., Cellobiose phosphorylase (EC 2.4.1.20) of *Cellulomonas:* Occurrence, induction, and its role in cellobiose metabolism, *Arch. Microbiol.* **135**:241–249.

Seifert, K., 1968, Zur Systematik der Holzfaulen, ihre chemischen und physikalischen Kennzeichen, *Holz Roh Werkstoff* **26**:208–215.

Selby, K., 1968, Mechanism of biodegradation of cellulose, in: *Biodeterioration of Materials* (A. H. Walters and J. J. Elphick, eds.), pp. 62–78, Elsevier, Amsterdam.

Sen, S., Abraham, T. K., and Chakrabarty, S. L., 1981, Characteristics of the cellulase produced by *Myceliophthora thermophila* D-14, *Can. J. Microbiol.* **28**:271–277.

Shane, B. S., Gouws, L., and Kistner, A., 1969, Cellulolytic bacteria occurring in the rumen of sheep conditioned to low-protein teff hay, *J. Gen. Microbiol.* **55**:445–457.

Shepherd, M. G., Tong, C. C., and Cole, A. L., 1981, Substrate specificity and mode of action of the cellulases from the thermophilic fungus *Thermoascus aurantiacus, J. Biochem.* **193**:67–74.

Sheth, K., and Alexander, J. K., 1969, Purification and properties of β-1,4-oligoglucan:orthophosphate glucosyltransferase from *Clostridium thermocellum, J. Biol. Chem.* **244**:457–464.

Shirling, E. B., and Gottlieb, D., 1966, Methods for characterization of *Streptomyces* species, *Int. J. Syst. Bacteriol.* **16**:313–340.

Shortle, W. C., Menge, J. A., and Cowling, E. B., 1978, Interaction of bacteria, decay fungi, and live sapwood in discoloration and decay of trees, *Eur. J. For. Pathol.* **8**:293–300.

Sijpesteijn, A. K., 1951, On *Ruminococcus flavefaciens,* a cellulose-decomposing bacterium from the rumen of sheep and cattle, *J. Gen. Microbiol.* **5**:869–879.

Sinha, R. N., and Ranganathan, B., 1983, Cellulolytic bacteria in buffalo rumen, *J. Appl. Bacteriol.* **54**:1–6.

Sinha, S. N., Ghosh, G. L., and Ghose, S. N., 1981, Detection of cellulase inhibitor in the wheat bran culture of *Aspergillus terreus, Can. J. Microbiol.* **27**:1334–1340.

Smith, W. R., Yu, I., and Hungate, R. E., 1973, Factors affecting cellulolysis by *Ruminoccus albus, J. Bacteriol.* **114**:729–737.

Sørensen, J., 1978, Capacity for denitrification and reduction of nitrate to ammonia in a coastal marine sediment, *Appl. Environ. Microbiol.* **35**:301–305.

Stack, R. J., and Hungate, R. E., 1984, Cellulose digestion by *Ruminococcus albus* strain 8, in: *Abstracts Annual Meeting,* American Society Microbiol., Abstract 033.

Stack, R. J., Hungate, R. E., and Opsahl, W. P., 1983, Phenylacetic acid stimulation of cellulose digestion by *Ruminococcus albus* 8, *Appl. Environ. Microbiol.* **46**:539–544.

Stackebrandt, E., and Kandler, O., 1974, Biochemisch-taxonomische Untersuchungen an der Gattung *Cellulomonas, Zentralbl. Bakteriol. Hyg. I Abt. Orig. A* **22**:128–135.

Stackebrandt, E., and Kandler, O., 1979, Taxonomy of the genus *Cellulomonas* based on phenotype characters and deoxyribonucleic acid–deoxyribonucleic acid homology and proposal of seven neotype strains, *Int. J. Syst. Bacteriol.* **29**:273–282.

Stackebrandt, E., and Kandler, O., 1980a, Fermentation pathway and redistribution of ^{14}C in specifically labelled glucose in *Cellulomonas, Zentralbl. Bakteriol. Hyg. I Abt. Orig. C* **1**:40–50.

Stackebrandt, E., and Kandler, O., 1980b, *Cellulomonas cartae* sp. nov., *Int. J. Syst. Bacteriol.* **30**:186–188.

Stackebrandt, E., Seiler, H., and Schleifer, K. H., 1982, Union of genera *Cellulomonas* Bergey *et al.* and *Oerskovia* Prauser *et al.* in a redefined genus *Cellulomonas, Zentralbl. Bakteriol. Hyg. I Abt. Orig. C* **3**:401–409.

Stanier, R. Y., 1942, The cytophaga group: A contribution to the biology of *Myxobacteria, Bacteriol. Rev.* **6**:143–196.

Stanton, T. B., and Canale-Parola, E., 1980, *Treponema bryantii* sp. nov., a rumen spirochete that interacts with cellulolytic bacteria, *Arch. Microbiol.* **127**:145–156.

Stapp, C., and Bortels, H., 1934, Mikrobiologische Untersuchungen uber die Zersetzung von Waldstreu, *Zentralbl. Bakeriol. Parasitenkd. Infektionskr. Hyg. Abt. II* **90**:28–66.

Stephens, G. R., and Heichel, G. H., 1975, Agricultural and forest products as sources of cellulose, *Biotechnol. Bioeng. Symp.* **5**:27–42.

Sternberg, D., and Mandels, G. R., 1979, Induction of celluloytic enzymes in *Trichoderma reesei* by sophorose, *J. Bacteriol.* **139**:761–769.

Sternberg, D., Vijayakumar, P., and Reese, E. T., 1977, β-Glucosidase: Microbial production and effect of enzymatic hydrolysis of cellulose, *Can. J. Microbiol.* **23**:139–147.

Stewart, B. J., and Leatherwood, J. M., 1976, Derepressed synthesis of cellulase by *Cellulomonas, J. Bacteriol.* **128**:609–615.

Stewart, C. S., Paniagua, C., Dinsdale, D., Cheng, K.-J., and Garrow, S. H., 1981, Selective isolation and characteristics of *Bacteroides succinogenes* from the rumen of a cow, *Appl. Environ. Microbiol.* **41**:504–510.

Stipanovic, A. J., and Sarko, A., 1976, Packing analysis of carbohydrates and polysaccharides. VI. Molecular and crystal structure of regenerated cellulose II, *Macromolecules* **9**:851–857.

Stoppok, W., Rapp, P., and Wayner, F., 1982, Formation, location and regulation of endo-1,4-β-glucanases and β-glucosidases from *Cellulomonas uda, Appl. Environ. Microbiol.* **44**:44–53.

Storvick, W. O., Cole, F. E., and King, K. W., 1963, Mode of action of a cellulase component from *Cellvibrio gilvus, Biochemistry* **2**:1106–1110.

Strayer, R. F., and Tiedje, J. M., 1978, Kinetic parameters of the conversion of methane precursors to methane in a hypereutrophic lake sediment, *Appl. Environ. Microbiol.* **36**:330–340.

Streamer, M., Eriksson, K.-E., Pettersson, B., 1975, Extracellular enzyme system utilized by the fungus *Sporotrichum pulverulentum (Chrysosporium lignorum)* for the breakdown of cellulose, *Eur. J. Biochem.* **59**:607–613.

Stutzenberger, F. J., 1971, Cellulase production by *Thermomonospora curvata* isolated from muncicipal waste compost, *Appl. Microbiol.* **22**:147–152.

Stutzenberger, F. J., 1972a, Cellulolytic activity of *Thermomonospora curvata:* Optimal assay conditions, partial purification, and product of the cellulase, *Appl. Microbiol.* **24**:83–90.

Stutzenberger, F. J., 1972b, Cellulolytic activity of *Thermomonospora curvata:* Nutritional requirements for cellulase production, *Appl. Microbiol.* **24**:77–82.

Su, T.-M., and Paulavicius, I., 1975, Enzymic saccharification of cellulose by thermophilic actinomyces, *Appl. Polym. Symp.* **28**:221–236.

Sundman, V., and Nase, L., 1971, A simple plate test for direct visualization of biological lignin degradation, *Paper Timber* **53**:67–71.

Swift, M. J., 1977, The ecology of wood decomposition, *Sci. Prog. Oxf.* **64**:175–199.

Swisher, E. J., Storvick, W. O., and King, K. W., 1964, Metabolic nonequivalence of the two glucose moieties of cellobiose in *Cellvibrio gilvus, J. Bacteriol.* **88**:817–820.

Szakács, G., Réczey, K., Hernádi, P., and Dobozi, M., 1981, *Penicillium verruculosum* WA30, a new source of cellulase, *Eur. J. Appl. Microbiol. Biotechnol.* **11**:120–124.

Tanaka, M., Taniguchi, M., Matsuno, R., and Kamikubo, T., 1981, Purification and properties of cellulases from *Eupenicillium javanicum, J. Ferment. Technol.* **59**:177–183.

Tanaka, T., Yamamoto, R., Oi, S., and Nevins, D. J., 1982, Purification and some properties of β-transglucosylases of *Sclerotinia libertiana,* having the ability to synthesize higher cell-oligosaccharides from cellotriose or cellotetraose, *Carbohyd. Res.* **106:**131–142.

Teather, R. M., and Wood, P. J., 1982, Use of Congo red–polysaccharide interactions in enumeration and characterization of cellulolytic bacteria from the bovine rumen, *Appl. Environ. Microbiol.* **43:**777–780.

Thakur, R. N., and Sastry, K. S. M., 1981, Leaf blight of *Mucuna prurita, Ind. Phytopathol.* **34:**394–395.

Thauer, R. K., Kaufer, B., and Scherer, P., 1975, The active species of "CO_2" utilized in ferredoxin-linked carboxylation reactions, *Arch. Microbiol.* **104:**237–240.

Thayer, D. W., 1978, Carboxymethylcellulase produced by facultative bacteria from the hind-gut of the termite *Reticulitermes hesperus, J. Gen. Microbiol.* **106:**13–18.

Thompson, N. S., 1983, Hemicellulose as a biomass resource, in: *Wood and Agricultural Residues* (E. J. Soltes, ed.), pp. 101–119, Academic Press, New York.

Tong, C. C., Cole, A. L., and Shepherd, M. G., 1980, Purification and properties of the cellulases from the thermophilic fungus *Thermoascus aurantiacus, Biochem. J.* **191:**83–94.

Tønnesen, B. A., and Ellefsen, O., 1971, Submicroscopical investigations, in: *Cellulose and Cellulose Derivatives* (N. M. Bikales and L. Segal, eds.), Vol. 5, Part IV, pp. 265–304, Wiley, New York.

Trivedi, L. S., and Rao, K. K., 1980, Factors influencing cellulase induction in *Fusarium* sp., *Curr. Microbiol.* **3:**219–224.

Trivedi, L. S., and Rao, K. K., 1981, Production of cellulolytic enzymes by *Fusarium* species, *Biotechnol. Lett.* **3:**281–284.

Ueda, K., Ishikawa, S., and Asai, T., 1952, Studies on the aerobic mesophilic cellulose-decomposing bacteria, part 5—2, Taxonomical study on genus *Pseudomonas, J. Agric. Chem. Soc. Japan* **26:**35–45.

Umezurike, G. M., 1979, The cellulolytic enzymes of *Botryodiplodia theobromae* Pat. Separation and characteriztion of cellulases and β-glucosidases, *Biochem. J.* **177:**9–19.

Umezurike, G. M., 1981, The β-glucosidase from *Botryodiplodia theobromae, Biochem. J.* **199:**203–209.

Urbanek, H., Zalewska-Sobczak, J., and Borowinska, A., 1978, Isolation and properties of extracellular cellulase–hemicellulase complex of *Phoma hibernica, Arch. Microbiol.* **118:**265–269.

Vaheri, M. P., 1982, Acidic degradation products of cellulose during enzymatic hydrolysis by *Trichoderma reesei, J. Appl. Biochem.* **4:**153–160.

Van den Berg, L., Patel, G.B., Clark, D. S., and Lentz, C. P., 1976, Factors affecting rate of methane formation from acetic acid by enriched methanogenic cultures, *Can. J. Microbiol.* **22:**1312–1319.

Van Gylswyk, N. O., 1970, A comparison of two techniques for counting cellulolytic rumen bacteria, *J. Gen. Microbiol.* **60:**191–197.

Van Gylswyk, N. O., 1976, Some aspects of the metabolism of *Butyrivibrio fibrisolvens, J. Gen. Microbiol.* **97:**105–111.

Van Glyswyk, N. O., and Hoffman, J. P. L., 1970, Characteristics of cellulolytic *Cillobacteria* from the rumens of sheep fed teff *(Eragrostis tef)* hay diets, *J. Gen. Microbiol.* **60:**381–386.

Van Gylswyk, N. O., and Labuschagne, J. P. L., 1971, Relative efficiency of pure cultures of different species of cellulolytic rumen bacteria in solubilizing cellulose *in vitro, J. Gen. Microbiol.* **66:**109–113.

Van Gylswyk, N. O., and Roche, C. E. G., 1970, Characteristics of *Ruminococcus* and cellulolytic *Butyrivibrio* species from the rumen of sheep fed different supplemented teff *(Eragrostis tef)* hay diets, *J. Gen. Microbiol.* **64:**11–17.

Varadi, J., 1972, The effect of aromatic compounds on cellulase and xylanase production of the fungi *Schizophyllum commune* and *Chaetomium globosum*, in: *Biodeterioration of Materials* (A. M. Walters and E. H. Hulchvan der Plas, eds.), pp. 129–135, Applied Science, London.

Varel, V. H., 1983, Characteristics of bacteria in thermophilic digesters and effect of antibiotics on methane production, in: *Fuel Gas Developments* (D. L. Wise, ed.), pp. 19–47, CRC Press, Boca Raton, Florida.

Varel, V. H., Fryda, S. J., and Robinson, I. M., 1984, Cellulolytic bacteria from pig large intestine, *Appl. Environ. Microbiol.* **47**:219–221.

Veiga, M., Esparis, A., and Fabregas, J., 1983, Isolation of cellulolytic actinomycetes from marine sediments, *Appl. Environ. Microbiol.* **46**:286–287.

Vian, B., 1982, Organized microfibril assembly in higher plant cells, in: *Cellulose and Other Natural Polymer Systems: Biogenesis, Structure and Degradation* (R. M. Brown, Jr., ed.), pp. 23–43, Plenum Press, New York.

Viljoen, J. A., Fred, E. B., and Peterson, W. H., 1926, The fermentation of cellulose by thermophilic bacteria, *J. Agric. Sci.* **16**:1–17.

Vogels, G. D., 1979, The global cycle of methane, *Antonie Leeuwenhoek J. Microbiol. Serol.* **45**:347–352.

Von Hofsten, B., Berg, B., and Beskow, S., 1971, Observations on bacteria occurring together with Sporcytophaga in aerobic enrichment cultures on cellulose, *Arch. Mikrobiol.* **79**:69–79.

Wellard, H. J., 1954, Variation in the lattice spacing of cellulose, *J. Polym. Sci.* **13**:471–476.

Westermark, U., and Eriksson, K.-E., 1974a, Carbohydrate-dependent enzymic quinone reduction during lignin degradation, *Acta Chem. Schand. B* **28**:204–208.

Westermark, U., and Eriksson, K.-E., 1974b, Cellobiose:quinone oxidoreductase, a new wood-degrading enzyme from white-rot fungi, *Acta Chem. Scand. B* **28**:209–214.

Whitaker, D. R., and Thomas, R., 1963, Improved procedures for preparation and characterization of *Myrothecium* cellulase. Part I. Production of enzyme, *Can. J. Biochem. Physiol.* **41**:667–670.

White, A. R., and Brown, Jr., R. M., 1981, Enzymatic hydrolysis of cellulose: Visual characterization of the process, *Proc. Natl. Acad. Sci. USA* **78**:1047–1051.

Whittaker, R. H., and Likens, G. E., 1973, Primary production: The biosphere and man, *Hum. Eco.* **1**:357–369.

Wicklow, D. T., Detroy, R. W., and Adams, S., 1980, Differential modification of the lignin and cellulose components in wheat straw by fungal colonists of ruminant dung: Ecological implications, *Mycologia* **72**:1065–1076.

Wiegel, J., and Dykstra, M., 1984, *Clostridium thermocellum:* Adhesion and sporulation while adhered to cellulose and hemicellulose, *Appl. Microbiol. Biotechnol.* **20**:59–65.

Wiegel, J., and Ljungdahl, L. G., 1979, Ethanol as fermentation product of extreme thermophilic, anaerobic bacteria, in: *Viertes Symposium Technische Mikrobiologie* (H. Dellweg, ed.), pp. 117–127, Verlag Versuchs- und Lehranstalt für Spiritusfabrikation und Fermentationstechnologie, Berlin.

Wiegel, J., and Ljungdahl, L. G., 1981, *Thermoanaerobacter ethanolicus* gen. nov., spec. nov., a new extreme thermophilic, anaerobic bacterium, *Arch. Microbiol.* **128**:343–348.

Wiegel, J., Ljungdahl, L. G., and Rawson, J. R., 1979, Isolation from soil and properties of the extreme thermophile *Clostridium thermohydrosulfuricum, J. Bacteriol.* **139**:800–810.

Wiegel, J., Braun, M., and Gottschalk, G., 1981, *Clostridium thermoautotrophicum* spec. nov. A thermophile producing acetate from molecular hydrogen and carbon dioxide, *Curr. Microbiol.* **5**:255–260.

Wilcox, W. W., 1970, Anatomical changes in wood cell walls attacked by fungi and bacteria, *Bot. Rev.* **36**:2–18.

Wilcox, W. W., 1973, Degradation in relation to wood structure, in: *Wood Deterioration and Its Prevention by Preservative Treatments,* Vol. 1 (D. D. Nicholas, ed.), pp. 107–148, Syracuse University Press, Syracuse, New York.

Winfrey, M. R., and Zeikus, J. G., 1979, Anaerobic metabolism of immediate methane precursors in Lake Mendota, *Appl. Environ. Microbiol.* **37**:244–253.

Winogradsky, S., 1929, Études sur la microbiologie du sol sur la degradation de la cellulose dans le sol, *Ann. Inst. Pasteur* **43**:549–633.

Winter, J. U., and Wolfe, R. S., 1980, Methane formation from fructose by syntrophic associations of *Acetobacterium woodii* and different strains of methanogens, *Arch. Microbiol.* **124**:73–79.

Wolin, M. J., 1979, The rumen fermentation: A model for microbial interactions in anaerobic ecosystems, in: *Advances in Microbial Ecology,* Vol. 3 (M. Alexander, ed.), pp. 49–77, Plenum Press, New York.

Wolin, M. J., and Miller, T. L., 1983, Interactions of microbial populations in cellulose fermentation, *Fed. Proc.* **42**:109–113.

Wood, H. G., Drake, H. L., and Hu, S.-I., 1982, Studies with *Clostridium thermoaceticum* and the resolution of the pathway used by acetogenic bacteria that grow on carbon monoxide or carbon dioxide and hydrogen, *Proc. Biochem. Symp.* **1982**:29–56.

Wood, T. M., 1968, Cellulolytic enzyme system of *Trichoderma koningii;* Separation of components attacking native cotton, *Biochem. J.* **109**:217–227.

Wood, T. M., 1980, Cooperative action between enzymes involved in the degradation of crystalline cellulose, in: *Colloque Cellulolyse Microbienne,* pp. 167–176, CNRS, Marseille.

Wood, T. M., 1983, Biochemistry of cellulase complex, in: *Proceedings from the Seminar, La Biomasse, Source d'Intermediaires Industriels,* ADEPRINA, Paris.

Wood, T. M., and McCrae, S. I., 1975, The cellulase complex of *Trichoderma koningii,* in *Symposium on Enzymatic Hydrolysis of Cellulose* (M. Barley *et al.,* eds.), pp. 231–254, Finland.

Wood, T. M., and McCrae, S. I., 1978, The cellulase of *Trichoderma koningii,* purification and properties of some endoglucanase components with special reference to their action on cellulose when acting alone and in synergisms with the cellobiohydrolase, *Biochem. J.* **171**:61–72.

Wood, T. M., and Phillips, D. R., 1969, Another source of cellulose, *Nature* **222**:986–987.

Wood, T. M., McCrae, S. I., and MacFarlane, C. C., 1980, The isolation, purification, and properties of the cellobiohydrolase component of *Penicillium funiculosum* cellulase, *Biochem. J.* **189**:51–55.

Wood, T. M., Wilson, C. A., and Stewart, C. S., 1982, Preparation of the cellulase from the cellulolytic anaerobic rumen bacterium *Ruminococcus albus* and its release from the bacterial cell wall, *Biochem. J.* **205**:129–137.

Yaguchi, M., Roy, C., Rollin, C. F., Paice, M. G., and Jurasek, L., 1983, A fungal cellulase shows sequence homology with the active site of hen-egg-white lysozyme, *Biochem. Biophys. Res. Commun.* **116**:408–411.

Yamane, K., Suzuki, H., and Nisizawa, K., 1970, Purification and properties of extracellular and cell-bound cellulase components of *Pseudomonas fluorescens* var. *cellulosa, J. Biochem.* **67**:19–35.

Yoshikawa, T., Suzuki, H., and Nisizawa, K., 1974, Biogenesis of multiple cellulase components of *Pseudomonas fluorescens* var. *cellulosa* I. Effects of culture conditions on the multiplicity of cellulase, *J. Biochem.* **75**:531–540.

Yu, I., and Hungate, R. E., 1979, The extracellular cellulases of *Ruminococcus albus, Ann. Rech. Vet.* **10**:251–254.

Zeikus, J. G., 1977, The biology of methanogenic bacteria, *Bacteriol. Rev.* **41**:514–541.

Zeikus, J. G., 1981, Lignin metabolism and the carbon cycle. Polymer biosynthesis, biodegradation and environmental recalcitrance, in: *Advances in Microbial Ecology,* Vol. 5 (M. Alexander, ed.), pp. 211–243, Plenum Press, New York.

Zhu, Y. S., Wu, Y. Q., Chen, W., Tan, C., Gao, J. H., Fei, J. X., and Shih, C. N., 1982, Induction and regulation of cellulase synthesis in *Trichoderma pseudokoningii* mutants EA_3-867 and N_2-78, *Enzyme Microb. Technol.* **4**:3–12.

Index